HOT SPOT OF
INVENTION

HOT SPOT OF
INVENTION

Charles Stark Draper,
MIT, and the Development
of Inertial Guidance
and Navigation

THOMAS WILDENBERG

NAVAL INSTITUTE PRESS
ANNAPOLIS, MARYLAND

Naval Institute Press
291 Wood Road
Annapolis, MD 21402

© 2019 by Thomas Wildenberg
All rights reserved. No part of this book may be reproduced or utilized in any form or by any means, electronic or mechanical, including photocopying and recording, or by any information storage and retrieval system, without permission in writing from the publisher.

Library of Congress Cataloging-in-Publication Data
Names: Wildenberg, Thomas, date, author.
Title: Hot spot of invention : Charles Stark Draper, MIT, and the development of inertial guidance and navigation / Thomas Wildenberg.
Description: Annapolis, Maryland : Naval Institute Press, [2019] | Includes bibliographical references and index.
Identifiers: LCCN 2019022492 | ISBN 9781682474693 (hardcover)
Subjects: LCSH: Draper, C. S. (Charles Stark) | Aeronautical engineers—United States—Biography. | Inertial navigation systems. | Gyroscopes.
Classification: LCC TL540.D67 W55 2019 | DDC 629.132/51 [B]— dc23
LC record available at https://lccn.loc.gov/2019022492

♾ Print editions meet the requirements of ANSI/NISO z39.48-1992 (Permanence of Paper).
Printed in the United States of America.

27 26 25 24 23 22 21 20 19   9 8 7 6 5 4 3 2 1
First printing

Contents

List of Illustrations — *vii*
Acknowledgments — *ix*
List of Abbreviations — *xi*
Author's Note — *xv*
Prologue   Hot Spot of Innovation and Invention — *xvii*

Chapter 1.	A Milestone in Aviation History	1
Chapter 2.	The Formative Years	8
Chapter 3.	Back to MIT	19
Chapter 4.	Aircraft Instruments and the Beginnings of the Instrumentation Lab	30
Chapter 5.	From Turn Indicator to Gunsight	46
Chapter 6.	War Work	59
Chapter 7.	Directors and Gun Fire-Control Systems	70
Chapter 8.	The A1-C(M) Gunsight	82
Chapter 9.	The "Immaculate Interception" and Other Air-Defense Activities	95
Chapter 10.	Inertial Navigation	103
Chapter 11.	Floated Gyros and SPIRE	116
Chapter 12.	SINS: The Submarine Inertial Navigation System	128
Chapter 13.	Professor, Prodigious Worker, Family Man	137
Chapter 14.	Inertial Guidance for Atlas and Thor	147
Chapter 15.	Titan, FLIMBAL, AIRS, and the MX/Peacekeeper	157
Chapter 16.	Polaris	164
Chapter 17.	Poseidon and Trident	174
Chapter 18.	Spy Satellites and Space Planes	189

Chapter 19.	To the Moon and Beyond	200
Chapter 20.	The Road to Divestiture	209
Chapter 21.	A Heterogeneous Engineer	224

Notes   *229*
Source Material   *259*
Index   *279*

Illustrations

Figures

1-1. SPIRE components	2
1-2. SPIRE system elevation view	5
1-3. SPIRE system plan view	6
5-1. Turn indicator patent	52
5-2. Doc's Shoebox	56
5-3. Superelevation	58
6-1. Mark 14 gunsight	64
6-2. Mark 14 gunsight in action	66
7-1. Mark 52 director	75
7-2. Gun Fire Control System Mark 63 Mod 2	78
8-1. Draper's notebook	86
8-2. A1-C diagram	89
8-3. A1-C in an F-86	91
11-1. Mueller Mechanical Integrating Accelerometer	120
11-2. SPIRE system diagram	122
11-3. Single-axis integrating gyro	124
18-1. Schematic of reconnaissance satellite	191
18-2. Agena spacecraft	193
18-3. Corona J-1 model camera	197

Table

16-1. Polaris Mark II Computer Characteristics	169

Acknowledgments

No scholarly study of this nature can be successfully achieved without the help and support of many people—from friends, colleagues, associates, archivists, and librarians to previous writers and historians who have laid the foundations on which the research for this book is based. That said, I am particularly indebted to a number of individuals who have made significant contributions to this work. A special thanks goes out to James Stark Draper, who was kind enough to provide a long personal interview that revealed many anecdotal aspects of his father's life and career, taking the time to show me through the family home and answering my numerous questions. I am also grateful to Martha Stark Draper Ditmeyer, "Doc's" daughter, now deceased, for her informative interview, as well as to Chip Collins, also deceased, for the same. I would also like to thank Catherine Granchelli for her help in arranging the aforementioned interviews, in setting up other interviews, and providing unfettered access to the historical material in the Charles Stark Draper Library Collection. I am also indebted to Debbie Douglas, director of the MIT Museum, for her well-thought-out comments and suggestions and for her assistance in pinning down some of the more obscure facts of Draper's career and the divestiture of the MIT Instrumentation Laboratory. Candyce Henry did a yeoman's work in her untiring reviews of the manuscript, as did Pelham Boyer in his editing.

Abbreviations

ABM	anti-ballistic missile
ADSEC	Air Defense Systems Engineering Committee
AFCRL	Air Force Cambridge Research Laboratories
AGC	Apollo Guidance Computer
AIRS	Advanced Inertial Reference Sphere
AMC	Air Material Command
APL	Applied Physics Laboratory
ARDC	Air Research and Development Command
BuOrd	Bureau of Ordnance
BuShips	Bureau of Ships
CCD	charged-coupled device
CEP	circular error of probability
CIA	Central Intelligence Agency
CID	Confidential Instrument Development (Laboratory)
CSDL	Charles Stark Draper Laboratory
DASO	Demonstration and Shakedown Operation
DOD	Department of Defense
DSKY	display keyboard (Apollo)
EA	electronics assembly
FBM	Fleet Ballistic Missile (program)
FCS	fire-control system
FLIMBAL	floating inertial measurement ball
FP	floated pendulum
GFCS	Gun Fire Control System
HIG	hermetically sealed integrating gyro
IAP	Improved Accuracy Program

IAS	Institute of Aeronautical Sciences
ICBM	intercontinental ballistic missile
IFOG	interferometric fiber-optic gyro
IL	Instrumentation Laboratory
IMU	inertial measurement unit
IRBM	intermediate-range ballistic missile
IRIG	inertial reference integrating gyro
JPL	Jet Propulsion Laboratory
LVDC	Launch Vehicle Digital Computer
MAST	Marine Stable System
MIRV	multiple independent reentry vehicles
MIT	Massachusetts Institute of Technology
MPMS	Missile Position Measurement System
MRAM	magnetoresistive memory
MRV	multiple reentry vehicle
NAC	November Action Committee
NACA	National Advisory Committee on Aeronautics
NASA	National Aeronautics and Space Administration
NDRC	National Defense Research Committee
nm	nautical mile
NSWC	Naval Surface Warfare Center
PIGA	pendulous integrating gyro accelerometer
PIPA	pulsed integration pendulum accelerometer
R&D	research and development
RCA	Radio Corporation of America
ROM	read-only memory
SAB	Scientific Advisory Board
SACC	Science Action Coordinating Committee
SAGE	Semi-Automatic Ground Environment System
SIBS	Stellar Inertial Bombing System
SINS	Submarine Inertial Navigation System
SLBM	submarine-launched ballistic missile
SPIRE	Space Inertial Reference Equipment
SPO	Special Projects Office
SSBN	nuclear-powered ballistic-missile submarine
SSPO	Strategic Systems Project Office

STG	Space Task Group
ULMS	Undersea Long-Range Missile System
USSR	Soviet Union (Union of Soviet Socialist Republics)
VLSI	very-large-scale integration
WDD	Western Development Division

Author's Note

To his family and close personal friends, Charles Stark Draper was known by his middle name. It was only after he received his doctorate degree in 1938 that he picked up "Doc," which was adopted by his colleagues, students, and professional associates as befitting Draper's character. Both appear frequently in the historical record that has been preserved in the writings and oral histories of those who knew and loved him. Although he was not called Doc until the 1940s, I feel that it represents the essence of the person who is the central theme of this work.

Prologue

Hot Spot of Innovation and Invention

> Our history is punctuated with hot spots of invention—areas where a critical mass of inventive people, networks, institutions, funding, and other resources converge and creativity flourishes.
>
> Dr. Arthur Molella[1]

Charles Stark Draper, often referred to as "the Father of Inertial Navigation," was the moving force behind the development of the floated gyroscope in the United States. He was an engineer, a scientist, and an inventor; an inspiring teacher; and a dynamic leader, responsible for creating the laboratory that brought inertial navigation to fruition for operational use in submarines, aircraft, and space vehicles. These factors alone make him worthy of study. But Draper also created and ran the famous laboratory, now bearing his name, that helped make the Massachusetts Institute of Technology (MIT) one of the nation's leading centers for government research. The story of Draper's life and his accomplishments cannot be separated from that of the Instrumentation Laboratory; they are one and the same. Thus, this biography of Charles Stark "Doc" Draper is also a chronological accounting of the MIT Instrumentation Laboratory and its contributions to the nation.

Draper's personality, drive, and intellectual curiosity were at the heart of the success of the MIT Instrumentation Laboratory. But Draper's success was also due to his association with MIT, which provided resources, funding, and an environment that enabled Draper to achieve greatness. The institute's engine laboratory and the research fellowship that drew him back to MIT to pursue a graduate degree laid the groundwork for his doctoral dissertation and for the development of both the engine indicator and the MIT-Sperry Apparatus for Measuring Vibration. Draper's early endeavors were collaborative efforts that involved his professors and fellow students, including such future luminaries as Harold E. Edgerton and Athelstan F. Spilhaus.

Draper was an astute student of human nature, which is not surprising given that he majored in psychology while attending Stanford University as an undergraduate. He was also a talented teacher who knew how to get the most out of students. He had a gift for attracting the best and the brightest, and he was adept at cultivating key decision makers in government, industry, and academia. Yet none of these traits would have led to greatness without the sponsorship of the MIT administration, which, starting with the tenure of Karl Taylor Compton, created an environment that promoted and facilitated research partnerships with the government.

For those who are interested in naval history, three of Draper's accomplishments stand out: the Mark 14 lead-computing gunsight, the Submarine Inertial Navigation System, and the inertial guidance systems designed and engineered by Draper's laboratory for the Polaris, Poseidon, and Trident ballistic missiles. The Mark 14 was the first of several Draper gunsights and directors that revolutionized antiaircraft gunnery in World War II. Close to 80 percent of all enemy aircraft shot down by the U.S. Navy ships in the Pacific between October 1944 and January 1945 were engaged by Draper-equipped antiaircraft guns.[2] Draper, the research institution bearing Charles Stark Draper's name that evolved from the MIT Instrumentation Laboratory, continues to be the Navy's sole source for Trident's Mark 6 guidance system.

1

A Milestone in Aviation History

The temperature was near freezing and a light rain was falling in the early morning hours of February 8, 1953, when Charles Stark Draper walked across the tarmac toward the Boeing B-29 bomber that was parked on the apron at Hanscom Field in Bedford, Massachusetts. Draper, chairman of the Aeronautical Engineering Department at the Massachusetts Institute of Technology (MIT), in the city of Cambridge across the river from Boston, was an expert in gyroscopic control. Today he was about to prove that an airplane could fly across the country relying solely on an inertial guidance platform for navigation. The B-29 that he was about to board was equipped with a 2,800-pound inertial guidance system called SPIRE (Space Inertial Reference Equipment), which Draper's MIT Instrumentation Laboratory had been perfecting for the past two years.[1] "Doc," as he was known to his close friends, fellow professors, and students, was an avid flyer and an expert in aircraft instrumentation. SPIRE was the culmination of a project that had originated in the late 1930s when Draper and his students began investigating the theoretical aspects of inertial navigation, along with the idea of creating a "closed box" navigator independent of external references.[2]

To construct SPIRE, the MIT Instrumentation Laboratory designed an inertial platform using three single-degree-of-freedom gyros that Draper's dedicated team of research engineers had developed for improved accuracy. The inertial platform was combined with a tracking unit containing two state-of-the-art accelerometers. These components were mounted on a series of instrument-supporting members on servo-driven gimbals.[3] In addition to the inertial platform, which was the heart of SPIRE, the system included a large electronic console, a navigation panel for position readouts, and a pilot's indicator showing which way the plane was heading. An analog computer was employed to convert from the inertial platform's polar coordinate system to the Great Circle coordinate system needed for navigation. Except for the pilot's indicator, mounted in the cockpit, all the components were in the bomb bay.

Digital computers were still in their infancy, microelectronics lay in the future, and gyros had yet to be miniaturized—factors that accounted for SPIRE's large size and weight. It must be remembered, however, that SPIRE was designed as a demonstrator to prove the feasibility of inertial navigation. It was put together "with little concern for small size or neatness," because Draper wanted to obtain practical results as soon as possible.[4]

SPIRE was an Air Force–sponsored research project initiated in September 1950 as a follow-on to an earlier program code-named FEBE. It was loaded on January 23, 1953, into a B-29 on loan from the U.S. Air

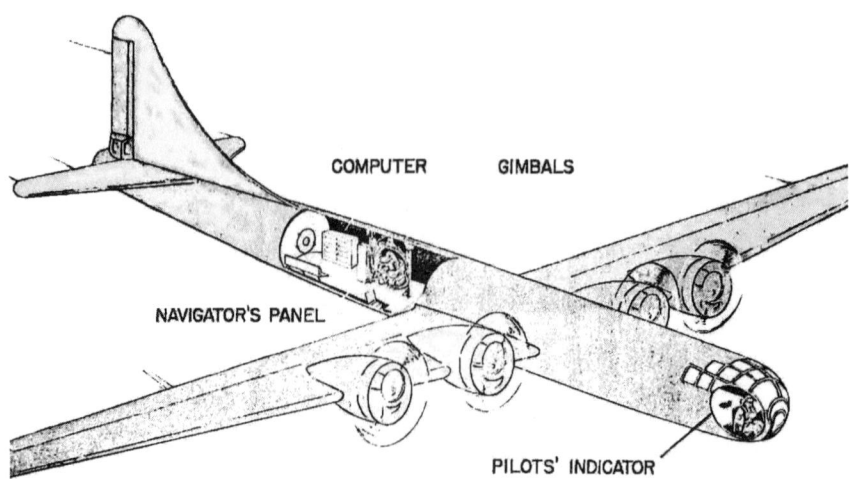

Fig. 1-1. SPIRE components located within the B-29 test vehicle. *CSDL Collection, MIT Museum*

Force and was given a one-hour shakedown flight on Friday, February 6.[5] Draper was so confident in its success that he secretly planned to demonstrate the system en route to a top-secret government-sponsored symposium on inertial navigation that was scheduled to begin the following Monday in Los Angeles. On Saturday, the day after SPIRE's shakedown, he told "Chip" Collins, chief pilot for the Instrumentation Lab's Flight Test Facility, to ready the B-29 for a coast-to-coast flight the next day.[6]

When Draper's B-29 took off at eight minutes past 8 o'clock, on February 8, Collins was at the controls, with copilot Dave Buxton and their flight engineers, Irv Levin and Charles Cameron, in the cockpit. In the bomb bay where SPIRE was situated were the other members of Draper's team: Roger Woodbury, John Hurse, and Joe Aronson, two Air Force observers, and a representative from the General Electric Company.

Once in the air, Collins headed west, brought the aircraft to a heading of about 270 degrees (which placed it on the Great Circle route to Los Angeles), climbed to their cruising altitude of ten thousand feet, and engaged the autopilot.[7] From then on, the flight was controlled solely by SPIRE, which had been programmed to keep the B-29 on course via signals to the autopilot. To avoid any aircraft that might interfere with the flight and force them to disengage the autopilot, the B-29's flight plan included a traveling "block" of airspace two miles ahead that would keep the airspace clear.

The flight was uneventful, and the navigation system seemed to be working fine, until they reached the Rocky Mountains just south of Denver and north of Colorado Springs. To clear the mountains, they climbed to 20,000 feet. The weather had been perfect until they reached the mountains, but there they ran into dense cloud cover. An hour or two from Denver they encountered unexpected air turbulence, and Collins suddenly noticed that the B-29 was turning ten to twelve degrees to the right.

"Collins," Draper exclaimed over the intercom, "what the hell's going on up there?"[8]

"Doc," answered Collins, "the system is commanding a turn to the right."

Draper and the crew monitoring the system knew that something was awry, because they could see the gimbal turning with respect to the aircraft. But they could not see what was happening to the rudder. There was a note of panic over the intercom, but Draper remained calm.

"Let's not do anything. Let's leave it alone. Let's see what it's going to do."

Unbeknownst to those on board the B-29, they had encountered a weather front and were being blown southward—that is, to the left. But SPIRE, sensing the wind drift, had adjusted the rudder so that the aircraft stayed on track. When they broke out of the cloud cover on the other side of the Sierras, they were right on course.

The aiming point for the flight was the intersection of the runways at the Los Angeles International Airport. An indicator light had been installed in the left side of the cockpit to show when they were over the aiming point. When it came on, Collins—according to his account, recorded years later—looked down and saw that they were passing over the apron area just 1,800 feet from the aiming point. Had this been true, the system's error over the twelve-hour, 2,600-mile flight would have been phenomenal, at one one-hundredth of a percent (0.00013). Over the years, Collins' enthusiasm for this accomplishment clouded his memory, for the actual error, based on the data recorded during the flight, was nine nautical miles.[9] Nevertheless, Draper was ecstatic with SPIRE's results, for they demonstrated beyond a shadow of a doubt that a completely self-contained system using sensors that relied solely on inertial principles could be successfully used to navigate over long distances.

As they had flown across the country, Draper, assisted by Roger Woodbury, had plotted the B-29's progress on a long roll of paper that showed the intended course as well as the actual course flown. The results were verified using photographs of prominent landmarks taken through the nose of the B-29.[10] After landing in Los Angeles, Draper and the rest of the SPIRE team stayed up all night putting together Lambert conformal maps showing the results achieved by the SPIRE system. The next morning they mounted the maps on a big board behind the seminar podium, using a brightly colored tape to show the exact track the B-29 had followed across the continent.

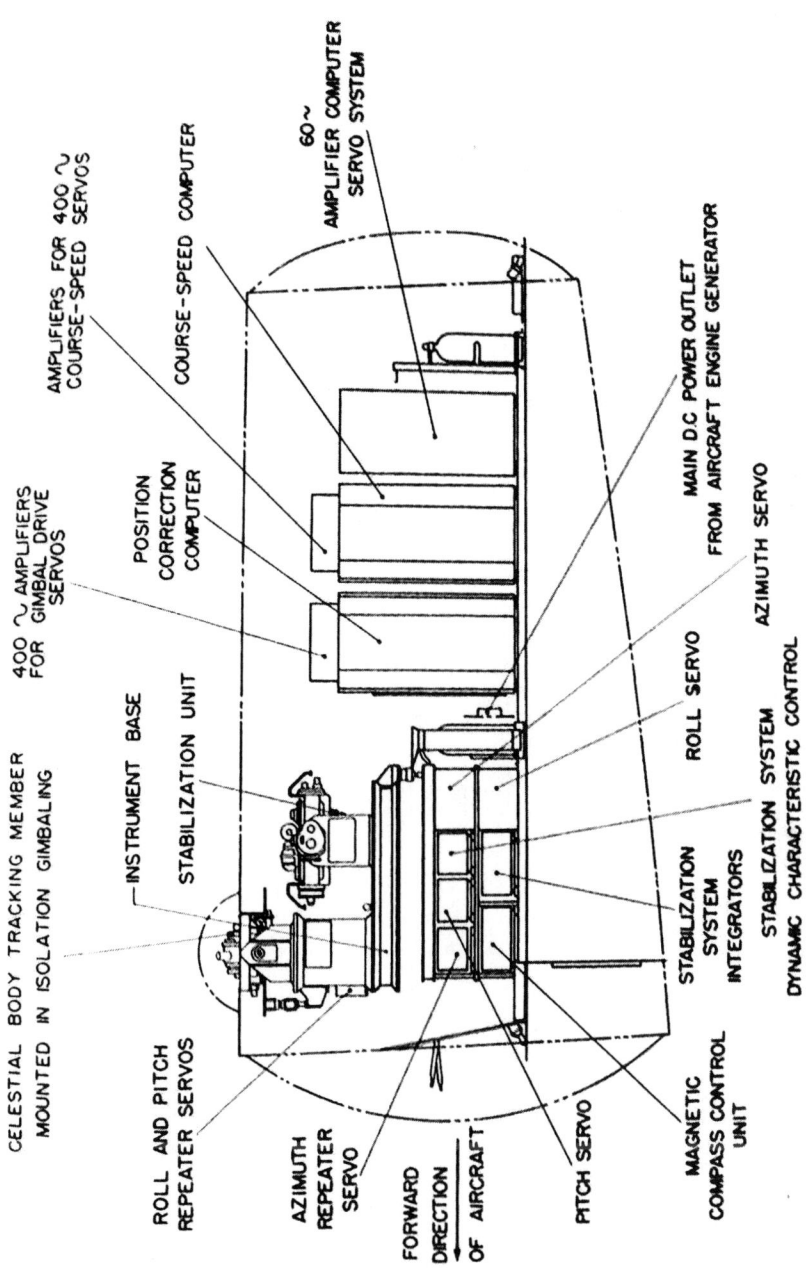

Fig. 1-2. Elevation view: SPIRE system installed in the aft pressurized compartment of a B-29. *CSDL Collection, MIT Museum*

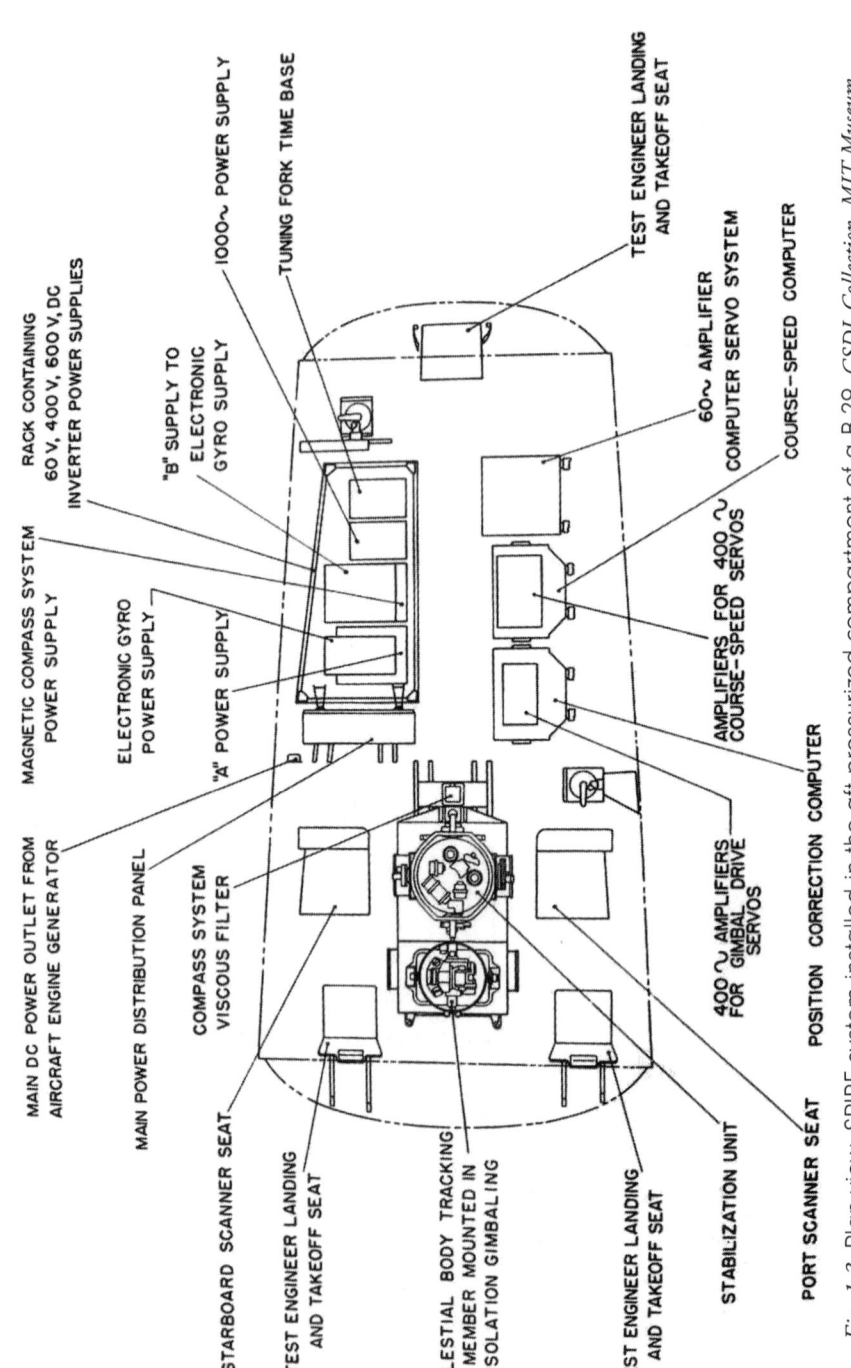

Fig. 1-3. Plan view: SPIRE system installed in the aft pressurized compartment of a B-29. *CSDL Collection, MIT Museum*

When the symposium began the next day, Draper was introduced as the first speaker. "Gentlemen," he began, "we have a system that works. We did it."[11] Then, to the astonishment of the attendees, he went on to describe the historic flight he had just made, "giving credibility to the enormous potential of inertial guidance."[12] It was a personal triumph for Draper as well as a technological achievement, one that showed his talent for garnering publicity—a talent that he successfully used throughout his career.

SPIRE was the forerunner of all modern navigation systems that depend on inertial guidance. It established the Instrumentation Laboratory as the leader in inertial navigation and guidance. SPIRE's success laid the foundation for the laboratory's future development of the guidance systems for the Thor, Polar, Titan, Poseidon, and Trident ballistic missiles.

2

The Formative Years

Charles Stark Draper was born on October 2, 1901, in Windsor, Missouri, a modest farming community typical of many throughout the country at the turn of the century.[1] Windsor, which is located twenty miles northwest of Clinton, is the county seat of Henry County. At the time of Draper's birth it had two schools, four churches, two newspapers, two banks, a flour mill, a feed mill, and a fruit evaporator.[2] His father, Dr. Charles A. Draper, was a dentist, his mother, Martha "Data" Stark Draper, a schoolteacher. She was the direct descendant of Gen. John Stark of Revolutionary War fame and a founding member of the Missouri chapter of the Daughters of the American Revolution. Martha was extremely proud of her heritage and made sure that this pride was passed on to her two sons.[3] Stories of General Stark's heroism at the battles of Bunker Hill and Bennington were undoubtedly told to Draper and his younger brother Clayton,* along with the general's often-quoted exhortation at Bunker Hill, "Live free or die!"[4] These stories, together with those extolling the exploits of his grandfathers, both of whom served in the Civil War, instilled in Draper a lifelong sense of patriotism and reverence for the military. As he recalled years later, "Grandpa Stark was a union Army captain and Grandpa Draper went the other way."[5] Needless to say Draper's youth was full of talk from both sides of the war.

---

* Clayton's first name was Ralph, but like his brother he went by his middle name.

While Draper was growing up in rural America, towns like Windsor, with a population of around 1,500 at the turn of the century, were rapidly changing as new technologies, such as the telephone, electric power, and the automobile, were introduced. The telephone came to Windsor in 1899, when the Missouri Union Telephone Company was formed. By the time Stark—as Draper preferred to be called—was ten, the town had become electrified. There is no record of when the first automobile appeared in Windsor—by 1916 there were 1.3 million cars on America's roads—but we know that Draper began working on automobiles when he was eleven or twelve years old. As he told Barton Hacker during an oral history interview in 1976, "Automobiles were just beginning to come in and the older people like my father and uncles, who had been associated with horses and mules all their lives, didn't feel at home with automobiles."[6] So at an early age he got the opportunity to drive automobiles and to work in garages, repairing cars when they needed fixing.

When the town's able-bodied men of military age went to war in 1917, Draper, who was only sixteen at the time, took on the responsibility of becoming the town plumber's assistant. He fixed faucets and toilets and installed plumbing fixtures. Draper also became the town electrician, stringing wires from poles and wiring houses. He shoveled coal in the power plant's boiler and did those things in the plant that an electrician would do when there wasn't anybody else to do it. He got three dollars a day and worked from seven o'clock in the morning to six o'clock in the afternoon, with thirty minutes off for lunch.

Draper never took any courses that taught him how to do these things. He just picked up what he could from local tradesmen and mechanics. What he couldn't learn from them, he had to figure out for himself. As Draper explained to Hacker, "I was working under situations where the customers got real unhappy if things didn't come out well."[7] This taught him the value of dealing with real-world situations and the need to come up with effective solutions to problems encountered. With very little help—most of the time there was nobody to show him what to do—Draper had to be extremely self-sufficient. This was an important personality trait, and he exhibited it throughout the rest of his life. Draper evidently cherished this type of activity, for in later years he frequently referred to himself as just a "greasy-thumb mechanic."[8] The work also laid the foundation for his future interest in the sciences.

In the fall of 1917, the Draper family relocated to Columbia, Missouri. Perhaps the family moved there so that Stark could attend the University of Missouri, in Columbia; Draper's schoolteacher mother felt strongly about the need for a good education. World War I was still going on when Draper enrolled in the university's School of Arts and Sciences. Feeling the need to follow in the footsteps of his patriotic ancestors, he duly joined the student Army Training Corps, was given a rifle and spent a couple of hours a day marching and drilling. This was his first taste of Army life. Draper never said very much about his first experience with the military, but it would not be his last.

Two years later, in 1919, the family moved again, this time to Palo Alto, California, and with the move came a major change in lifestyle. Draper's father was no longer earning a living as a dentist in a small Missouri farm town; he had become an independent businessman. His first venture was the Windsor Flour Mills, which he acquired with R. F. Williams in 1910.[9] Somewhere along the way he traded the old mill for a quarter section of farmland near El Dorado, Kansas.[10] This fortuitous transaction would have an immense impact on the Draper family's fortunes, for in it a huge oil field had been discovered in October 1915. The El Dorado field, which encompassed thirty-four square miles in Butler County, soon became the largest producing oil field in the United States.

Dr. Draper struck it rich in mid-December 1916, when the first of many wells that would be drilled on his property found oil at a depth of 2,482 feet.[11] Within weeks, at least eleven oil wells had been "spudded in" on the Draper farm, each of which was expected to yield between six and eight hundred barrels a day. According to the local press, Dr. Draper was entitled to receive one-eighth of the oil and one cent per thousand cubic feet of gas found on his land.[12] With crude oil selling for about a dollar per barrel, each well would be earning the Draper family about a hundred dollars a day. This was a considerable sum of money in 1916, equivalent in value today (2019) to a bit over $2,300 per day per well. The papers reported that sixty-four wells were planned for the property. As one put it, Dr. Draper "came floating westward on the tide of oil prosperity."[13]

After moving to Palo Alto with his family, Draper transferred to Stanford University. He decided to major in psychology, because he was curious about things of the mind.[14] He joined a fraternity, played tennis, and became the staff photographer for the *Stanford Pictorial*. He also drove a Stutz Bearcat, an expensive, high-powered sports car. As the writer Stephen Birmingham notes, it "was the glamorous symbol of the Flapper Era, when in its snappy bucket seats, fair-haired youths in raccoon coats with Prohibition flasks in their pockets tore across the landscape and through the pages of Scot [F. Scott] Fitzgerald novels."[15] Draper's son James claims that his father always did things with "pizzazz and dash."[16] The Stutz provides an intriguing suggestion of Draper's lifestyle while attending Stanford University, as does the bootlegging story that follows.

On January 20, 1919, during Draper's first year at Stanford, the Volstead Act outlawing the sale of alcohol went into effect, ushering in the era of Prohibition. Beer drinking had been a regular form of celebrating and socializing by male students on many campuses before the act passed. Now, alcohol flowed more freely than ever, and a majority of students—both male and female—"developed an appetite for synthetic gin, raw hooch, or anything" they could get their hands on.[17] "Speakeasies" became the rendezvous for many collegiate "whoopee parties" and fraternity blowouts. Bootleggers turned limousines into mobile bars, parking them outside football stadiums before big games and on Saturday nights did a booming business on Fraternity Row. If the bootlegging story Draper told James many years later is true, then Stark too was an active participant in these shenanigans.

According to the tale, Draper drove the Stutz to Half Moon Bay one day with at least one or two fraternity brothers to pick up a load of bootlegged liquor that was being landed on the beach. But the cops showed up, and he went roaring down the beach in the Bearcat with "a rumble seat full of cases of hooch and the fraternity brothers jumping off the car."[18] Of course, it was his car, and he couldn't abandon it. So he got caught, was tossed out of the fraternity, and was placed on probation by the university. Whether the story is true or not cannot be verified at this late date. Draper was known to be an avid drinker throughout his life, and the account seems plausible based on his affinity for taking direct action when needed to solve a problem. Earlier he had driven the Stutz up the front steps of his fraternity house. His fraternity brothers had been discussing the porch-climbing abilities of the various automobiles then on display at a "gas wagon" exhibit being held at the university, and someone suggested that Draper try it in his roadster.[19] After several unsuccessful attempts, Doc succeeded in making the ascent, to a round of applause from the delighted onlookers.

Draper's mother was bound and determined to turn her son into a doctor, but he was too young to go on to medical school when he graduated from Stanford in June 1922.[20] His birthday was in October, and he had not reached the twenty-one-year minimum age that was then a requirement of most medical schools. That summer he began taking a course that would prepare him for becoming a ship's radio operator, which he was thinking of doing. This plan was sidelined when the opportunity arose to take a cross-country trip with a fraternity brother who was on his way to Harvard University. Doc went along just for the ride, but when they crossed the Harvard Bridge across the Charles River from Boston to Cambridge, he sighted a group of buildings that would become the center of his life for the next five decades. Half a century later Draper described what happened next:

> I saw this collection of white buildings over on the right-hand side. And I just was curious so I went over and asked them what it was. And they said, well, it was a place they called MIT. And I asked them if they had a catalogue, and they said yes. So I registered more for curiosity than anything else. . . . [T]hey let me in on the strength of a bachelor's degree from Stanford [and the $250 tuition to fill an empty slot in enrollment].[21]

There is more to the story of why he entered the Massachusetts Institute of Technology than just his being a little curious. Doc had an inquisitive intellect; he wanted to know how and why things worked. Curiosity about the mind's workings had brought him to psychology; now he wanted to know the science behind the gadgets he had fixed at Stanford. While attending classes there, he had taken a part-time job in the psychology department as Professor Frank Angell's apparatus man. "Most of the people in psychology were female types . . . not interested in gadgetry and he was the one guy" who was.[22] His job was to fix all the contraptions used in the class experiments—something he was well qualified to do, given his experience fixing everything from automobiles to toilets in Windsor. He did not think these contraptions were very good: the quality of the information they provided was poor. Draper's inquisitive mind was so concerned with the principles of nature and how things worked that he wanted to find out why these devices didn't work. MIT, which was an engineering school, seemed to be a good place to find out. Draper decided to major in electrochemical engineering,* for the same reason he had majored in psychology—that is, he was simply curious about it.[23]

Draper registered in the fall of 1922 as a freshman, and the next two years, as he later remarked in his oral history, "were pretty much of a nightmare." He had never had any courses in mathematics, geometry, chemistry, or physics—all prerequisites for the undergraduate work at MIT.[24] He had been admitted with the understanding that he would make up these missing prerequisites during the next summer session. In the meantime, Doc struggled to keep up with his freshman classmates who had been better prepared. He told Barton Hacker that he "was so god damn busy trying to stay ahead of the devil that I never had a chance to even look around and see how close the devil was." Doc's "first summer at MIT was a highly stressed affair," according to his son James.[25] He took chemistry and mechanisms in the first half of the summer term and algebra, solid geometry, and trigonometry in the second half. Things began to get a little better in his third year at MIT, but he still had a problem with grades.

In a letter to his parents dated April 2, 1925, he told them he had taken a vacation for a couple of days, in which that he "saw a lot of shows—ate a lot of expensive food and otherwise threw away thoughts of school."[26] Draper should have been spending his vacation time studying, but he did not. "With every possible thing in my favor to start with," he wrote, "I have managed to get myself in a very low scholastic condition to such an extent that even my best friend among the professors expressed himself as being disgusted with me." Draper overcame this lapse in focus on his schoolwork and managed to graduate on schedule after four years with a bachelor of science degree.

---

* The scientific and technical specialization dealing with problems in the study, development, and application of instruments for measuring.

Not much is known of Draper's extracurricular activities as an undergraduate, but there is evidence that he was an enthusiastic pugilist, if not a very successful one; he claims to have had his nose broken five or six times.[27] It has been reported that the first man to break Draper's nose was his boxing instructor, Napeen Boutilier, a local firefighter who was an outstanding amateur boxer.

While an undergraduate at MIT Draper enrolled in the Reserve Officer Training Corps program. When he graduated on June 8, 1926, he received a commission as a second lieutenant in the U.S. Army Reserve in addition to his diploma. What Draper actually intended to do with his life after graduating remains unknown; there is no record of his thoughts on this subject. However, the fact that he remained in Cambridge after graduating to work on Professor Donald C. Stockbarger's calcium fluoride furnace suggests that he meant to continue his studies at MIT.[28]

A month after Draper's graduation, on July 2, Congress passed the Air Corps Act.[29] The new law interrupted any plans that Draper may have had with regard to graduate school by offering him an opportunity to learn how to fly. He had been interested in flying ever since his first flight in a Curtiss JN-4 as a student at the University of Missouri.[30] The trip had lasted an hour and was the beginning of his curiosity about and interest in becoming a pilot.[31] After that, Draper flew whenever he got the chance.

In addition to creating the Air Corps, the act provided for a five-year expansion program that would increase the number of commissioned officers and aviation cadets. To do so, the Air Corps Training Center at Brooks Field, near San Antonio, Texas, had to expand its output of flyers. This in turn required an increase in the number of students accepted for flight training. That, finally, opened the door for Draper, who was not in the regular Army and would not otherwise have been considered for such training.

On August 12, 1926, orders were cut directing him to proceed to Brooks Field to attend primary flying school, reporting on September 11.[32] When Draper arrived at the training center he would have been subjected to a physical examination and a personality study to determine his suitability for flying.[33] In one of the well-publicized physiological tests given all prospective pilots in that era, he was placed in a chair and spun around until he was dizzy. The test was designed to observe his eye movements, reaction time, and coordination.

Draper undoubtedly also had to undergo the rigors of the Ruggles Orientator. This primitive simulator tested a potential airman's aptitude for flying. It consisted of a small cockpit-like enclosure suspended within a gimbal ring assembly that allowed it to rotate in three axes. The motion in each direction was regulated by small electric motors that could be controlled by both the student riding in the cockpit and the instructor outside. The instructor turned the cockpit on its axes, and as he did so the student tried to keep the simulator horizontal or level. If the instructor's actions were not promptly countered, the student would quickly be spinning upside down, over and over, or around and around in circles.[34] All flight school candidates were put through this test, which was considered to an effective tool in determining a student's inherent flying abilities.[35]

Draper never said much about his experience at Brooks Field, except that he was taught to fly in San Antonio by the Army. He never mentioned the Ruggles Orientator, but Robert Duffy's memoir claims that Doc's sessions in it revealed a tendency toward airsickness.[36] Perhaps Doc was embarrassed to admit this;[37] he just told people that he had done poorly. This, in effect, was true: he "was [found] professionally disqualified for further training" by the school's Faculty Board on November 8, 1926.[38]

Draper asked to remain in the Air Corps Reserve. His request was granted, and he was placed in Brooks Field's Engineering Division. He was later reassigned to the Power Plant Branch of the experimental Engineering Section at Wilbur Wright Field, near Riverside, Ohio. Nevertheless, "washing out" was a huge setback for him. It was a "shattering experience" that had a marked impact upon his temperament, for it gave him "a viewpoint of sympathy and understanding for students in scholastic trouble."[39] It engendered a strong effort on his part to be a good teacher and "above all to accept the responsibility for effectively transferring information to those who find difficulty in learning."

After flunking out of the Army's flight training program, Draper, disappointed at his failure, joined up with another washed-out student and drove back to the Bay Area in a four-cylinder Dodge.[40] He rested at home for a while before going to New York City to work for a research and development laboratory recently established by Reginald E. Gillmor.[41]

Gillmor, a former naval officer with a degree in electrical engineering, was well connected with the Navy and in touch with its technological needs. As an ensign, he had been responsible for the installation and operation of the first Sperry gyrocompass produced for the Navy, on the battleship *Delaware* in 1911.[42] Gillmor resigned from the Navy in the fall of 1912 to join the Sperry Gyroscope Company, which Elmer Sperry had founded in 1910, and was sent to England to run its London branch. He returned to active duty during World War I, serving as flag secretary to Adm. William Sims, commander of U.S. Naval Forces in European Waters.[43] In the fall of 1918, after war ended, Gillmor returned to the United States and rejoined the Sperry company, this time as its vice president and general manager.

After leaving Sperry again in 1926, possibly for health reasons, Gillmor managed to secure from the Navy a small research contract to demonstrate the feasibility of using infrared radiation for communication.[44] Since infrared light cannot be detected by the human eye, it potentially represented a secure method of communication between ships and shore installations. Gillmor hired Draper to work on this contract on the basis of a recommendation from Peter Hume, the son of a naval officer who had served with Gillmor.[45] Hume had been a classmate of Draper at MIT and had taken the same electrical engineering course; Draper had helped Hume get through the laboratory part of the course, for which Hume felt indebted: Hume apparently told Gillmor that Draper "was a pretty good guy." It helped that Draper had an undergraduate degree in electrochemical engineering and had worked on radiation measurements with Donald C. Stockdale, an instructor in MIT's physics department.[46]

To fulfill the contract, Gillmor set Draper up in a sparsely equipped, one-room laboratory at 83 Fourth Avenue in Manhattan.[47] After several months of hard work, Draper assembled a primitive signaling device using a locomotive headlight reflector and a Thalofide infrared-sensing photoelectric cell invented by Theodore Case. The apparatus responded to infrared radiation by sounding a buzzer.

The equipment worked fairly well when tested on the Potomac River, proving the validity of the concept and suggesting the likelihood that infrared receivers could be developed into practical instruments. But as Doc later reported, the Navy "saw no foreseeable applications for infrared" and did not renew Gillmor's contract. This put Draper out of a job, but the time he had spent working on Gillmor's infrared contract with the Navy had more value than the outcome would suggest: it had created a personal relationship with Gillmor that would pay dividends in years to come when Gillmor returned to the Sperry Gyroscope Company as its president. It had also provided Draper with the opportunity to conduct independent research and introduced him to the nuances and pitfalls of Navy contracts. The lessons he learned were undoubtedly helpful in securing support for the many research projects that were to follow.

In the meantime, another opportunity had presented itself. In the first week of May 1927, Draper received a letter from Professor Henry M. Goodwin, head of the electrochemical engineering program of MIT's physics department, asking if he would be interested in returning to Cambridge as a graduate student on an automotive engineering fellowship with a stipend of a thousand dollars.[48] Draper accepted the offer and returned to MIT in the fall intent on studying the spectrum of engine-fuel flames, as Goodwin had suggested in his letter. Professor Goodwin had, according to Draper's son James, played an important part in Draper's whole career at MIT, and he was to be instrumental in urging Draper to go for his doctorate.[49]

"Detonation"—or engine knock, as it is generally known—was then a very troublesome issue for the automotive industry. To investigate this phenomenon, Alfred P. Sloan Jr. and Henry M. Crane of the General Motors Corporation were willing to provide MIT with funds to do research on the combustion process within the cylinders of an internal combustion engine. The faculty was looking for somebody with a physics or chemical background to work on this project. Draper, argued Professor Goodwin, was the ideal candidate.

The Crane Automotive Engineering Fellowship Draper held was funded by a three-thousand-dollar gift to MIT made by Henry Crane, who was a well-known automotive engineer and had graduated from MIT in 1896 with degrees in mechanical and electrical engineering.[50] He had been hired by a classmate, Alfred Sloan, now president of General Motors, to be his special assistant, advising on engineering matters.[51]

The individuals involved, the external source of funding made available to MIT, and the type or research for which Draper's fellowship was awarded are indicative of how small was the community of industrial researchers in the United States at this time. It also illustrates the closeness of the MIT community and the strength of the university's connections with leaders of industry. Edward Roberts, MIT's David Sarnoff Professor of Management Technology, characterizes these relationships "as the essence of an institution devoted to binding mind and hand, *'mens et manus.'*"[52]

Although Draper would start his graduate work studying the combustion process of the internal combustion engine, the opportunity to work on the cutting edge of aircraft instrumentation and the beginnings of a working relationship with the Sperry Gyroscope Company were just around the corner.

3

Back to MIT

Draper returned to MIT in the fall of 1927. He intended to study the spectroscopy of fuel flames in combustion engines, in accordance with Goodwin's fellowship offer, while completing the requirements for a graduate degree. For Doc it was the opportunity of a lifetime. He could pursue his interest in physics while working in Charles Fayette "Fay" Taylor's Aeronautical Engine Laboratory, run by Taylor's brother. Edward S. Taylor had joined MIT a year earlier as an associate professor of aeronautical engineering after working for the Wright Aeronautical Corporation, where he had been in charge of engine design and development.[1] While at Wright he had been instrumental in developing the company's revolutionary Whirlwind radial engine, which provided a greater power-to-weight ratio than conventional in-line engines.

C. Fayette Taylor had been enticed to MIT by the course in aeronautical engineering that had been instituted in the fall of 1926 (and perhaps by the promise of leading it). As the writers of *Aerospace Engineering during the First Century of Flight* note, "Taylor had a long and distinguished career at the institute and is best known as a pioneer in the development of the internal combustion engine.... His research and teaching at MIT formed the basis of scientific design and operation still in use today. It established MIT as an internationally renowned center in this field."[2] On receiving his appointment to MIT's faculty, Professor Taylor invited his brother Edward to join him on the aeronautical engineering staff.

"Together they built a teaching and research program on airplane and other types of internal combustion engines."[3]

Draper's attempt to investigate the fuel flame combustion process immediately ran into trouble when the physics department refused to move one of its spectrographs to the engine lab. The mechanical engineering department was not about to move an engine to the spectrography lab, either.[4] Without a spectrograph to work with, Draper could not study the fuel flame as originally intended. While studying everything he could about the theory of the internal combustion engine in preparation for his spectroscopy studies, he had become intrigued with the pinging noise made when an engine "knocked." This phenomenon occurs when the unburned "end gas," under increasing pressure and heat (from the normal progressive burning process and hot combustion-chamber metals) spontaneously combusts, producing a sharp pressure spike.[5] To investigate this phenomenon further, he began to make small microphones to pick up the noise.[6] Instead of investigating the flame process within an internal combustion engine, Draper decided to focus his research on the knocking process. He began working on a pressure-sensing diaphragm connected to a carbon-granule microphone element that could be used to collect pressure data within the cylinder of an internal combustion engine. Draper built the sensor in the engine lab, then housed in a little building "about twice the size of a [railroad] boxcar" between Vassar Street and the railroad tracks.[7] He intended to use the project as the basis for his master's thesis required for the master of science degree, which he expected to receive in June.

That spring, Draper was approached by C. Fayette Taylor. The professor wanted Doc to stay at MIT so that he could continue to study engine knock. He offered Doc an appointment as a salaried research assistant in the Aero Engineering Laboratory; that would allow Doc to continue his research on detonation. As a member of MIT's staff, Doc would not have to teach any courses or grade papers, and he would be able take courses without paying tuition. This would enable Draper to pursue the doctorate degree that both Taylor and Professor Goodwin were recommending. Doc accepted Taylor's enticing offer and graduated on June 5, 1928 (his thesis was titled "A Method for Detecting Detonation Waves in the Internal Combustion Engine"). His master's degree in hand, he traveled to Palo Alto for a family visit and returned to Cambridge after the summer break.

When the fall semester started Draper enrolled as a doctoral candidate in physics, with a minor in math. He continued to work on the detonation problem, improving the pressure sensor and adding an oscillograph to analyze its output.[8] Doc's research into engine knock was now being conducted in the newly completed Aeronautical Engine Laboratory, run by Edward Taylor. This state-of-the-art facility had been constructed through the generosity of Alfred P. Sloan Jr., who had donated $65,000 ($950,000 today) for it.[9] The one-story building was located just east of the Guggenheim Aeronautic Laboratory, another new building (completed in 1928), financed by a gift from the Daniel Guggenheim Fund for the Promotion of Aeronautics.

Draper continued to work on his engine indicator while taking courses in the physics department, which he considered a good environment. He was very involved with the department's faculty and thought that it was "a very pleasant kind of outfit."[10] As a student, Draper was more concerned with trying to understand the principles contained within the textbooks used in his course work than in working out the academic problems placed at the end of most chapters. Being able "to deal with real problems of the real world" was already becoming the guiding principle that fueled his quest for knowledge.

Draper's interest in engines and the close working relationship that he established with Edward Taylor paid off handsomely the following year, when he was appointed a research assistant in aeronautical engineering for nine months, with a salary of $1,900 plus tuition. While the former paid his living expenses, the latter allowed him to continue working on his doctorate in physics without having to ask his father for money. After the appointment was confirmed, Draper was assigned to the engine laboratory, where he became Taylor's assistant. He ran engines, took data, corrected reports, and did additional research on detonation.

Work as a research assistant in the Aeronautical Engineering Laboratory was a plum assignment for Draper, who maintained an avid interest in flying despite failing to qualify as an Army pilot. He now enrolled in the Curtiss Wright Flying School in Boston to perfect his flying skills.[11] He took his first training flight in a Curtiss Fledgling on October 12, 1927, in a commercial pilot's course. By December he was doing rolls, spins, and loops. He graduated shortly thereafter, although there is no evidence that he applied for a commercial pilot's license.

In September 1929, Draper purchased a Curtiss Robin in partnership (later he would take over sole ownership) with Assistant Professor Daniel C. Sayre, who taught aeronautical engineering in the mechanical engineering department.[12] They paid only $750 for the plane, according to Draper, much less than the reported selling price of four thousand dollars.[13]

The Robin was a high-wing monoplane with seats for a pilot and two passengers. It was powered by a ninety-horsepower OX-5 engine. According to his wife Ivy, Draper sold some shares of American Telephone and Telegraph to pay for his share of the plane.[14] A banker advised against selling the stock, which was then selling for more than three hundred dollars a share, but Doc went ahead anyway. A month later the stock market crashed and AT&T "tanked" along with the rest of the market, reaching a low of around $165 a share.

Draper could have taken his doctoral exams in the spring of 1930 if he had wished and would likely have passed. But he was in no hurry to get his doctorate. He had already secured a full-time position as a research associate starting in the fall semester, and there seemed no need to rush ahead. Instead of spending the summer studying for his exams, Draper headed off to Europe secure in the knowledge that when he returned to Cambridge he would have a paying position within the institute that would allow him to continue to pursue his degree. He figured that the three months he would spend in Europe would not affect his path to a doctorate. When he returned to Cambridge, however, the physics department, in Draper's words, "had been annihilated."[15]

Since its founding in 1861, MIT had been a technology-driven college. When Draper entered the institute in 1922, its main function was to serve as a training ground for industrial workers, not to be a first-rate university dedicated to advanced research in science and technology. By the time Doc was ready to take the doctoral exam, the institute's physics department, according to MIT historian Philip Alexander, "suffered from narrowness and tunnel vision, its prime goal being to drill engineering students in professionally relevant principles."[16] To rectify this situation, Karl T. Compton, MIT's new president, called on theoretical physicist John C. Slater to take over the department, replacing Charles Norton as chairman.[17]

A year earlier, MIT's Board of Trustees had approached Compton, then head of the physics department of Princeton University, and offered him the institute's presidency, to take up the task of revitalizing MIT.[18] Compton was brought in as a reformer, a man who believed that "the fundamental sciences must be made the backbone of the Institute."[19] After becoming president, he transformed MIT's administrative and academic structure, strengthened the scientific curriculum, and developed a new approach to education in science and engineering. Compton was convinced that the future of MIT depended upon strong science departments, especially physics, which he wanted to be the best in the country.[20] The institute's goal, he explained in his report for the academic year 1930–31, "was to offer a 'technological education,'" by which he meant an "education in the fundamental principles," along with "a training in the application to important basic processes and problems," rather than a mere "technical education."[21]

The new physics chairman, Slater, as noted in the department's online history, "was already an accomplished theorist and a participant in the development of what would become known as quantum mechanics."[22] He shared Compton's belief in the importance of basic science and had the latter's ear. Together they revamped the physics curriculum and brought in a coterie of up-and-coming physicists. The advanced courses were all made electives, and a list of subjects was prepared that covered the important developments in experimental and theoretical physics.[23]

This was bad news for Draper, whose previous doctoral work, in applied physics, became irrelevant under the revised curriculum. He would have to start all over again. Professor Philip M. Morse, one of the new physicists hired by Slater,* told Draper that "his years of endeavor of passing courses in physics had produced nothing that would be of any use in getting a doctor's degree under the new regime."[24] His recommendation was to forget the whole thing. Draper was not dissuaded; he kept his position as a research associate, which allowed him to take courses at little or no cost. In Draper's mind, he was still being paid to work toward his doctorate, and he continued with his research on the internal combustion engine.

---

* Morse joined the faculty in the fall of 1931.

In the spring of 1931, Draper was asked to teach a course on aircraft instruments, filling a vacancy created when Professor William G. Brown left MIT in 1929.[25] The Taylor brothers were well aware that Draper liked gadgets, owned an airplane, and was making his own aircraft instruments. He was the best candidate for the job, and C. Fayette Taylor, acting head of Aeronautical Engineering Section within the Department of Mechanical Engineering, was in the ideal position to recommend him. It was a great opportunity for Doc; he began teaching the course—his first—in the fall semester.

For teaching aids he was given a carton of World War I instruments and several boxes of slides.[26] He also set up a small laboratory to study instrumentation in a little corner room in the engine laboratory. That, as Doc recalled in 1976, was how he got into the instrument business: "Anybody who wanted to do anything could get lab space. The one thing that there wasn't any of, and you didn't waste any time trying to get any of it, was money. So all the stuff I made I got out of some automobile junkyard. I had a whole instrumentation laboratory, which was made up completely and entirely out of stuff that cost nothing. Either stuff that I'd chiseled or got free some other way. I even bought the machine tools that we used to make the parts."[27] Draper soon had a pair of students helping him. They worked for free, doing thesis work related to aircraft instruments. In the meantime, Doc continued to devote his principal research efforts toward the development of engine indicators and the study of the combustion process.[28]

Draper's first paper on the subject, coauthored with Edward Taylor (who was the lead author), was published in March 1933. The article, which discussed a new high-speed engine indicator, appeared in the March issue of *Mechanical Engineering*.[29] The apparatus it described was the culmination of the work that Doc had begun in 1928 for his master's thesis. In the interim, Draper, with the help of the MIT academic community, had made significant improvements. A number of professors and fellow students provided advice and support. Professor Arthur C. Hardy, an expert on electro-optics who had just perfected the first recording spectrophotometer, suggested that he reduce the size of the apparatus and modify the oscillograph to use motion-picture film, which could then be projected and enlarged or traced with a pencil.[30] Professor Louis H. Young helped with the vibration recorder.[31] Harold E. Edgerton, who would later win fame as an inventor and entrepreneur but was then a fellow graduate student, helped with the oscillograph.[32] C. Fayette Taylor provided advice on inserting the pressure sensor, and Edward Taylor, gave valuable aid "in the form of practical suggestions."[33] Supported by the Taylor brothers and all the others, Doc designed and built an "engine indicator," based on his improved pressure sensor, a vacuum-tube amplifier, and a special oscillograph. The latter was capable of resolving frequencies of over 10,000 cycles per second and a number of successive engine cycles. The oscilloscope had been developed in conjunction with fellow doctoral candidate, David G. C. Luck.[34]

The oscillograph in itself was significant enough to warrant a paper of its own, in the August issue of the *Review of Scientific Instruments*.[35] Because conventional oscillographs used large amounts of expensive photographic film, Draper and his fellow researcher redesigned the optics so that one second of activity was captured on twenty-five feet of film instead of the 160 feet that would have been needed in the conventional instrument. As Michael Dennis points out in his doctoral thesis, "Not only did this save money, but Draper's machine used conventional 35mm film and standard light bulbs rather than the more expensive motion picture film and projection bulbs."[36]

The engine indicator project conducted under Edward Taylor's tutelage was a milestone in Doc's career. He was on his way to making a name for himself. Taylor's brother considered Draper's achievement important enough to warrant a mention in the aeronautical engineering section of the 1933 *President's Report*.[37] While working on the engine indicator, Draper continued to study for his doctoral exam in physics; he took and passed it in the fall of 1934. "Twelve tried, four flunked and Stark led the rest in both written and oral exams by a large margin," wrote his younger brother in a letter to their parents.[38]

Draper's research on the internal combustion engine went far beyond his original interest in instrumentation. By the end of 1934 he had accumulated enough knowledge on the subject to prepare a special report on it for the National Advisory Committee on Aeronautics (NACA).[39] Draper's work was a tour de force as a study of the physical process that caused knocking in an internal-combustion engine. He based his findings on the experimental data acquired from laboratory apparatus he had created to measure pressure-wave frequencies generated during the combustion cycle. In an impressive demonstration of his mathematical abilities, Draper applied the calculation of resonant pressure-wave frequencies to prove that his empirical results agreed with the mathematical theories of sound.

NACA was a federal agency founded on March 3, 1915, to undertake, promote, and institutionalize aeronautical research. As Michael Dennis rightly points out, C. Fayette Taylor's membership on the NACA power-plant committee "undoubtedly assisted in the publication of Draper's work."[40] An important role might also have been played by Jerome C. Hunsaker, "a one-man clearinghouse for American aeronautical engineering" with strong connections with the Navy, NACA, and industry.[41]

Hunsaker, an international recognized authority on aeronautics, was head of the Department of Mechanical Engineering at MIT, one of the new faculty recruited by Karl Compton in the early part of 1933. Hunsaker had agreed to take over the department with the understanding that he would be given responsibility for instruction in aeronautical engineering.[42] C. Fayette Taylor was not pleased with his resulting removal as head of the aeronautical engineering program, but Vannevar Bush, vice president of MIT and dean of the School of Engineering, soothed his feelings by promising him exclusive control over all of the department's power plant work. Hunsaker's selection as president of the Institute of Aeronautical Sciences (IAS) during its first official meeting, at the Yale Club in New York on October 17, 1932, is indicative of the scope of Hunsaker's renown and influence within the aeronautical community. At the time of its founding the IAS was a cross between a club for gentlemen engineers and a professional society that would enable elite specialists to interact with colleagues in other disciplines.[43] As Bill Trimble writes in Hunsaker's biography: "He [Hunsaker] saw the institute casting a wide net, telling the press at the time that it 'will bring in for the discussion of aeronautical problems, scientific experts in many fields.'"[44]

Before Hunsaker's arrival, C. Fayette Taylor had commented on the difficulty "of securing publication of papers through ordinary channels."[45] Hunsaker's prestige and extensive contacts outside of academia undoubtedly helped to reverse this situation, making it easier for the likes of Draper to get published—he may even have given Draper a nudge or two.

By the beginning of Hunsaker's second year as head of the Department of Mechanical Engineering, Doc and his instrument laboratory had impressed Hunsaker so much that he felt compelled to laud Doc's accomplishments. When the *President's Report* for 1933–34 appeared, in October 1934, Hunsaker's summary of aeronautical engineering included this passage: "The Aeronautical Instrument [sic] Laboratory, under the immediate direction of Mr. Draper, has made notable contributions in the field during the last two years and is attracting wide attention. New types of automotive engine indicators and vibration indicators have been developed. Progress has been made in the rationalization of instrument behavior through the analysis by means of instrument theory and the experimental evaluation of the magnetic compass."[46]

Clearly, Hunsaker recognized Draper's scientific expertise, for in the following year, Doc—still *without* a doctoral degree in hand—was promoted from instructor to assistant professor. This was an extraordinary appointment, given the overwhelming importance of the PhD, and it is indicative of the high esteem in which Hunsaker and the institute's administration held him.

Draper's advancement to assistant professor coincided with the formal inauguration of the Instrumentation Laboratory under his direction in 1935.[47] He described the new laboratory during one of the technical sessions conducted during the Fourth Annual Meeting of the Institute of Aeronautical Sciences, held at the Pupin Physics Laboratory of Columbia University on January 29–31, 1936.[48] It had special equipment, he told the audience, that enabled researchers to study the manner in which instruments responded to changes in an activating force under actual operation conditions.

Instead of the calibration equipment used in conventional laboratories for routine test work, Draper had installed apparatus to study the fundamental problems of instrumentation. These devices allowed Doc and his students to verify the theoretical models of various instruments, which was the focus of the work being done in the Instrumentation Laboratory. Draper sought to define instrument design with a set of differential equations that engineers could use to understand and improve existing measuring devices.

Draper's mathematics relied on four characteristics uniquely defined by the work conducted in the Instrumentation Laboratory:

1. The equations relied on nonnumerical coefficients determined by actual experimentation.
2. The analysis made via the study of the laboratory's instruments made extensive use of dimensionless quantities.
3. The solutions to Draper's equations yielded graphs that could be used to solve a variety of design problems.
4. Draper developed a new form of notation and a lexicon for use with the differential equations developed to analyze how instruments functioned in the real world. Not only did this terminology distinguish Draper's methodology from others', but its adoption would implicitly acknowledge the superiority of the methods of the MIT Instrumentation Lab.[49]

However, gaining that adoption—which meant translating Draper's framework into a vocabulary readily understood by others—proved difficult. Draper tried to gain acceptance within the scientific community of his equations and their unique notation (for details of which see chapter 12, page 129) with three detailed articles and numerous presentations at several professional society meetings. As Michael Dennis notes, Draper's vocabulary, which was "rooted in the [instrumentation] laboratory[,] . . . was at one with the laboratory's pedagogical mission, [but] only those trained within its walls understood what went on within."[50] Robert Duffy deems Draper's self-defining mathematical notation "a noble experiment which never quite captured the hearts and minds of either the educator or the educatees."[51]

In 1935, Draper's chances to obtain the coveted PhD were greatly improved when Professor Slater announced that in the near future the physics department would be offering a doctor of science degree in applied physics.[52] This option had been brought about by the Visiting Committee—composed of distinguished scholars, MIT graduates, and members of the MIT Corporation, operating as an advisory group—which pointed out the industry's need for men trained in applied physics. The committee had convinced Slater "that the present preliminary Doctor's examinations required more knowledge of advanced theoretical and atomic physics than was reasonable to expect of a man specializing in applied physics, and it was consequently decided in the future to offer an alternative preliminary examination in applied physics, of equal difficulty with the examination in pure physics but dealing more with those parts of physics and related subjects which are important in the application of physics."[53]

In 1938, the Department of Physics, which had heretofore declined to accept any dissertation based solely on applied science, finally relented in Draper's case and agreed to award him the PhD in science based on his work on engine knock. By then it was obvious to Doc that the submission of his dissertation would be just a formality. He was already well established as a faculty member and had worked with the members of the examining committee for years. He duly produced one but guessed correctly that no one on the committee would bother to wade through the 529-page tome on "The Physical Processes Accompanying Detonation in the Internal Combustion Engine." Draper, with his fondness for Scotch and his proclivity to a roguish sense of humor, decided to test his scholarly inquisitors. He placed a numbered voucher in the back pages of each volume of his dissertation authorizing the reader one bottle of Scotch whisky.[54] None were cashed in. Professor Morse—who had originally advised Draper to "forget the whole thing"—signed off as professor in charge.

Legend has it that Draper took more courses at MIT than anyone else ever has. In addition to the ninety-five regular courses required for his undergraduate degree, Draper was required to take five more during the summer between his freshman and sophomore year to make up for the prerequisites he was missing. No one knows exactly how many graduate courses Doc took, but he appears to have taken as many as he could. Fortunately, he left his class notes behind, and from them his son has counted seventy. Which of these courses were required for his master's and doctorate degrees has not been determined. It is highly doubtful that anyone else has come anywhere near the 170 courses taken by Doc during his student years at MIT.[55]

4

Aircraft Instruments and the Beginnings of the Instrumentation Lab

As mentioned in earlier pages, Draper's interest in flying paved the way for his appointment as an instructor to teach the course in aircraft instruments, an assignment that helped solidify his staff position as a research associate within MIT's aeronautical engineering program. It also led to an increasing interest in the physical and mechanical limitations affecting the accuracy and usage of instrumentation used for flight. After his appointment, Draper's flying experiences continued to spark his interest in the behavior of various instruments critical to piloting an aircraft and awakened him to the problems of aerial navigation.

By the spring of 1931 Draper had enough time and confidence in flying the Robin to fly it across the country for a visit to family and friends in Palo Alto.[1] "Don't worry," he wrote his mother before he left, "I've got my parachute."[2] Draper was only kidding, but the comment shows both his humorous side and his confidence in his flying abilities.

On June 2, 1931, he took off, heading for Syracuse, New York, the first of two dozen stops. To avoid going over the Rockies, Draper chose to fly the southern route, via El Paso. The flight plan took him from New York across Pennsylvania, Ohio, Illinois, Missouri, Oklahoma, Texas, New Mexico, and Arizona to the California border.[3] From there he flew to Palo Alto. He was accompanied on the journey by Winthrop H. "Win" Towner, an MIT undergraduate studying aeronautical engineering. Towner, a qualified pilot in the Army Reserve who had graduated from the Air Corps advanced flying school in 1929, served as relief pilot.[4] It took seven days for them to reach Palo Alto, using a Rand-McNally map for navigation. Along the way they stopped at airports for fuel, overnight accommodations, or engine repairs. They were forced down twice.[5] The first incident occurred near Sunbury, Ohio, when a water line broke. Then, between Tucson and Phoenix, the Robin's engine started missing, causing them to lose altitude. It took a while to find a safe landing place, but they made it down safely. The problem turned out to be a fouled spark plug caused by graphite that someone had advised them to add to the oil to help lubricate the valve stems. The spark plug was easily replaced, however, and they were soon on their way again.

The OX-5 engine, in Doc's opinion, was not very reliable.[6] Several years later, on a flight to San Francisco, the Robin's radiator fell apart near St. Louis. Draper reached St. Louis safely, but once there he had two choices: he could stay in St. Louis hoping that "some day a new radiator could be found[,] or ride a bus to California."[7] Doc decided to sell the Robin and take the bus to his home in Palo Alto.

From the endless hours he had spent in the Robin, Doc understood the problems associated with the instruments then in use for flight control. He was particularly concerned with the reliability and accuracy of the artificial horizon and the magnetic compass, both of which were essential for aerial navigation. It was not long before he began to study the details of how these mechanisms worked, looking for ways in which they might be improved. He started by investigating the mechanism that powered the artificial horizon, the turn indicator,* and directional gyro; Draper considered these three instruments collectively "an essential part of the equipment of any well-equipped airplane."[8] All three relied on the input of an air-driven gyroscope powered by suction generated by a venturi.

The venturi was a low-cost, mechanically simple, and easily implemented solution to the air-supply problem. The problem with this approach was that the venturi could freeze up in bad weather, just when the instrument was needed the most. The solution, in Draper's view, was to replace the venturi with a positive-displacement pump that would run the gyro under positive air pressure. This approach was more expensive and

---

* Also called the turn-and-bank indicator.

mechanically complicated than the venturi. If Draper expected instrument manufacturers to replace the venturi with the positive-displacement pump, he realized, he would have to demonstrate scientifically its advantages. He would begin by determining investigating the physical characteristics of the venturi method.

Draper could have attacked this problem himself, but he would have had to fit it in between his teaching duties and the ongoing research on engine knock. Instead, Draper recruited Athelstan F. Spilhaus, a twenty-year-old graduate student in aeronautical engineering from South Africa. He framed Spilhaus' assignment in such a way that it could fulfill the thesis requirements for his graduate degree. This approach to problem solving and people became one of the hallmarks of Draper's performance as a laboratory administrator—that is, using "live" projects as it became an instrumental teaching tool that prepared his students for the real-world challenges they would face once they left academia.

Under Draper's guidance, Spilhaus measured the speed of the gyro rotors and estimated the amount of suction needed to drive the instruments.[9] Together they computed the horsepower that it took to turn the gyros, then computed the drag of the venturis in the wind tunnel. They found that it took about eight horsepower to drive the little gyros.

After Spilhaus received his master of science degree in June 1933, Draper arranged for him to spend the summer working for the Sperry Gyroscope Company. The company had recently been divested by General Motors, which had acquired it in 1929 when GM bought North American Aviation Corporation from Clement Keys. The new company, christened the Sperry Corporation, was incorporated in Delaware on April 13, 1933, with Thomas A. Morgan as president.[10] The realigned corporation included the Sperry Gyroscope Company and the Ford Instrument Company. Only 10 percent of its business was in aeronautics at the time, though Morgan, who was "high" on aviation, committed a substantial portion of its research-and-development budget to the aircraft business.

Reginald Gillmor, the ex–naval officer who had been hired by Elmer Sperry in 1913 to run the company's London office, was placed in charge of the Sperry Gyroscope subsidiary and made a vice president of the corporation.[11] What Gillmor did between 1927, when Draper worked at the Electro-Physical Laboratories, and 1933 has not been established. But it is probable that Doc's prior association with the new head of Sperry Gyroscope facilitated Spilhaus' placement for summer employment.

Working as an engineer for Sperry during the summer of 1933 was a new experience for Spilhaus; his only previous employment having been as an apprentice mechanic. Thanks to Doc, however, he "knew probably as much about the mathematics and aerodynamics of gyro instruments as anybody there."[12] His thesis, "Air Suction Methods in driving of Gyroscopic Instruments for Aircraft," made him an ideal candidate for the summer job as a research engineer.

Spilhaus spent an interesting summer in Sperry's Brooklyn plant, according to his oral history. One day he was called into Elmer Sperry Jr.'s office, where Leslie F. Carter, one of Sperry's engineers, was proudly showing off the "beautiful" double venturi that he had devised to power Sperry's aircraft instruments. Preston Bassett, the chief engineer of the Sperry Gyroscope Company, was also present. Draper was not there, but Spilhaus had the presence of mind to tell Elmer "the whole story about this ridiculous waste of horsepower to drive the little gyro instruments. . . . I would use pumps," he said. "Tiny little pumps off the engine."[13] It was pretty dramatic, according to Spilhaus, especially when Sperry bent the double venturi over his knee and told Bassett, "Pres [Preston,] I think we've got to get away from venturis. Let's forget about the darn thing."[14] As far as Sperry was concerned, the venturi was out.

After spending the summer in Brooklyn, Spilhaus returned to MIT to continue toward a PhD in aeronautics. He moved in with Draper and his roommates, sharing an apartment across the street from the Guggenheim Building for aeronautical engineering. "We could walk across in our bedroom slippers and go to work, which we did very often." They chipped in and hired a cook: "He was a fine old black man who came in and cooked our dinner."[15]

Spilhaus thought of Draper as a tremendous scholar, a great influence on his own studies. Another of the roommates dubbed Draper "Droopy Drawers," because he would walk around in the apartment wearing slippers, a shirt, and his underpants. He wore a green eyeshade when studying and, as Spilhaus recollects, was "thoroughly comfortable" in this attire. Also, in Spilhaus' view, Doc "was an aeronautical nut," because he loved to fly whenever he could and ignored the minor accidents that seemed to happen at the end of his flights.[16] At one point his roommates began calling him "Snowdrift" Draper: the meteorological division's airplane, in which he was an observer, turned upside down in a snowdrift. Although Doc was not piloting the aircraft, his roommates could not resist the opportunity of ribbing him.

Draper's aeronautical prowess made a lasting impression on Julius A. Stratton, an assistant professor in the physics department who had taught one of the many physics courses taken by Draper. For any pilot heavily involved in mastering blind flying, the need to prevent stalling followed by spinning was a serious issue. After describing the nature of this problem to Stratton, Draper volunteered to give his former teacher a practical demonstration of this complicated aspect of flying. "Jay," as he was known to his friends and close associates, politely declined. Then, one Sunday, as Stratton later related to Howard W. Johnson, MIT's president, he ran out of excuses. Lacking the simple courage to say no, Stratton took off with Draper in a small, open-cockpit airplane:

> It was a beautiful morning: my confidence ebbed slowly back as we flew serenely over Concord and then headed toward Boston. I sat behind Stark as he lectured with shouts and gestures on the idiosyncrasies of instruments. Then, at a point directly over the harbor, it happened. The nose went sharply up. For a moment we hung on the propeller, as if suspended in the stillness of space. Suddenly the plane pitched forward; I gazed down on the water far below. We began to pick up speed. I became aware of a strange phenomenon slowly, then more rapidly the Boston skyline began to rotate about the plane. The Customs House tower moved in a majestic circle. It occurred to me that I had left a good many things undone and that the Department would be hard put to find someone to teach mechanics. The plane leveled out, rolled over, and settled down on a quiet course. I said nothing. No student but Draper has ever done such a thing to me before or after.[17]

Stratton was duly impressed by the inadequacy of the instrumentation and with Draper's ideas about needed improvements, but he never flew with Draper again.

In mid-May 1933, just before Spilhaus was scheduled to graduate, Draper began to conduct test flights in the meteorological division's airplane to study the performance of various flight instruments.[18] The meteorological station, at the East Boston airport, had been established in November 1931 under the direction of Professor Carl G. Rossby. Daniel Sayre, one of Draper's roommates, piloted the Cessna that was used during the first flights to collect aerological data. Sayre had earned his pilot's license while still a student and had then gone on to organize the Boston Airport Corporation, which established the first passenger service between Boston and Nantucket.[19] The daily flights made by Sayre, who was an assistant professor of aeronautics in the meteorological division, were part of a research program designed to gain new knowledge to aid in weather forecasting, measuring the variation of temperature and moisture at various altitudes in different air currents. To collect this data, the meteorology group within the Department of Aeronautics needed a data recorder that could function under the environmental conditions encountered at 17,000 feet. Pen-and-ink markings were unreliable, both because moisture absorbed by the paper would cause it to expand, distorting the record and because the ink might freeze. To solve this problem, Draper designed and installed a data recorder that used a metal stylus attached to a thin sheet of smoked aluminum on a revolving drum.[20] At the end of each flight the recorder was removed from the aircraft and taken to the meteorological group, where the data was transcribed for later use and study.

Draper began conducting his own experiments during these flights. By then the Cessna had been replaced with a Curtiss Robin fitted with a Challenger engine. The Challenger, which was rated at 185 horsepower, produced more than twice the power of the OX-5 in Draper's plane. The larger engine allowed the craft to reach 17,000 feet with two persons on board: the pilot and an observer, who could concentrate on the instruments recording atmospheric conditions at various altitudes. The craft was equipped with a special glass-enclosed cabin and new equipment, making it a real "flying laboratory."

Draper used the flights to make manometric measurements of the suction venturi used to drive the gyroscopically controlled aircraft instruments that he wanted to improve. To determine the optimum position of the venturi, he placed the device in various locations; after a number of trials, Draper concluded that the upper side of the wing near the outer portion of the slipstream gave the best results.[21]

Draper also wanted to collect experimental data on the horsepower needed to drive the air pumps that Spilhaus had studied in order to compare them with Spilhaus' calculated, theoretical results. He purchased a Deslauriers propeller-driven pump for this purpose and had it installed on the meteorology group's Robin so he could study the pump's operating characteristics and ability to power an air-driven gyro. Draper encountered unexpected mechanical difficulties with the pump and was not able to collect enough data for a meaningful comparison. Despite its problems, however, Draper retained the Deslauriers pump as an emergency backup in case the venturi froze.

In addition to his work on the venturi and other methods of supplying air to air-powered instruments, Draper tested the performance of various flight instruments in extreme flying positions and different altitudes, using a motion-picture camera to record their behavior. He also experimented with radio antennas, to determine which type functioned best and what was the most advantage position for them.

After his summer sojourn in Brooklyn, Spilhaus, having completed all the courses available in aeronautics, took the introductory course in meteorology. He became so intrigued with the subject that he decided to "go upstairs" to the meteorological group, where he became so engrossed in meteorology that he changed his major to it. (Spilhaus went on to become a well-known meteorologist, geophysicist, and inventor. He is remembered for inventing the bathythermograph and launching the weekly science-oriented comic strip *Our New Age,* syndicated in more than a hundred newspapers around the world from 1957 to 1973.)

With Spilhaus out of the picture, Draper recruited Walter McKay and George R. Stuart, undergraduates in the aeronautical engineering program, to undertake another one of his "pet" projects: a study of the physical constants of the magnetic compass, one of the essential flight instruments found on all aircraft. To indicate compass heading this instrument employs a magnetic float attached to a compass card supported by a pivot assembly. To prevent the magnetic float from swinging wildly, the instrument case is filled with a damping fluid that restrains the float. This reduces the effects of aircraft vibration so that the heading may be read more easily through the glass face mounted on one side of the case. An aircraft's magnetic compass is typically a self-contained unit, requiring no external power. It is relatively inexpensive to manufacture and highly reliable. This makes it extremely useful as standby or emergency backup for the gyrocompass, which depends on an uninterrupted external air supply.

The problem with the magnetic compass of Draper's day was that it gave erroneous readings during banked turns and airspeed changes. The error induced during a banked turn was caused by a phenomenon known as "magnetic dip," the tendency for the magnetic float to align itself with the magnetic forces surrounding the Earth. This phenomenon does not pose a problem near the equator, where the magnetic lines of force are parallel to the surface of the earth. It does cause a problem, however, the closer you get to the poles, where magnetic lines of force become perpendicular to the surface of the Earth. In the Northern Hemisphere the compass needle will point downward (i.e., "positive dip") and in the Southern Hemisphere, upward ("negative dip").

The compass card is mounted so that its center of gravity is well below the pivot point of the pedestal. It moves in response to the centrifugal force acting on it; in a banking turn, that causes the compass to dip to the low side. The resulting error is most apparent when turning through headings close to north and south. In fact, when the aircraft makes a turn from a heading of north, the compass briefly indicates a turn in the opposite direction—what flyers refer to as the "northerly turning error." When the aircraft makes a turn from a heading of south, the compass indicates a turn in the correct direction but at a considerably faster rate than is actually occurring—the "southerly turning error."

Magnetic dip, along with the forces of inertia acting on the float, also causes errors in the compass reading when the aircraft is accelerating or decelerating on easterly or westerly headings. Because of its pendulous-type mounting, the after end of the compass card tilts upward during acceleration (increasing airspeed) and downward in deceleration (reducing airspeed). When the aircraft is accelerating on either an easterly or westerly heading, the error appears as a turn indication toward the north. When it is decelerating on either side of these headings, the compass indicates a turn toward the south.[22] In order to fly a straight course under these circumstances, a pilot would watch his other instruments, such as the turn indicator and the directional gyro.

The inherent errors in the magnetic compass made it an ideal choice for study. The two undergraduates assigned to this task were charged with determining the physical constants of the magnetic compass so that it could be accurately modeled mathematically.[23] Draper, in conjunction with McKay, would later use this information and his (Draper's) growing knowledge of instrument dynamics to patent an improved magnetic compass, one that was "designed in order to be free of northerly turning error and thus enable a pilot to fly a straight course solely by reference to his magnetic compass and without the aid of other instruments."[24]

By the summer of 1934, Doc had accumulated enough data on the performance of aircraft instruments to come up with a generalized mathematical model of the dynamic errors that occurred during actual flight conditions. As Draper described in a draft paper that he forwarded to Elmer Sperry Jr. for review on August 19, 1934,

> Aircraft flight instruments such as the magnetic compass, altimeter, rate of climb meter, gyro-horizon, etc. have been so improved that calibration errors are now practically insignificant. However, these instruments are accurate only when conditions are undisturbed or static. Under varying conditions, such as vibration, airplane motions, and when the actuating external force is changing, the instruments have errors caused by the dynamic characteristics of the indicating mechanism. These errors have long been known to exist and to have been designated as lag, oscillations, etc. but have never been mathematically determined or scientifically evaluated.[25]

Draper's paper went on to explain how that three experimentally determined coefficients—mass, damping (or friction), and the elastic (or restoring) force—were sufficient to account for the dynamic errors in any given instrument. They were related by the following expression:

(mass coefficient) (acceleration) + (friction coefficient) (velocity) + (elastic coefficient) (displacement) = (externally applied actuating force)

He then showed that the formula could be translated into a mathematical equation:

$$m \frac{d^2x}{d^2t} + b \frac{dx}{dt} + kx = \text{externally applied force}$$

where

m = effective mass of the system
b = effective friction coefficient
k = effective elastic coefficient, which tends to return the indicating element to its zero position
x = displacement of the indicating element from its equilibrium position
t = time

Draper's mathematical model was a scholarly achievement in and of itself, but there was a practical aspect to it too. In addition to explaining the dynamics of instruments, it could be used to determine the extent to which a given instrument could be improved. As William G. Denhard, a longtime alumnus of MIT's Instrumentation Laboratory declared, this was the beginning of "Doc's engineering philosophy that analysis could lead to better instrument design."[26] This was contrary to current practice, by which pilots and most people in the aviation industry believed in the "try it and fly it" approach to change. To these folks the idea of using analysis to aid in designing instruments was looked upon as "frothy, intellectual, and mind ramblings." But not for Elmer A. Sperry Jr., whose father had been a famous inventor noted for both his scientific acumen and his ability to hire talented assistants. Sperry read Draper's draft paper, no doubt, with great interest, as it provided tools to improve the design of the company's instruments. Elmer quickly returned one of the copies, with suggested changes as Draper had requested, and kept the other.[27]

How and when Sperry had first become aware of Draper's capabilities is not known for certain. Perhaps the two men met at an annual meeting of the Institute of Aeronautical Sciences. As for Draper's paper, it is likely that Sperry, having recently been elected to the governing council of the IAS and being generally aware of Draper's growing stature within the MIT community, had suggested that Draper prepare it for presentation at the IAS annual meeting.[28] By then Sperry wanted Draper to come to work for him; Sperry "made no bones about saying he was the best equipped man in the country" to work on the sort of things created at his company.[29]

In November 1934, Draper and his brother Clayton, now a New York stockbroker, had dinner with Elmer A. Sperry Jr. Clayton, in a letter home, described the event to their parents: "He [Sperry] couldn't possibly have praised your eldest son higher. . . . [T]hey built trick gadgets, then came to Stark to find out how they worked or rather why."[30] The letter is important because it shows the nature of the personal relationship that Draper had already established with Sperry and his company, which was one of the country's leading suppliers of gyroscopes and aircraft instruments. Draper's ongoing connection with the Sperry Gyroscope Company would later prove critical to the successful development of the Mark 14 gunsight, which catapulted Draper and his laboratory onto the cutting-edge of fire-control research and development for the U.S. Navy's Bureau of Ordnance.

By 1934 Draper had established a burgeoning consulting business in his "spare" time, doing work for the United Shoe Company, the Warner Brothers Paving Company, and the Palmer Engine Company. To help with his consulting, which the MIT administration encouraged, he hired Edward Gugger, who had previously worked as a constructor of apparatus in the Aero Department.

Consulting had become fully accepted as a legitimate activity at MIT in the early 1930s, when the institute's committee on consulting invented what became known as the "one-fifth rule."[31] The rule specified that professors might spend one day a week as industrial consultants, with no questions asked; the rest of the week, they were obligated to work for the university. The rule said nothing about the weekend. Presumably, professors could work as a consultant three days out of seven if they wished. More importantly, a professor who worked part-time for industry was encouraged to find work there for his students while they pursued their degrees. This became the standard academic model at MIT, and it explains why Draper sought help from his students.

When Draper returned to the MIT campus in the fall of 1934, he was about to be thrust into a new project that would firmly establish his reputation within MIT and the aeronautical industry. At the time, the rapidly increasing horsepower of aircraft engines and the introduction of controllable-pitch propellers resulted in torsional vibration—the periodic motion of a propeller shaft twisting about its axis, first in one direction and then in the other—that was creating structural failures in the propeller shafts of the bigger engines.[32] Many other parts of an airplane—of necessity a light, flexible structure—are also prone to failure if subjected to a high level of vibration at certain frequencies over a period of time. The problem of torsional vibration in radial engines raised its ugly head when controllable-pitch propellers, which were heavier, replaced fixed-pitch propellers. When the Wright R-1820 engine began breaking propeller shafts, the U.S. Navy turned to MIT's prestigious Aeronautical Engine Laboratory to solve the problem. Hunsaker—still well connected with the U.S. Navy—negotiated a contract with Cdr. Ralph C. Weyerbacker, head of the Material Division in the Navy's Bureau of Aeronautics, to conduct a joint program to develop equipment for recording vibrations in aeronautical crankshafts.[33] Once the project was under way, Edward Taylor, who was initially in charge of the project, came up with an undamped-vibration damper that fixed the crankshaft breakage problem. After that he shifted his attention to improving engine designs, leaving Draper to take over the vibration monitoring system.[34]

As noted earlier, Draper was already well regarded by Hunsaker and had worked alongside Edward Taylor on the engine indicator paper. Taylor obviously had enough confidence in Draper to give him the responsibility for completing what was a very prestigious and important project for Taylor's laboratory, given the limited amount of government funding available during the 1930s.

As before, Draper, heavily occupied by his teaching duties, still working on the detonation problem, and deeply immersed in the generalized study of instrument dynamics, looked to one of his talented graduate students for help. George C. Bentley was the ideal candidate. He had just completed his master's thesis on a self-contained instrument for recording vibration. Applying well-known engineering principles and off-the-shelf equipment, the two men worked together to construct a rugged, self-contained device to collect data on the vibration in various parts of aircraft while it was in the air. The apparatus used two electro-magnetic pickups (one for linear vibrations and one for torsional vibrations) connected to a compact photographic oscillograph via a special vacuum-tube integrating amplifier.[35] The finished instrument provided a visual record of the linear and torsional vibrations experienced by airplane structures and power plants. With this device they were able to collect vibration data on the propeller, crankshaft, engine, and structure for a number of different aircraft—the first time that such data had been obtained in flight.[36] This was an important accomplishment, for it provided a new tool for analyzing the unanticipated problems of the new types of aircraft that were being developed by the nascent aircraft industry. The two men presented their work at the annual meeting of the IAS in February 1936.[37] The Sperry Gyroscope Company thought that Draper's apparatus had profit-making potential and negotiated a contract with MIT for an exclusive license to sell a commercial version of it.

Research projects and consulting were essential parts of MIT's education and financial structure. The relevant policy stems from the "Technology Plan" established in 1918 to encourage large corporations to become continuing sources of financial aid to make up for the loss of state funding after a Massachusetts constitutional convention announced that the state could support only institutions under state control.[38] Under the plan, participating industrial firms paid an annual fee for special privileges in using the results of the research done by teachers at the institute. To manage corporate contracts and solicit new research partners, the administration created the Division of Industrial Cooperation and Research. The Technology Plan had resuscitated MIT during the early 1920s, bringing in funds from a wide range of industrial corporations, but it had fallen into disuse by the time Compton took over MIT's presidency.[39] Compton reaffirmed the administration's commitment to close ties with industry. Both he and Bush considered service to industry to be a central tenet of the institute's mission, but they wanted industrial collaboration to advance, and not hinder, teaching and research.[40] Thus every research agreement had to be approved by the director of the Division of Industrial Cooperation and Research, who, under Compton's direction, made sure that the contract gave publication rights to the research staff and respected the autonomy of the institute.

When companies needed outside help with technical problems, faculty members were encouraged, as noted above, to provide consulting services, provided that they did not compete with outside consulting firms. Compton, in keeping with his promise that the fundamental sciences would constitute the backbone of the institute, levied a 50 percent "tax" on all outside consulting fees to ensure that the faculty accepted only intellectually valuable assignments.[41] The proceeds of the tax were disbursed to faculty members according to the intellectual merit of their work, as determined by a special committee.

On August 1, 1936, the Sperry Gyroscope Company entered an agreement with MIT to commercialize Draper's work, paying one dollar, a variable annual fee, and a 5 percent royalty on the selling price of each instrument.[42] In addition to turning over all of his drawings, the agreement stipulated that Doc would act as a consultant as the company worked to bring the device to market; the institute's variable fee would be determined by the amount of work Doc performed. MIT would own any future patents that Sperry might apply for, in return for an exclusive license to manufacture the MIT-Sperry Apparatus for Measuring Vibration. The agreement was financially rewarding to both parties. Within six months, Sperry had sold $20,000 worth of MIT-Sperry vibration equipment to the U.S. Navy, which was a considerable sum of money (over $360,000 in current dollars) at the time.

The vibration-measuring project was another milestone in Doc's career. It marked the unofficial beginning of the MIT Instrumentation Laboratory, which was given the name by Jerome Hunsaker in October 1934 when he mentioned for the first time in an MIT document the achievements of the "Aeronautical Instrument Laboratory" in the aeronautical engineering section of the annual *President's Report*.[43] The project for the Navy was also the first MIT research contract completed by the instrument lab under Draper's leadership. Draper's second research project for MIT was a study of the theory of the acoustic altimeter funded by NACA. Draper completed the project during the 1936–37 academic year.[44] His findings were published in August 1937, as NACA Technical Note No. 611.[45]

In the spring of 1937 Draper turned his attention once again to engine knock. C. Fayette Taylor had been urging NACA and the Bureau of Aeronautics to support the development of a flight-ready detonation detector, arguing that it was one of the most urgent needs of military and civil aviation.[46] Taylor, emboldened by the success of MIT's vibration apparatus, persuaded MIT to seek funding to develop a knock meter from Draper's engine indicator on the basis of studies that had already been done in the Sloan Laboratory for Aircraft and Automotive Engines.

The engine indicator was based on the pressure-sensing pickup that Draper had begun working on for his master's thesis. The "Knockmeter," as it became known, was developed with the help of the Taylor brothers and C. L. Williams. To C. Fayette Taylor went the honor of describing the new device to measure the pressure change in an engine cylinder when the team's paper was read at a meeting of the Society of Automotive Engineers in January 1934.[47] Taylor's name appeared first in the paper's list of authors, as befitting his status as director of the Sloan Laboratory. Taylor provided the institutional and intellectual framework for this project as well as for Draper's continuing study of the combustion process. As Michael Dennis makes clear in his thesis, "Draper's existence traded upon Taylor's power and authority, both of which flowed from the ICL [internal combustion engine laboratory] and the constellation of institutions and technical problems Taylor built around internal combustion engines. Teaching as well as research would play a major role as Draper worked to develop a similar constellation of technical problems and institutions linked to precision measurement."[48]

Funding to develop the detonation detector was obtained from the Navy and NACA, each of which contributed $1,200. Sperry contributed another $5,000 toward Draper's research, and the Wright Aeronautical Corporation supplied an engine valued at $5,000. The aircraft instrumentation business at this time accounted for about half of Sperry's Gyroscope's sales.[49]

In April 1938, after the engine-knock project had been under way for nearly a year, the Navy, anxious to obtain a detonation detector for its own laboratory needs, wrote to Draper requesting information on the project's status. By then Draper had completed the basic research and assembled a laboratory prototype. Working with Joseph H. Lancor, an electrical engineer hired to work on the project, he had by now developed a set of sensors that measured cylinder's vibration and internal pressure—both of which were needed to reveal the presence of detonation.[50]

Draper referred the Navy's request to Vannevar Bush. Bush responded after meeting with Draper and Hugh Willis, chief research engineer for the Sperry Gyroscope Company, to discuss the status of the project. In a letter to Capt. Sydney Kraus, head of the Material Division in the Bureau of Aeronautics, Bush explained that Draper's work had progressed to the point that it was now possible to produce a detonation indicator for "quantitative measurements of detonation intensities in aircraft engines."[51] Further, Draper could build such indicators for the Navy, because it was a suitable continuation of the research project that the Navy had helped fund. The cost to the Navy would be $7,500 for one device and $10,000 for two. Bush assured Kraus that Professor Draper would cooperate closely with the Navy and assist in testing.

Willis had wanted Sperry itself to manufacture the laboratory equipment the Navy was seeking, but Bush considered the indicators to be still a research project, one whose results could include the development of new instruments. Dennis' thesis concludes that "Bush's strategy created a future for Draper's research in two direct ways. First, if the Navy wanted a detonation indicator, they would have to support the Instruments Laboratory and with it both students and mechanics. Second, even if the Navy did not accept Bush's offer for the development of an indicator, Draper's laboratory would still have an important relationship with Sperry."[52]

The arrangement worked out by Bush satisfied all parties: Draper would build the devices for the Navy, by which the laboratory would gain further practical experience; the Navy would get the laboratory detonation indicators it wanted; and Sperry would get the fruits of Doc's labors and the commercial rights to the MIT-Sperry detonation indicator that was to be supplied to the military services during World War II.

5

From Turn Indicator to Gunsight

As a pilot who was well versed in the practical use and limitations of aircraft instrumentation, Draper was fascinated with the subject of blind flying. He owned a well-marked-up copy of Howard C. Stark's book, considered the authoritative work on blind flying when it was published in 1931. Draper was so absorbed with the topic that he even took time from his busy schedule to write an article on it for MIT's student publication, *Tech Engineering News,* in December 1936.[1] The editors of *Science Digest* considered it worthy to be reprinted in their own publication. Draper's long-term interest in blind flying had been sparked in part by Jimmy Doolittle's history-making flight of September 29, 1929, when he took off from Mitchel Field on Long Island, flew out fifteen miles, returned to his starting point, and landed again, relying solely on instruments. The most important of these aids were the artificial horizon and directional gyro developed by the Sperry Gyroscope Company. (Doolittle's flight had a direct bearing also on the beginnings of the MIT Instrument Laboratory—which became officially the MIT Instrumentation Laboratory only in the 1940s—for it required the services of Professor William G. Brown, whose departure from MIT opened the door for Draper to take charge of the aircraft instruments course.)

In the summer of 1938, while on his annual excursion to the West Coast to visit his family, Draper traveled to Oakland, California, to take the instrument-flying course at the Boeing School of Aeronautics.[2]

Although Boeing's course was intended to prepare private pilots for the instrument rating now needed to fly the radio beam on U.S. airways, Draper appears to have been more interested in trying out the new Link Trainer that the Boeing school used in the course. Shaped vaguely like an airplane with short, stubby wings, the Link Trainer was a ground simulator equipped with a hooded cockpit that allowed a pilot to practice instrument flying without leaving the ground. Draper spent about a hundred hours in the Link Trainer and of actual flight time under the hood of the Boeing 203 biplane trainer.[3] In return, he paid the tuition and gave some lectures on aircraft instruments. Draper never found the time to practice for or take the exam that would have given him an instrument rating, but as he wrote years later, "I did develop some opinions about much needed improvements for the bank and turn indicator."[4]

The turn-and-bank indicator was one of the three instruments Draper considered essential for flying. It was the most important instrument in Howard Stark's "1-2-3 order" method of blind flying, the others being the altimeter, airspeed indicator, and climb indicator.[5] Stark's second method, the "direct method," relied on the Sperry artificial horizon and gyro compass in place of the turn indicator and climb meter. When Stark's book was published in 1931, pilots had to be familiar with both methods, as well as with various combinations of the two, "for it is quite possible to have one or more instruments out of commission, even if both complete sets are available, and both groups should be available to ensure safety in instrument flying." Howard Stark advised "that while it is simple to fly by the Horizon and Directional Gyro alone, a better job can be done if other instruments are also referred to." Stark made the need to have two sets of instruments abundantly clear: "If one of these instruments should go out of commission, the parachute may be the only solution if he [the pilot] lacks the other instruments or is unable to use them."[6]

Clearly the turn-and-bank indicator was an important instrument, one that Draper considered worthy of improvement. While at Boeing's school he discussed his ideas for the design of a new turn indicator with Raymond J. Stephan, one of the school's instructors. He also discussed the idea with Hugh Willis, Sperry's chief research engineer, whom he visited in Seattle at the end of August.[7] Draper then returned to Palo Alto for a few days before heading to Dayton, Ohio, by train to join his fiancée, Ivy Willard.

The couple had first met in the fall of 1930 after Draper had returned from his summer-long study of aviation in Europe. Ivy, the twenty-year-old daughter of a Canadian lumber baron, had been working as a secretary in the MIT physics department when she became acquainted with Draper's younger brother, then a student at the Harvard Business School. It was Clayton who introduced them.[8] Ivy had been born in Montreal, Canada, but had grown up in St. Johnsbury, Vermont. Her studies at the University of Vermont were interrupted when she came down with polio in 1928. The disease sent her to Boston for treatment. "I spent a whole year," she told a reporter for a Boston paper in 1962, "most of it on my back, in the Massachusetts Memorial Hospitals getting over the effects and trying to put some life back into this right leg."[9]

Ivy and Doc began dating while Draper was working on his doctoral dissertation. Ivy, or so the story goes, was still using a leg brace the doctors had prescribed. Draper didn't like it. He insisted that she take it off and try walking with him to the Harvard Bridge. When they got there, she threw the leg iron into the Charles River. It's still there, "unless some poor fish is using it to keep its drooping fins up."[10] Draper, according to Ivy's later recollection, was so bound and determined to learn everything he could that he did not stop to bother with his doctorate or getting married until he was good and ready. At thirty-seven years of age, with his PhD in hand and a promotion to associate professor, there was nothing to hold him back from marriage. Ivy Willard and Charles Stark Draper were married in the Air Corps chapel at Wright Field near Dayton on September 7, 1938. According to Ivy, the location was "sort of half-way between the Atlantic and the Pacific to make it convenient for relatives and friends to attend."[11] None of their parents were at the wedding, however.[12]

Draper had a number of professional friends at Dayton, and had stopped there on the way back from California to consult with Fred Dent, a protégé who was now head of the Dynamics Branch of the Experimental Engineering Division at Wright Field.[13] When Draper returned to the MIT campus at the end of the second week of September, there was a letter on his desk from George G. Brady, chief engineer for the Curtiss Propeller Division of the Curtiss-Wright Corporation, offering a consulting arrangement.[14] Dent had recommended Draper, telling Brady that as far as he knew Draper was not linked to any airplane manufacturer.[15]

In a follow-on note addressed to "Dear Stark," Dent suggested that in view of his approaching marriage and "continual increased expenses" he might be interested in consulting for Curtiss-Wright, on propeller control and allied problems. Draper accepted and became a consultant, at a fee of three hundred dollars per quarter. This does not sound like much today, but in 1938 a new car cost about eight hundred dollars, the average monthly rent for a house was twenty-seven dollars, and a loaf of bread cost nine cents. Draper gained additional income as a consultant on an altitude and rate-of-climb meter for Pioneer Instrument Company and on an electronic balance for the Amtorg Trading Company.

The growth of his personal income aside, Draper was becoming increasingly frustrated with his inability to secure funding for the Instrument Laboratory. Lacking institutional sources, he was obliged to bear the burden of supporting the cost of new equipment and his personal research, and is capacity to do so was continually dependent on his ability to secure consulting or contract jobs from the outside. Despite repeated requests to the administration, he failed completely to attract any support for "his instrument laboratory" or to get an assistant assigned to serve as his understudy.[16] Discouraged by what appeared to be the institute's lack of interest in his laboratory, he began to look for opportunities elsewhere.

A chance to break away from MIT arose in the early part of 1936, when Frederick C. Dumaine, president of the Waltham Watch Company (located about seven miles west), requested the names of faculty members who would be able to help his company. The request was passed to President Compton, who recommended Draper. After an initial meeting with Dumaine, Draper asked his stockbroker brother to find out what he could about the firm. The Waltham Watch Company, according to Clayton's analysis, was making "real money" and was financially sound.[17] Clayton wondered if the company might be considering enlarging the scope of its business to include instrument work. Clayton thought that a "shot at doing commercial work . . . might prove very valuable training."

A month later, Draper reached an agreement with Dumaine on terms of future employment with the Waltham Watch Company.[18] Draper would remain at MIT for the coming year. He would not take on any new consulting positions and would use the interim to complete all consulting work in progress. In a year's time he would be ready to devote all his consulting time to the Waltham Watch Company, for which he would receive seven hundred dollars a month. Within three years he would either work himself into a permanent position there or withdraw entirely. There were other conditions too, including the requirement that Waltham hire George Bentley as his assistant. Draper, stifled by the lack of funding needed to expand the activity of the Instrument Laboratory, perceived the agreement with Waltham as a means of establishing his own laboratory free from the constraints of academia. It appears too that he thought he was being pushed out by the administration.[19] There was also the possibility of making some real money, especially if he could get Waltham into military contracting. During his studies of the watch business, Draper had learned that the Army Air Corps purchased close to 1,500 timepieces each year.

Most of the watches used for aerial navigation at that time were manufactured in and imported from Switzerland, because of their excellent quality and accuracy of their jewel movements. Now, however, with the increasing prospect of war, the supply of foreign watches was likely to dry up. In wartime the requirement for accurate timepieces and precision time devices for fuzes would increase dramatically, providing the Waltham Watch Company an excellent opportunity to capitalize on demand.

Draper spent most of the following summer at Waltham studying the company's research-and-development needs and preparing a report, "The Problem of a Research Division."[20] By the beginning of September he was looking forward to working for the company. Draper wrote Dumaine on September 9, "I feel now that I can definitely be of service to Waltham. It will be a pleasure to finish up the next school year so my entire time can be devoted to Waltham's problems."

But Draper was about to change his mind, and what caused him to do so has never been made explicitly clear. Sometime between his letter to Dumaine on September 9 and a meeting with Karl Compton on or about the 30th, he was unexpectedly promoted to full professor.[21] Before the meeting Draper thought that the administration was trying to ease him out of MIT.[22] Compton listened sympathetically to Draper's concerns but made no firm commitment with regard to future support, explaining the budgetary constraints under which he himself had been operating. However, he seemed open to the possibility of finding funds for the assistant Draper wanted to hire.

With a full professorship in hand and Compton's reassurance, Draper decided to stay at MIT after all. He left his agreement with Waltham unconsummated and returned his attention to his teaching, management of the Instrumentation Laboratory, and his studies of aircraft instruments and aerial navigation. Freed from Waltham's consulting he resumed his interest in the aircraft turn-and-bank indicator and convinced Sperry to provide $1,500 for his effort to improve it.[23]

Sperry's turn indicator and those of its competitors then on the market relied on an air jet–driven gyroscopic rotor to provide a stable reference point. The gyro was supported by a spring-restrained, single-degree-of-freedom gimbal carried by ball bearings. When the airplane turned, the precession of the gyroscope acted on the spring, causing the instrument's indicator to tip away from zero in the direction of the turn. During his investigations, Draper found that engine vibrations produced small dents in the ball-bearing races. These caused the gimbal to hang up during small turns, causing jerky motion of the indicating needle that made smooth flying difficult. Draper was sure that "replacing the ball bearings by spring suspensions and a proper amount of viscous damping would eliminate the erratic effects."[24]

To construct a working prototype, Draper recruited Harry Ashworth, a precision machinist and instrument expert from the Ames Aircraft Company. Ames, located at the Boston Municipal Airport in East Boston, was in the business of repairing aircraft instruments, propellers, magnetos, carburetors, starters, etc. Working out of Draper's "one room shop," Ashworth built two turn indicators that had suspended springs and damped gyros. He then mounted them on a custom-built panel that could be installed in any airplane. One of the instruments gave rate of turn about the yaw axis, the other provided a standard readout. Draper made a number of flights with this unique arrangement and with it "not only found flying easy and pleasant, but acquired reasonable effectiveness in carrying out a wide range of turning maneuvers."[25]

In May 1940, Draper took the two turn indicators to the Sperry plant in New York for flight testing in the company's Lockheed Hudson.[26] Draper was not invited along for the tests, which were conducted that morning. He did not find out the results until lunch with the pilots and engineers who had evaluated the new indicators. Draper was disappointed to learn that although Sperry's engineers thought the new turn indicator was an advance over the current instrument, they felt there was no reason to make any changes in the current model, which was selling quite well.

What happened next was a pivotal moment in Draper's career. He later described the episode in an article published by the American Institute of Aeronautics and Astronautics in 1980. "I was disappointed" with the results

Turn Indicator U.S. Patent 2,291,612

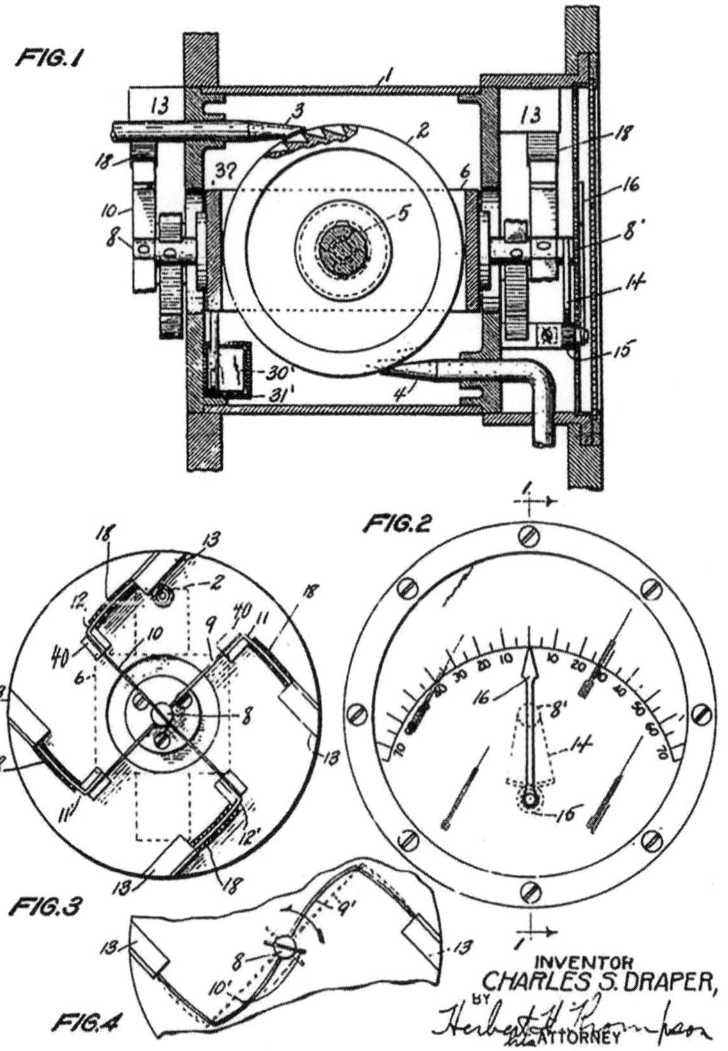

Fig. 5-1. Turn indicator patent. *U.S. Patent Office*

of Sperry's evaluation, he wrote, so "I searched my mind for ideas that would allow me to continue to work toward realizing instruments for sensing rotation with respect to inertial space."[27] As he ate lunch with the Sperry personnel, somebody brought in the latest newspapers, with the news of the German invasion of France. Among the headlines was a story about the difficulties faced by French gunners trying to hit fast-moving German tanks with their 75-mm artillery pieces. Draper immediately realized that the same principles used in the turn indicator could be applied to a gunsight. Instead of indicating turning rate, the gyroscope, which measured the angular rate of change over time, could be used to follow a moving target. Draper perceived that "gyroscopic instruments could be mounted on guns and connected to tilt levers for offsetting an optical index by the angles required to correct for target motion during the flight of the projectile."

Upon his return to MIT, Draper borrowed every book he could get his hands on about ballistics and fire control to make himself completely knowledgeable in the subject. With Ashworth's help, he began building an experimental gyroscopic gunsight. By chance, two of Draper's Navy students, Lt. Horacio Rivero Jr. and Lt. Lloyd M. Mustin, had just completed a jointly authored thesis addressing the short-range antiaircraft problem. Both men had taken Draper's advanced course in instruments, which included considerable work on the theory and application of the gyroscope, and their thesis had been required for their master's degrees. Mustin, who had served as an antiaircraft gunnery officer in the cruiser *Augusta* (CA 31), thought that gyros would provide a way of computing lead angle. He and Rivero analyzed attacks on ships from airplanes, especially divebomb, strafe, and torpedo. "As far as is known," wrote the two lieutenants, "no control device for the short-range problem has been developed anywhere which pretends to solve the three-dimensional problem involved."[28]

Together the two naval officers had set out to analyze how a gyroscopic device based on a commercially available aircraft turn indicator could be used to predict the path of an incoming airplane and thus act as a computing fire-control system for a machine gun. When they approached Draper for help, he shut up like a clam. Draper did not discourage the two students but was evasive about the problem of reducing the bearing friction in a gyroscope so that it would not precess when extraneous forces were applied. Draper had solved it in the turn indicator, but was reluctant to disclose information while the patent was still pending.[29] He was in the midst of deriving the theory behind the gyro gunsight's operation and of developing its basic design criteria and was not about to divulge any of its secrets yet to the outside world. Nevertheless, by the time he had worked out all the details, Draper's interest (perhaps influenced by his students' paper) had turned from killing tanks to bringing down aircraft.

In late May or early June Draper presented his idea on the gyro gunsight to Hugh Willis, suggesting that a small project be set up to test the concept with a laboratory model.[30] Draper had worked closely with Willis on the Sperry-MIT vibration apparatus, and it was Willis to whom he had turned with the idea for the turn indicator. Willis, with the backing of Preston Bassett, Sperry's chief engineer, approved Draper's proposal, despite the fact that Sperry's fire-control group almost unanimously disapproved of it. With the support of Karl Compton and the assistance of Nathaniel McLean Sage, director of MIT's Division of Industrial Cooperation, Draper negotiated a contract with the Sperry Gyroscope Company to transform the turn indicator into a working model of a gunsight for fire control against moving targets.[31] The agreement was confirmed by Willis in a letter to Sage on June 19, 1940. The project would be a joint research effort, with Sperry paying for salaries and material and MIT contributing restricted laboratory space, the equipment of Draper's machine shop, and procuring outside help in manufacturing key components.[32] Sperry would be given the opportunity to patent any devices Draper developed, with the understanding that because of the potential importance to national defense, the matter of licensing for the public good would be jointly decided by the officers of Sperry and MIT.

Sperry's funding allowed Draper to acquire space on the second floor of the Guggenheim Aeronautical Laboratory building and turn it into a laboratory. It would soon become known as the Confidential Instrument Development Laboratory—abbreviated "CID Lab." Draper moved Harry Ashworth with Edward P. Bentley (not to be confused with George C. Bentley) into the new space to help develop a prototype.[33] Edward Bentley, another of Draper's talented students, had just obtained his doctorate in science and had remained at MIT as the Quondam Research Assistant in Aeronautical Engineering. More assistants for Draper were acquired through the MIT placement office, which recruited MIT alumni Edmund B. Hammond (BS Aero 1940) and Victor C. Smith (ScD 1930).

By the time he moved into his new space, Draper had worked out the theoretical basis for the sight and had begun to experiment with hardware.[34] After analyzing the ballistic problem, he conceived the idea of using two gyros (similar to the one used in the turn indicator) to sense, respectively, the horizontal (traverse) and vertical (elevation) angular velocities of the lines of sight. While analyzing physical relationships embodied in the gunsight's mechanisms, Draper discovered an important principle of gyroscopic theory, one that would have a profound effect on the effectiveness of Draper's lead-computing gunsight. It was also a milestone in the ongoing development of the single-degree-of-freedom gyro, which was to become the hallmark of Draper's research.

What Draper found was that for the gyroscope to perform properly in a fire-control application, its "characteristic time"—the time required to reach equilibrium when subjected to a sudden change in angular velocity—had to be "substantially equal to, but not less than, the time of flight"; otherwise the gyroscope would tend to oscillate, and a smooth tracking rate could not be obtained.[35] Using the mathematical model that he devised for this phenomenon, Draper proved that the characteristic gyro time could be changed by varying the damping factor. This was an important finding, and it verified the approach that he had used in his turn indicator to dampen the spring stiffeners.

Ivan A. Getting, one of World War II's scientific luminaries, provides the best and most easily understood description of how Draper's lead-computing sight worked: "In the sight were mirrors positioned by the gyro, which were processed by springs. The operator tracked the target keeping the reticle image on the target. The greater the angular rate of the target, the greater the stretch of the springs, the greater the deflection of the mirrors, and the larger the lead angle. The gyro provided a spatial reference frame, and if the angular rate were multiplied by a function of the range (by varying a spring constant) the lead angle resulted with the sight mount pointing at the future target position."[36] To compensate for the effects of gravity, Draper attached a small weight to the elevation gyro to provide "superelevation"—the amount of elevation needed to compensate for gravity.

A one-dimensional (i.e., single-gyro) working model of the sight, "about the size of a typewriter," produced in the CID Lab was ready for testing in December 1940.[37] Draper took it to MIT's rifle range for testing; there it was fastened to a single-shot, .22-caliber rifle mounted on a pipe stand. A moving target was provided by an image of an airplane imprinted on a laboratory towel carried by a clothesline running over two motorized pulleys.[38] Draper used the gunsight to point the rifle at the target, which was about seventy-five feet away, tracked it until the optics settled, then pulled the trigger.

While Draper was testing the working model, Professor Sir Ralph Howard Fowler arrived at MIT to discuss the gunsight project.[39] Fowler, formerly Plummer Professor of Theoretical Physics at the University of Cambridge and now British scientific liaison officer for North America, had arrived in the United States in September as a member of the British Technical and Scientific Mission to the United States. The technical mission, organized by Sir Henry Tizard, was to exchange secret information on the scientific progress being made in a variety of new weapons being developed in Great Britain. Fowler was head of the committee on antiaircraft, fire control, directors, and searchlights, which was one of

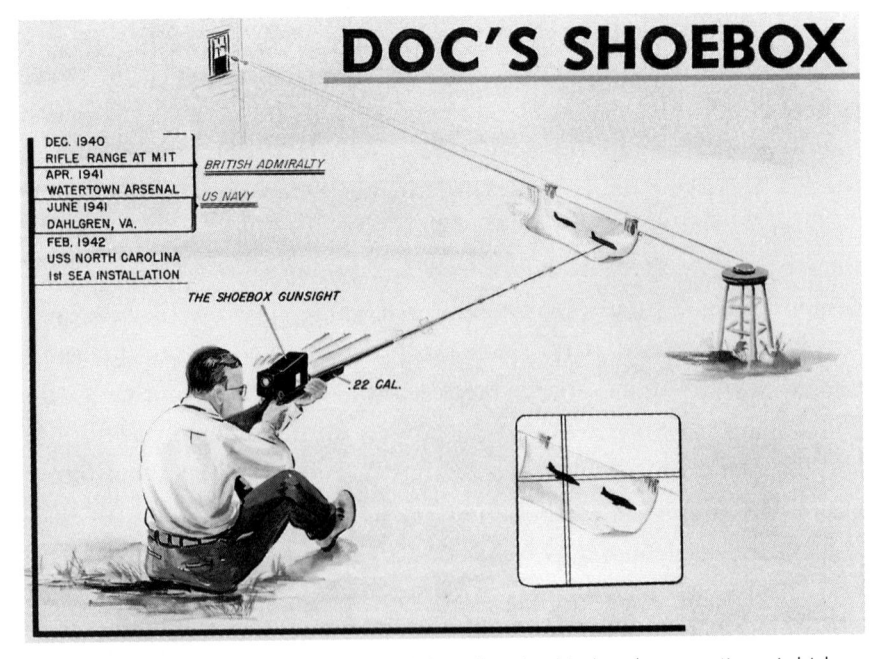

Fig. 5-2. Charles Stark Draper tested the idea for his lead-computing sight by mounting a prototype on a .22-caliber rifle that he then fired against moving targets set up at the Watertown Arsenal. *Charles Stark Draper Laboratory, reprinted with permission*

eight specialized groups set up by Tizard.[40] He visited the Sperry Gyroscope Company twice in October in company with fellow members of the Tizard Mission, as the British Technical and Scientific Mission was known: John Cockroft, F. C. Wallace, and E. G. Bowen.[41] It was probably during one of these visits that Fowler learned of Draper's research.

Fowler "showed great interest in the gyro gunsight" during a follow-up visit to Cambridge (the one in Massachusetts, of course, not Cambridgeshire) and insisted on trying it out for himself.[42] He used the "shoebox" sight to shoot at the moving target a number of times and was so impressed with the results that he asked Draper for the names of contacts at Sperry with whom he could discuss commercial terms. As a result of Fowler's unqualified endorsement, the Admiralty in London contracted with the Sperry Gyroscope Company to produce four of Draper's lead-computing sights for the Royal Navy.[43] Sperry, which was unable to manufacture the sight's gyroscopes, had to order them from Draper's laboratory, via the Division of Industrial Cooperation.

The Royal Navy gunsight project proved a debacle, and what happened remains unclear. There are indications that Sperry was ultimately unable to produce a workable unit. Draper hinted at this in 1969, recalling that "firing tests [of the Sperry units for the Royal Navy] showed indifferent results."[44] Even more damning is the handwritten draft of a letter from Draper to Bassett months afterward complaining about the procedures Sperry was using to produce the Mark 14 gunsight (the U.S. Navy's version of Draper's design) and its lack of qualified engineering personnel.[45]

Meanwhile, Draper continued to refine the sight's design. In February 1941, Lt. Leighton I. "Lee" Davis, then a graduate student in aeronautical engineering on sabbatical form the U.S. Army, arranged for Draper to use the range at the Army's Watertown Arsenal, in Watertown, New York, for further testing of the sight.[46] By then Draper's team had produced a second prototype having two gyros, as Draper had originally conceived and as described above. The testing of the second prototype at Watertown was conducted in a big concrete building called the "skull cracker."

As Draper recalled years later, a number of demonstrations of the gunsight at both the MIT range and the Watertown Arsenal "aroused no real interest in either the Navy, Army, Air Force, or the Sperry Gyroscope Company, which was doing quite well selling conventional fire control systems."[47] So the gyro gunsight, according to Draper, was put "on the back of a dark shelf and effectively forgotten"—that is, until the end of May, when Horacio Rivero, now assigned to the Bureau of Ordnance in Washington, D.C., showed up in Draper's office.

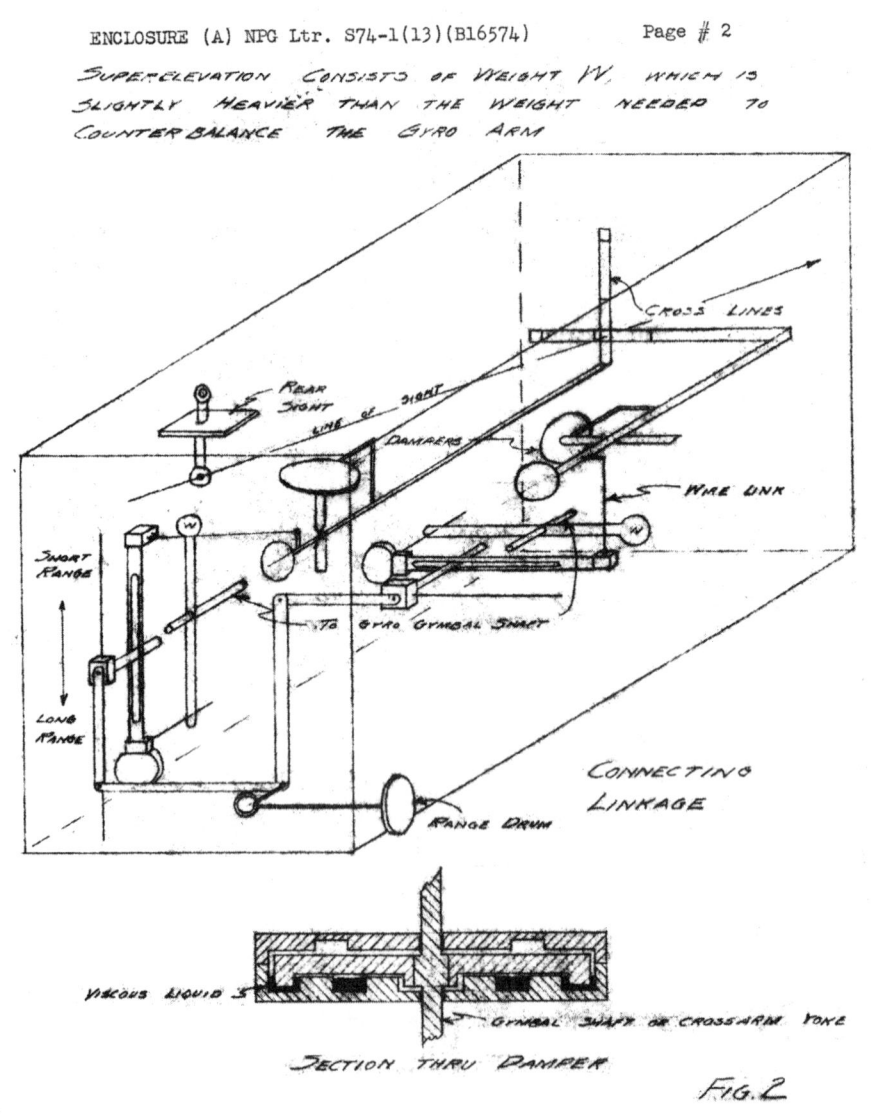

Fig. 5-3. Early design detail for Draper's lead-computing gunsight showing a section through the gyro damper and the weight added to provide superelevation to compensate for the effects of gravity on a projectile's trajectory. *CSDL Collection, MIT Museum*

6

War Work

After graduating from MIT, Horacio Rivero served brief tours of duty at the Naval Gun Factory in Washington, D.C., and the Naval Proving Ground at Dahlgren, Virginia, before being assigned to the Bureau of Ordnance (BuOrd). When he arrived, BuOrd's personnel were having a tough time trying to keep up with the ordnance needs of the fleet, which was expanding at breakneck speed in response to President Franklin Delano Roosevelt's plan for a "Two Ocean Navy." Bureau personnel were so overburdened with producing existing weapons that they were unable to carry out any development work. To help alleviate these and other problems in providing antiaircraft defenses for the fleet, Rear Adm. William H. P. "Spike" Blandy, chief of BuOrd, reorganized his bureau, the Navy's premier shore establishment, creating three divisions, for production, maintenance, and research and development (R&D). The last of these was to conduct fundamental studies, design systems, and engineer production runs of new weapons and fire-control systems.

When the reorganization took effect, on April 10, 1941, Lieutenant Rivero was assigned to the Antiaircraft Production Section, with additional duty as head of the radar desk in the Fire Control Section of the R&D Division.[1] The two biggest problems facing Rivero and the Fire Control Section were the development of a director for the fleet's light antiaircraft guns and the implementation of a fire-control radar, already in the design stage, for gun laying and fire control.

Although a number of firms and the Naval Gun Factory were working to develop a director for use with the fleet's antiaircraft machine guns, none of their designs appeared to lend themselves to quantity production. The questions of when and how Rivero became aware of Draper's lead-computing gunsight at MIT remain unanswered, but he could have done so from any number of sources within Ford, Sperry, or MIT with whom Rivero would have come in contact in the course of his work. It certainly would have piqued his interest, given his MIT thesis and his knowledge of the director problems facing BuOrd.

A month after the reorganization, Rivero obtained clearance from the Sperry Gyroscope Company to inspect Draper's secret project.[2] On May 28, Rivero, in company with Lt. Cdr. Marion E. Murphy, head of the Fire Control Section, traveled to MIT to inspect a new radar being developed by its Radiation Laboratory, then the center for U.S. radar research. After that Lieutenant Rivero and Lieutenant Commander Murphy went to Draper's office, where they discussed the development of his gyroscopic sight, then known as the "black box." Upon their return to Washington Murphy reported their findings about Draper's lead-computing sight to Capt. Garret L. Schuyler, director of the R&D Division. Schuyler, intrigued, set out to compare Draper's lead-computing antiaircraft sights with those being developed by the Bendix Aviation Corporation and the General Electric Company. Draper's sight, he discovered, had several advantages over them, not least its innovative use of viscous damping to increase stability in order to overcome the detrimental effects of vibration. Schuyler sent Murphy back to Draper's lab in company with Cdr. Ernest E. Hermann, the officer in charge of coordinating the bureau's antiaircraft program, to find out more about Draper's gyro sight. In the days that followed other naval officers, including Lt. Cdr. Marion E. Miles, a member of the Navy Department's Antiaircraft Board, arrived at Draper's door to study the gunsight, and arrangements were made to test it at the Naval Proving Ground at Dahlgren, Virginia.

Murphy and an entourage of officers from the Bureau of Ordnance and the Navy Department's Fleet Training Division—a coterie that included no less than six future admirals—traveled to Dahlgren on July 24 to witness tests of the laboratory model of Draper's lead-computing sight jury rigged to a 20-mm Oerlikon gun.[3] The free-swinging 20-mm gun was controlled by the body movements of the gun's operator through a set of handlebars and shoulder rests. The gun had been originally designed to be used with conventional open-ring sights; hitting a fast-moving target with this type of sight depended on the gunner's ability to lead the target in both traverse and elevation. To assist the gunner in this rather difficult endeavor, every fifth round in the ammunition drum was usually a tracer. These rounds provided a visual indication of where the rounds were heading, enabling the gunner to adjust his aim as needed. But much depended upon the individual gunner's skill. If Draper's lead-computing sight worked, it would make it much easier for the gunner to lead the target, greatly increasing his ability to obtain hits.

Six firing runs were conducted against a thirty-foot towed target sleeve that day. Although no hits were made; visual observation of the tracer pattern showed that the sight generated good deflection (horizontal lead) but was low in elevation. The sight functioned properly, without any adjustments, despite the fact that it fell off the mount twice due to vibration. Nevertheless, the inspector in charge of ordnance submitted an unfavorable report, based on a number of practical problems that he perceived in Draper's design. Others in attendance that day, including Murphy, correctly judged that these faults could be readily set right.

The next day, Murphy telephoned Omar B. Whitaker at Sperry Gyroscope and told him that the bureau's personnel were impressed with the sight. He asked Sperry to provide descriptive specifications and an approximate price for fifty sights: thirty for the Oerlikon and twenty for a remote-control gun director then being considered within BuOrd. The Navy wanted Draper's sight as soon as possible. "If this pilot lot of sights comes up to [our] anticipated expectations," he told Whitaker, "there would undoubtedly be a demand for them in large quantities."[4]

Although the engineers in Sperry's fire-control group were still skeptical of Draper's sight, Sperry officials were persuaded to cooperate with the Navy and agreed to supply twelve prototypes for further tests.[5] Once again, the Sperry engineering staff was unable to produce the prototype sights specified by the Navy and turned to Draper for assistance. On September 5, Sperry sent Draper a standard subcontract for the production of twelve sights. By now BuOrd had come to recognize that the sight, if satisfactory, could also serve as the heart of a small gun director needed to control the 40-mm Bofors antiaircraft gun, which was about to enter service with the fleet. Only seven of the twelve of the prototype sights ordered from the confidential instrument lab were to be mounted directly on Oerlikons, the other five would be used in the development of a new director for the Bofors (see p. 68). The delivery schedule was more than Draper and his assistants could handle by themselves, so they hired a local manufacturing company to assist them.[6]

Another version of Draper's homemade sight—possibly one produced for the Admiralty—was tested at Dahlgren on September 11 and 12. Although no record of these tests has been located, the results must have been excellent, for the Bureau of Ordnance issued to Sperry on October 7, 1941, a letter of intent to produce 2,500 sights for the U.S. Navy. Sperry, according to the recollections of Robert C. Seamans Jr., one of Draper's graduate students hired to work in the CID Lab, claimed that Draper's lead-computing sight was impractical to manufacture.[7] As Seamans noted, Sperry was developing its own, in-house sight, which its engineers thought would do a better job. The Navy, Seamans later recalled, told Draper that if he wanted the project to move ahead, he had to demonstrate that his sight could be manufactured in quantity. It was the gyroscope that Sperry was having difficulty producing. To prove that they could be reliably manufactured, Draper had a local machine ship in Newton, Massachusetts, owned by Fred MacCloud and John Sattlemyer, manufacture fifty damped gyroscopes to order. The company had no name, so Draper christened it Duolcam, which is "MacCloud" spelled backward. Duolcam wound up supplying all of the gunsight gyros to Sperry, which, as the prime contractor, was then responsible for putting together the finished product.

At the end of October, Draper and his "gyro gang" traveled to the deserted fields of the Dam Neck antiaircraft range near Virginia Beach, Virginia, to observe the final testing of a production prototype of the lead-computing sight. During the demonstration that followed, Lt. Cdr. Ernest M. Eller, assistant production officer in the BuOrd Production Division and a tracer expert with more than 100,000 rounds of target experience, tested the first model. "Stepping from the gun after firing its first rounds, he turned to Draper and said, 'We'll buy it.'"[8] When "Nat" Sage (head of Sperry's industrial cooperation division) learned that the

Navy had accepted the gunsight he went "dancing down the corridor" to report the news to James Killian, Compton's executive assistant, acting as Compton's surrogate while he was directing radar research for the Office of Scientific Research and Development.[9]

On December 27, the Navy, having officially accepted Draper's gunsight, designated it the "Gun Sight Mark 14." Sperry was given a contract and began production of the first 2,500 units ordered by the Navy. Draper's MIT CID Lab was subcontracted for design and testing.[10] Draper was not happy with the initial approach being taken by Sperry's engineers, however. The personnel in charge of the project, in his opinion, did not understand the essential problems involved; suggestions to improve the design based on practical experience in field trials by the Navy and testing at the CID Lab had been consistently ignored. Failure to correct these problems, he believed, would "seriously endanger the success of the gyroscopic sight."[11] In a note to Preston Bassett, Sperry's chief engineer, Draper identified four critical problems that he felt had to be solved if the sight was to be successful:

1. Temperature control of the damping fluid
2. Production of the variable-spring systems for range adjustment
3. Selection of thermostat units
4. Selection of bearings for the gyroscopic rotors.

Unlike the turn indicator, which relied on a simple, elementary form of viscous damping, such as an oil-soaked sponge, the Mark 14 had a liquid-filled cylindrical damping device fitted to the shafts on each gyro. The liquid had to be maintained at a uniform temperature to ensure that the viscosity did not change, which would have negatively affected the sight's performance. This necessitated a thermostatically controlled heating element. Draper believed that all four problems could be "solved by a systematic attack." He advised Bassett that "a good man should be placed full time" on each of these problems.

Draper's concerns must have been resolved, for when the initial deliveries of the Mark 14 were tried out at Dam Neck, they were enthusiastically received by the gunners. The sight was knocking down three out of every five target sleeves fired upon and was "outshooting tracer control by a huge

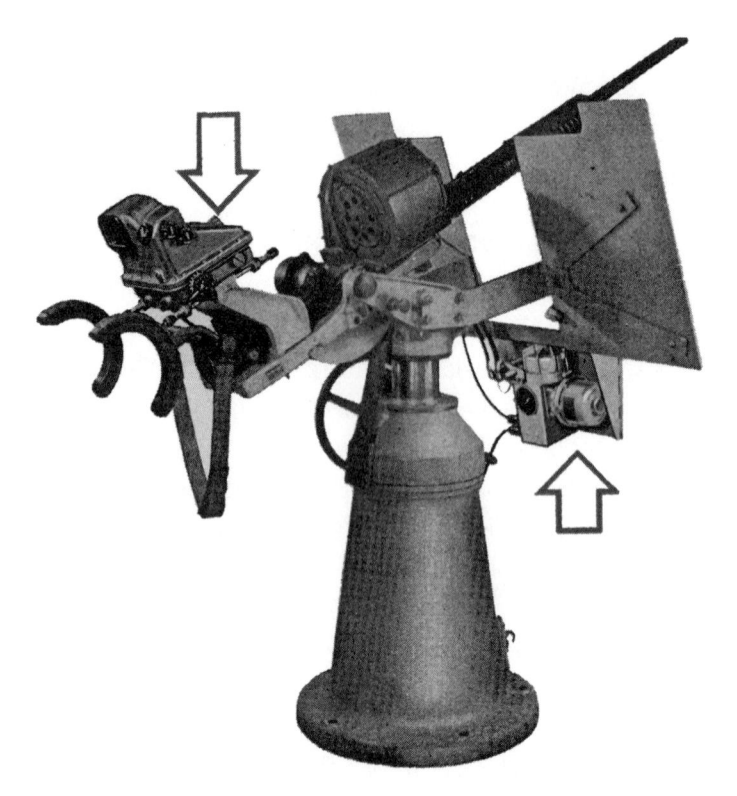

Fig. 6-1. The Mark 14 gunsight, shown above mounted on a 20-mm Oerlikon machine cannon, consisted of the sight itself and a power unit that supplied clean, dry air to spin the gyros. *U.S. Navy*

margin."[12] By July 1942 the Sperry-Draper sight had begun to appear in the fleet (although it would be months before the sight began to be produced in significant numbers).[13] It was used in action for the first time on August 24, 1942, when at least one sight, installed on a 20-mm gun on the battleship *North Carolina* (BB 55), engaged enemy aircraft during the battle of Stewart Island.[14]

David Mindell, Dibner Professor of History of Engineering at MIT, states that unlike the Navy's existing directors, which relied on the Ford rangekeeper for computing,

> the Mark 14 succeeded not because of the quality or precision of its computation but rather because of its compromises. Estimating range provided the most significant shortcut. Rather than using a

bulky and slow rangefinder, the operator merely estimated range by eye and then dialed it in by hand—a rough approximation, but the range to an attacking airplane was likely changing rapidly anyway. Moreover, because the device employed polar coordinates, such errors diminished in significance as the target got closer (in contrast, the prewar Cartesian directors exacerbated the errors at close range).[15]

The Mark 14 gunsight proved so useful that close to 85,000 of them were produced during World War II, at an average price of $1,300 per unit.[16] The huge success of Sperry's Mark 14 was a financial boom for MIT, Draper, and also Edward Bentley, who was listed as a coinventor on the patent, filed April 17, 1942. By the end of 1943 MIT had received $250,000 for the 26,000 Mark 14 gunsights delivered during the first two years of the war.[17] On December 20, 1943, Killian advised Draper that the executive committee had authorized the coinventors 7 percent of the income derived from the Navy's contract. Draper received $8,750 (50 percent) for his contribution to the project, Bentley received $6,562.50 (37.5 percent), with the remainder (12.5 percent) going to Joseph J. Jarosh—a 1931 alumni with a bachelor of science in mechanical engineering—and John Ashworth.[18] These were considerable sums of money, considering that the typical prewar salary for an assistant professor at MIT was only $3,000 per annum.

The success of the Mark 14 gunsight was due in large part by the transformation that had taken place when MIT created the Confidential Instrument Development Laboratory, with Draper as its director. The CID Lab was destined to become the de facto engineering design shop for BuOrd's antiaircraft-fire-control requirements. As Draper made clear years later in an interview with Barton Hacker, his laboratory "did everything": "We conceived the stuff; we did the mathematics; we did the design; we did the making of the parts; we did the assembly of the parts; and we did the testing. . . . We did everything till we came out with a piece of working hardware, in which case we then transferred the information to whatever company was going to manufacture the material."[19]

Once the officers in BuOrd became familiar with it, it quickly became apparent to them that Draper's lead-computing gunsight could solve

Fig. 6-2. The illustration, taken from a U.S. Navy training manual, shows how the Mark 14 gunsight allowed the gunner to lead the target.

another urgent problem facing the Fire Control Division: providing a suitable gun director for the 1.1-inch antiaircraft gun. Several designs for a suitable director were under development by various suppliers; none of them seemed promising. The solution was to incorporate Draper's lead-computing sight in a stand-alone director remotely connected to the mount. This would alleviate the problem of vibration due to the gun's recoil and provide a computing device less complicated and cheaper to manufacture than the rangekeepers that had heretofore been the centerpieces of the Navy's antiaircraft directors. It was this thinking that had led to the decision to allocate the five experimental sights that the Confidential Instrument Development Laboratory fabricated under subcontract to Sperry for the Navy's remote-control director.

In addition to the twelve Mark 14 experimental sights contained in Sperry's initial 1941 contract, the Navy insisted that Sperry supply four pedestal-type mounts.[20] Before the year was out, Sperry had a prototype, known as the "barber chair," ready for testing on 1.1-inch weapon. It was sent to Dahlgren along with one of the Mark 14 gunsights fabricated by the CID Lab, modified for the ballistics of the 1.1-inch gun.[21] The tests conducted at Dahlgren revealed a number of serious defects in the design of the "barber chair" that Sperry was unable to correct.

This was a serious setback for BuOrd's antiaircraft program, one that required drastic action on the part of Lieutenant Commander Murphy and Commander Hermann. The two traveled to Sperry's Brooklyn, New York, offices immediately after the New Year holidays to confer with Sperry's engineers. At a meeting on January 3, 1942, they decided to abandon the "barber chair" and start anew on a design, to be designated as the "Gun Director Mark 51."[22] Because of the wartime urgency, the two naval officers decided to assume direct responsibility for the new design, taking the unconventional step of recommending that a pilot production order for a thousand Mark 51 directors be authorized in advance of a satisfactory test. Their recommendation was formally approved by Admiral Blandy on January 16, 1942.

In the meantime, a preliminary design conference was held at the Sperry plant on January 13 to discuss how best to proceed with the design of the Mark 51 director. The scheme that emerged called for the development of a simple "dummy gun" director mounting a Mark 14 sight that would be capable of transmitting train and elevation orders remotely to the gun mounts. A second design conference, with Sperry and its subcontractor the Multiscope Company in attendance, was held at BuOrd's Washington, D.C., offices on January 23. What followed, according to BuOrd's documentation, was "a concentrated effort by various agencies to expedite the design and production facilities for the Mark 51 director."[23] Although not directly involved in the design of the pedestal mount, the CID Lab was responsible for ensuring that the Mark 14 gunsight would function successfully when installed on the Mark 51 director. This required design changes to accommodate the different ballistic characteristics of 1.1-inch and 40-mm guns for which the Mark 51 was intended.

While overseeing these modifications, Draper took time to collaborate with Gordon S. Brown, the director of MIT's Servomechanisms Laboratory, on the servomechanisms needed for remote control of the gun mounts that would use the Mark 51 director. The Servomechanisms Laboratory, established in 1940, had grown out of the Department of Electrical Engineering's response to a request made a year earlier by the U.S. Navy for a special course on fire control for naval officers assigned to MIT's graduate program.[24] While working on improving servomechanisms to control large guns, Brown developed an increasingly broad working relationship with the Sperry Gyroscope Company, which soon led to an agreement between Sperry and MIT to design servomechanisms for the Navy.[25]

Toward the end of March 1942, Hollis C. Walter, one of Sperry's engineers, suggested that Draper concentrate on the sights needed for the new director and not concern himself with the manufacturing problems associated with the production version of the Mark 14.[26] Draper refused to get out of the picture, however, because of problems that were likely to pop up with the Mark 14. As he explained in his response to Walter, "The hundreds of hours I have spent working on the practical difficulties of design and construction should enable me to speed up the changes which will be required to produce satisfactorily operating equipment. I will do anything within my power to help matters along."[27]

In April, the Confidential Instrument Laboratory shipped the eleventh Mark 14 gunsight (i.e., designated No. 11), assembled with ballistic corrections for the 40-mm gun, to the Naval Gun Factory.[28] From there it was sent to Dahlgren, where it was installed on the first Mark 51 director completed by Sperry for testing with the remotely controlled, power-driven, 40-mm quadruple antiaircraft gun. Tests of the No. 11 sight, along with the No. 1 sight, which had been set up to handle the ballistics of the 1.1-inch antiaircraft gun, were conducted between May 4 and 7, 1942.[29] Ed Bentley, project manager for the Mark 14, was there to represent MIT and provide whatever assistance was needed.

The Mark 51 controlled the gun mounts to which it was "slaved" via remote-control servomechanisms. Sighting was similar to the Mark 14, except the operator positioned the sight using a set of handlebars that moved the sight in elevation and traverse. The director's single operator tracked the target using the Mark 14 reticle, keeping the director pointed in the predicted direction of the target. The director's movements were then slaved to an 1.1-inch or 40-mm gun mount to which it was connected; the gun moved along with the Mark 51 director. Range was estimated by the operator at a fixed value, so that the lead angles would be approximately correct for at least one moment in time when the gunner fired at an incoming target. The results of the testing were mixed.

The No. 1 sight performed poorly, exhibiting erratic deflection. When Bentley tested it afterward, he found that it had excessive "dead space" (in effect, "slack") in deflection. Fortunately, this could be attributed to "considerable handling by personnel unfamiliar with its use" after the unit had been shipped to the Naval Gun Factory for instructional purposes at the Fire Control School there.[30] Capt. David I. Hedrick, inspector of ordnance in charge of the Dahlgren Proving Ground, did not consider this performance "indicative of the results to be expected with a new and properly adjusted sight"—an interesting turnaround for a man who had disapproved of Draper's sight when it was first tested at Dam Neck.

As for the No. 11 sight, with 40-mm ballistics, it gave a "creditable" performance at ranges less than a thousand yards. Hedrick believed that better performance would be obtained with better training and the development of a satisfactory technique for use of the sight. Notwithstanding any reservations, however, he recommended the adoption of the Mark 51 director, in light of the preliminary test results and "the urgent need for a simple and quickly produced director." Production ramped up quickly. The first production models of the Mark 51 director for installation in the fleet began arriving in June 1942.

7

Directors and Gun Fire-Control Systems

During the first year of World War II, the U.S. Navy relied increasingly on the new 20-mm Oerlikon and 40-mm Bofors guns to supplement the 5-inch guns that had previously been the fleet's main defense against aircraft. The Mark 14 sight and the Mark 51 director were important additions to the fleet, greatly improving the lethality of both systems. During this period the fleet had also begun the first tentative use of radar to direct the fire of its 5-inch, 38-caliber dual-purpose (i.e., surface and air) guns. The "5-inch/38," in conjunction with the Mark 37 director, was considered the primary antiaircraft weapon in the U.S. Navy. It had been placed on all new capital ships, destroyers, and naval auxiliaries starting in the mid-1930s.

In 1942, the Navy lost four aircraft carriers in five major engagements in the Pacific. At the end of the year, *Enterprise* (CV 6) was the only U.S. carrier afloat that had experienced and survived multiple attacks by Japanese aircraft. As the first year of the war drew to a close, her commanding officer, Capt. Osborne B. Hardison, sent a memorandum to the chief of the Bureau of Ordnance addressing the effectiveness of the ship's antiaircraft guns, directors, and fire-control radar.[1] The memo laid out their good points and the faults that had been revealed in action, and it offered suggestions for correction.

First to be discussed were the *Enterprise*'s 20-mm Oerlikons. The ship's forty-four single-mount 20-mm guns were chiefly responsible for the large number of aircraft that had been destroyed by the ship's antiaircraft batteries. Hardison considered the 20-mm an excellent weapon, despite its limited range and striking power. The heavy volume of fire put out by the weapons of this type—they were installed in the galleries along the flight deck—was thought to be in itself an effective deterrent to accurate bombing; *Enterprise*'s 20-mms were believed responsible for the low percentage of bombs dropped against the ship to hit. Although the ship's gunnery officers considered it an excellent piece of equipment, the gun's Mark 14 Mod 2 sight had a number of defects that needed to be corrected before it could be considered truly effective. In their opinion, the sight's window was too small and susceptible to fogging; its mechanism was sensitive to vibration and susceptible to failure; and the sight could not be rapidly shifted from one close-in target to another without upsetting the gyro. Most of the gunners on the *Enterprise*, according to the ship's gunnery officers, still believed they could obtain more hits using tracer control.

Some of the aforementioned defects were addressed in an upgraded version of the sight designated the "Mark 14 Mod 3," but not all of them. Keeping the reticle (the cross hairs in the eyepiece) on target continued to be a problem for most gunners. As they attempted to follow the target, the reticle often lagged behind. This meant that it had to be moved ahead more rapidly. Lt. Robert E. Wallace, a former gunnery instructor and the officer in charge of the automatic weapons on board the battleship *Idaho* (BB 42) would later recall "that the reticle was indeed a jumpy sort of thing. It seemed to bounce all around inside the eyepiece."[2] To be effective, however, the Mark 14 had to be moved steadily and, above all, smoothly. "Those two gyroscopes" Wallace was sure, "knew exactly what they were doing and they tried their best to respond to the directions they received. Given smooth and steady movement, they worked like a charm."

Gunnery officers charged with training personnel in the use of the Mark 14 quickly realized that its effectiveness could be improved with more training and better understanding of the sight's operation. A training manual produced later in the war emphatically pointed out, in capital letters, that a Mark 14 user's "toughest job is TRACKING SMOOTHLY. . . . YOU CAN DO IT IF YOU PRACTICE."[3] The manual recommended that the operator "move the gun with smooth, steady motion and avoid jerky or uneven action."

Next to be reviewed in Hardison's memorandum were the newly installed 40-mm Bofors guns. "The 40-mm gun," he reported, "appears to be an excellent intermediate caliber weapon and excellent results were obtained with it in our last action."[4] Its greater range, relative to the 20-mm, enabled gunners to start firing accurately at dive-bombers as soon as they could be seen. Like any new weapon, however, it had a number of teething problems that had to be fixed to reduce jamming, improve firing, and provide for low-angle shooting.

Hardison did not mention the Mark 51 in connection with the 40-mm. But the Mark 51s in use on board *Enterprise* must have performed well, for he suggested that they should be modified for use with the 5-inch/38 guns as well. The highly complex Mark 33 director* that, with its crew of several men, controlled the 5-inch/38s was proving ineffective against the dive-bomber. The most urgent need—according to *Enterprise*'s gunnery experts—was some means of "one-man control" to replace the separate Mark 33's pointer and trainer. It was impractical, they reported, "for two men with no means of communication to bring the gun on the same target, let alone use proper lead, when there are several targets in [the] same general area."[5] The use of the Mark 51 director would allow each 5-inch/38 to be directed independently so that a large number of attacking planes could be fired on simultaneously. A service representative from General Electric (manufacturer of the Mark 51) had been interviewed with regard to the possibility of installing Mark 51 director to control the pair of 5-inch/38 guns mounted in each of the four sponsons strategically located around the flight deck. The installation, according to the service "rep," would be simply a matter of wiring. "Heartily concur," wrote an unidentified reader in the margins of BuOrd's copy.[6]

The last item covered on Hardison's report was the Mark 33's radar, then designated the "FD" radar. "At no time since its installation has [the FD radar] been useful," wrote Hardison in the after-action report.[7] No one on board fully understood either its function or operation, and its electronic transmitters and receivers had been improperly sited. Information about the problems with the Mark 33 director and its associated radar was passed to Vannevar Bush, who had taken leave from MIT to

---

* The Mark 33 director predated the Mark 37 but was essentially the same in function.

run the wartime Office of Scientific Research and Development (OSRD). Bush contacted Merle Tuve, head of the National Defense Research Committee's (NDRC's) Section T, which had developed the proximity fuze. Development work on the fuze was now winding down, and Tuve was looking for a new project for his scientists. Fire control was a natural fit for the section, which had formed a close working relationship with BuOrd personnel. Bush on January 11, 1942, requested in writing that Tuve meet with Harold Hazen, head of NDRC Division 7, to examine "the possibilities of a self-contained fire-control equipment for naval 5-inch antiaircraft guns."[8] Tuve and Hazen duly exchanged visits during the next few weeks. The two men called on Draper while in Cambridge, where Hazen's office was located, to discuss the status of his continuing work for the Navy.

At the time of their visit, Draper was heavily involved in completing the first four models of the Mark 15 gunsight, under development for BuOrd. The new sight was a more rugged version of the Mark 14, with a telescopic sighting system for more accurate tracking at greater ranges.[9] It could also correct for drift and ballistic wind and compute a dead time for fuze setting. The Mark 15 sight was part of a new, lightweight, inexpensive director, the "Mark 52," that was designed to control [existing guns] longer-range antiaircraft guns. Such a director was desperately needed to control the 3-inch/50-caliber deck guns being fitted to destroyer escorts rapidly nearing completion. As of then these ships lacked antiaircraft directors and could ill afford the weight or space of the tacheometric directors like the Mark 33 and Mark 37 already in wide use. Unlike Draper's sights, which provided a lead angle by measuring target-position rate of change in the horizontal and vertical planes, tacheometric directors—highly complex instruments that depended on analog computers—generated target position, speed, direction, and rate of target-range change on the basis of observed target data. This information was used to calculate the elevation and bearing of the antiaircraft guns required to hit the target, given its predicted movement. The Mark 37 required a seven-man crew, two officers and five enlisted men.

The Gun Director Mark 52 consisted of a Mark 15 sight installed on an improved Mark 51 pedestal stand and a radar that was capable of supplying range information but lacked sufficient angular accuracy to permit radar-controlled fire at unseen targets—"blind firing," in naval parlance.

Continuous range information, read by a radar operator below decks, was electronically differentiated to determine range rate, which was transmitted electronically to the Mark 52 gunsight, along with the range. Both pieces of data were used to compute fuze time for transmission to the guns, which in turn fired time-fuzed projectiles.

Once the Navy decided to go ahead with the Mark 52—or any other Draper fire-control device, for that matter—dozens of draftsmen and engineers from Sperry would suddenly show up at MIT.[10] The administration would clear out a couple of classrooms, make them collectively a restricted area, and set up drawing boards where engineers and designers would convert Draper's ideas into production models. Robert Seamans describes what occurred when Draper saw a production model: he "would be fuming at what happened to his designs when they went into production. They obviously had to make changes in the design, in order to mass produce, and in so doing, Doc—and he was always very vocal about this—felt that they had really jeopardized the design to some extent, or its accuracy."[11]

Work on the Mark 52 director proceeded rapidly. By the end of October 1943, the first fourteen units were ready for delivery to the Navy for shipboard and training use. In service, the medium-range Mark 52 director proved successful in controlling the 3- and 5-inch guns on destroyer escorts and auxiliaries. Seven hundred thirty-nine of the directors were delivered before the end of 1945.

While Draper was working to perfect the Mark 15 and its director, Tuve and Hazen continued to investigate the status of antiaircraft fire control with an eye toward the development of a new, radar-controlled director for the 5-inch/38, one that could be put into service quickly. In early March 1943 both men concluded that the most expeditious approach was to adapt components of fire control that were in existence or under development. A variant of the director based on Draper's Mark 15 sight seemed the most promising. "[Draper] has the entire development rather well thought out in a primary way," Hazen reported, "and has a considerable part of it reduced to operating models."[12] Section T, which had already begun to work on the project, was considering a variation of the Draper's disturbed-line-of-sight principle using radar inputs to assist the operator.[13]

Fig. 7-1. The Mark 52 director consisted of a Mark 15 sight installed on an improved Mark 51 pedestal stand mounting a radar capable of supplying range information. *U.S. Navy*

Hazen was very willing to provide assistance to Section T and readily agreed to furnish Tuve with pertinent results of Division 7's work on fire control. Not everyone was happy with this arrangement. There were those in Division 7, such as Duncan J. Stewart, who resented the incursion of Tuve's Section T on their "turf." Dr. Tuve and Dr. Robert B. Brode (a physicist then working with Tuve), he complained, "have no more business messing around with a director design than I have with proximity fuses."[14] Others, such as Warren Weaver—for personal reasons—wanted "nothing to do with Tuve."[15]

By April 1943, Section T's extremely close relationship with the Navy and its intrusion into the domain of Division 7 had "sparked a violent debate" on the nature of the relationship between Section T and the rest of the civilian research establishment.[16] Regardless of where Section T appeared on the OSRD organization chart, most of the scientists in Division 7 believed that it was really part of the Navy. This notion can be attributed to the fact that Section T's activities were by this time under the umbrella of the Applied Physics Laboratory (APL) of the Johns Hopkins University, which in turn was funded primarily through Navy contracts overseen by Capt. Samuel Shumaker. The rumbles of dissent from Division 7 forced Bush to explain the rationale for selecting Section T to solve the 5-inch/38 fire-control problem: "The Navy, through Admiral Blandy, has specifically requested that Section T be maintained as a unit so that it will be immediately available in case emergency work is needed in the field for which it was originally created [i.e., the proximity fuze]. At the same time the Navy specifically requested that Section T be authorized to undertake the development of fire-control equipment for the Naval 5-inch antiaircraft guns."[17] Bush had accordingly authorized Section T to aid Draper in completing the development of the Mark 52 director as rapidly as possible and to undertake the independent development of a new director suitable for use with the 5-inch/38.

In August 1943, while work on the Mark 52 director was progressing, APL began discussions with the Navy on an entirely new director for 40-mm antiaircraft guns, one with capability for blind firing at night and in poor visibility.[18] Blind firing was urgently needed in the fleet to combat the growing threat of night attack by torpedo planes. The new director was initially designated the "Gun Director Mark 57." Later, as more components were added, the Mark 51 "handlebar" director became one part of an integrated fire-control system, the "Gun Fire Control System (GFCS) Mark 57." Although the Mark 57 director resembled the Mark 51 in appearance, it operated on the line-of-sight principle, in which the tracking head was always pointed at the target and not along the line of fire as in Draper's sights.

While Tuve's APL group was developing the Mark 57 GFCS, Draper began work on his own version of a blind-firing director, to be called the "GFCS Mark 58."[19] The two teams worked in parallel: Tuve concentrating on the line-of-sight gyro system, Draper sticking with the combat-proven "disturbed reticle" system he had invented. Because blind firing required a much narrower radar beam, Draper was forced to replace the Mark 52's ten-centimeter Mark 26 radar with the more advanced three-centimeter Mark 29 Mod 1, which had been developed for APL's Mark 57 GFCS. In order to keep the radar on the target, Draper's team had to design a means of displacing the forty-five-inch antenna from alignment with the director axis so that it could remain parallel to the line of sight. They accomplished this using electrical pick-offs from the gyros to transmit line-of-sight displacements to the antenna mount. Target data from the radar was projected into the Mark 15 sight's field of vision, allowing the operator to stay on the target. The Mark 58's handlebar director was not powered, however, and wind loading on the antenna made it difficult for the pointer to maintain control. To solve this problem Draper's team switched to the Mark 34 radar and moved the antenna to the gun mount. The new arrangement was designated as "Gun Fire Control System Mark 63."

When in June 1944 the time came to conduct trials of the Mark 63 prototype at Dam Neck, Draper was there, along with Bob Seamans to ensure that everything went smoothly. They stayed in Virginia Beach, at a place right next to the ocean called the Gay Manor. They did not think much of the place, which they glibly referred to as the "Gay Manure."[20] Seamans shared a room with Draper for several weeks while the new director was being tested.[21] Each day they were up before dawn, then drove in "a gangster type black Cadillac" to a local hash house for breakfast, and arrived at the base by seven o'clock for a full day's work.[22] Evenings were spent discussing improvements that might be made to the equipment under test. "Creativity was improved by the consumption of

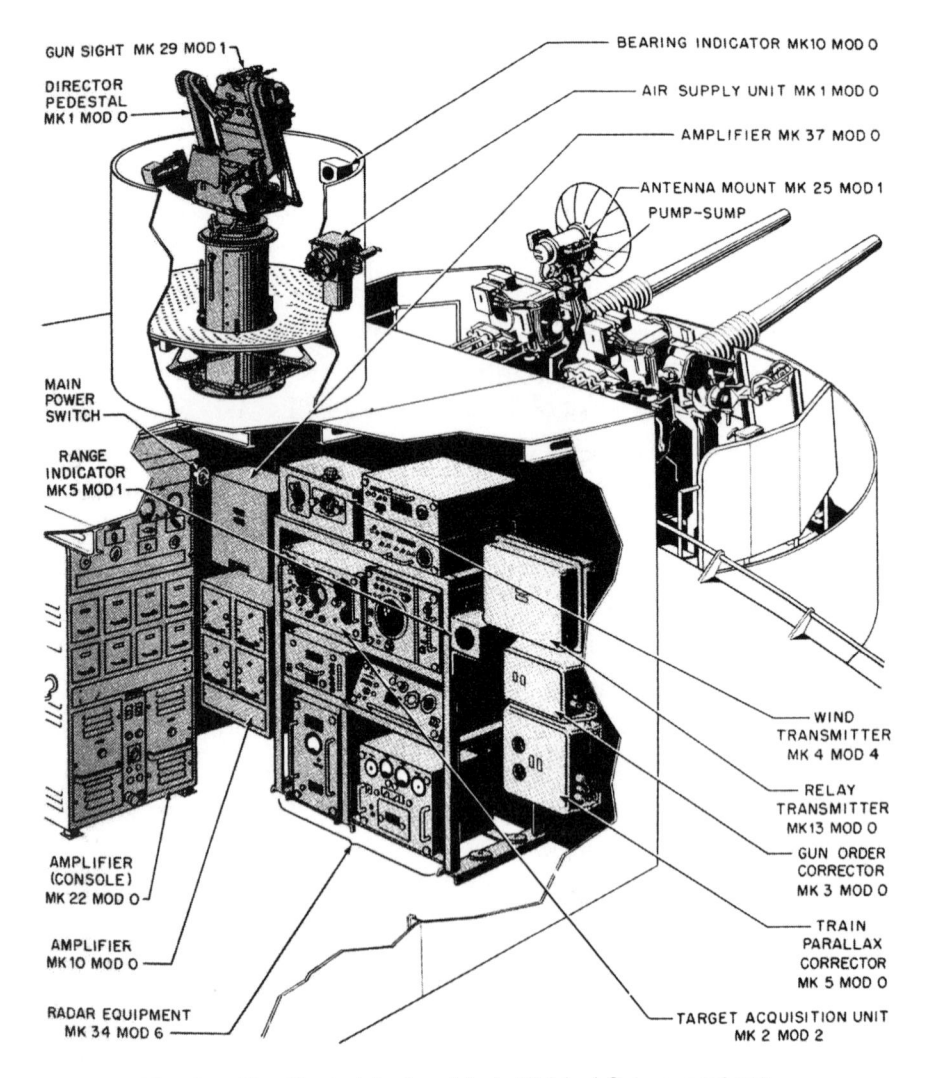

Fig. 7-2. The Gun Fire Control System Mark 63 Mod 2, in a postwar configuration controlling the fire of a dual 3-inch/70 gun mount. *U.S. Navy*

'soothing syrup,'" Draper's euphemism for the alcoholic beverages that he consumed on a regular basis and cherished as a social lubricant.

Virginia was then a "dry" state, but that did not keep Draper from his evening cocktail. A number of months earlier, John B. Nugent, an employee of the instrument laboratory and commodore of a yacht club near Cambridge, had asked Draper to address a gathering of its members. As an honorarium, Draper received a case of liquor.[23] Because the MIT campus was alcohol-free, he had the box marked the "John B. Nugent Medicinal Aid Foundation." While Draper was in Virginia the John B. Nugent Medicinal Aid Foundation supplied "nonlinear damping fluid" to Dam Neck.[24]

Seamans recalled many years later never having a problem getting to sleep during their stay in Virginia, but that was not the case with Draper, who had to read a book to take his mind off the day's activities in order to sleep. As he explained to Seamans at the time, "One thing you never want to do is bring along a very good book, because you'll tend to read it and stay awake too long and then you'll be tired the next day. You want to have a book that's so bad the you'll fall asleep on about the third page."[25] Seamans tells us that Draper would start reading "and all of a sudden he'd just conk out and the book would fall on his chest." The next day they would be up at six and do it all again. In the evening Draper would have drinks and socialize with the naval officers involved in hopes of finding out their reactions to his equipment and the day's testing.

By the time the firing trials of the Mark 63 got under way, the words "Confidential" and "Development" had been dropped from the lab's name. The problem was that at a distance, the *D*—as in "CID Laboratory" printed on identification tags—looked like an *O*.[26] This made "CID" look too much like "CIO," the then-well-known acronym for the Congress of Industrial Organizations, a union. MIT's administration changed the name to simply the "Instrumentation Laboratory," which it remained until MIT divested itself of the laboratory years later.

The first of the 402 Mark 63 GFCSs delivered to the Navy during World War II went to the aircraft carrier *Bon Homme Richard* (CV 31) in November 1944. The Mark 63 was the last fire-control system designed by the Instrumentation Laboratory to enter production during the war. After the war ended, the Navy terminated virtually all work on fire control.

The laboratory's staff dropped from between thirty and forty people to just one or two until 1946, when the Navy awarded a contract for the development of an advanced fire-control system based on a new design, the "Mark 64," on which work had started before the war ended.[27] It was soon named "Gunar," and the Mark 64 designation would be abandoned.

Gunar was ultimately developed to control the fire of 3-inch/50, 3-inch/70, and 5-inch/54 guns against high-speed aerial targets.[28] Like all of the previous fire-control systems developed by the Instrumentation Laboratory under Draper's supervision, it relied on the disturbed-line-of-sight principle used on the Mark 14 gunsight. Gunar was engineered as an on-mount antiaircraft fire-control system with stabilization and automatic tracking.[29] Its components were able to handle greater ranges and faster changes in range-rate than had previous systems. Further refinements to ballistic corrections were added to ensure accuracy at longer ranges. Also, while signals from Gunar's radar controlled the gun mount in the automatic mode, the system could be operated in a secondary mode using an optical telescope.

The finished design that emerged from the MIT Instrumentation Laboratory was packaged in two versions: an on-mount system known as Gunar Marks 1, 2, and 3 Mod 1; and a stand-alone director, the Mark 69 Mod 1 GFCS.[30] Both versions were capable of tracking a target at speeds up to 1,200 knots at ranges out to 40,000 yards. Modifications ("mods") continued over the next ten to twelve years. At that point, because its interest in air defense had shifted away from gunfire toward radar-guided missiles, the Navy made an across-the-board decision not to upgrade shipboard antiaircraft guns.[31]

Gunar was never fitted on board a Navy ship. The system was sold to the Canadian navy, however, which installed it on *St. Laurent*–class destroyers. Though Gunar achieved only limited impact, it provided Draper with the inspiration for the "floated gyro," which was to become the critical element in the future success of the Instrumentation Laboratory. Bob Seamans remembers carrying out preliminary design work for the Gunar system in conjunction with Draper and a couple of designers working on a conceptual layout of the hardware required. The instruments they built included a very large gyroscope, so large that Draper put one in a glass case for display in his office. "He said it was very helpful to look at something that was really a bad design, so that every time he felt that he was really doing something very well, he'd look at that and realize, it might be as bad as the design that he and I came up with."[32] The huge gyro proved to Draper that the precision of a gyro could not be improved by making it bigger. Seamans concluded that "[Draper] had to take [a] new approach, and that's when he started the idea of a floated gyro."

8

The A1-C(M) Gunsight

In March 1943, while Draper was still heavily involved in developing antiaircraft directors for the U.S. Navy, he received a visit from Lee Davis, who, now a lieutenant colonel, was about to take command of an Army Air Forces dive-bomber squadron.[1] Draper had first come in contact with his young protégé four years earlier when Davis was an instructor in the Department of Natural and Experimental Philosophy—which we would now call "physics"—at the U.S. Military Academy at West Point, New York.

Davis, himself a West Point graduate, had been posted to the Military Academy in January 1939, after serving a two-year tour of duty as a pilot with the 6th Pursuit Squadron at Wheeler Field, Hawaii.[2] The squadron was equipped with the vintage Boeing P-12 and the outdated Boeing P-26. These open-cockpit biplanes with fixed landing gear were the primary fighters in the Air Corps inventory when Davis joined the squadron in 1936. Davis had originally been slated to attend the Air Corps Tactical School at Maxwell Field, Alabama, but his orders were changed when someone in charge of assignments discovered that Davis had achieved good grades in the mechanics course while a cadet at West Point. This made him the ideal candidate to fill the vacancy left when Maj. John M. Weikert, who had been teaching the course, had been reassigned.[3]

Davis' teaching duties included supervision of the thermodynamics laboratory, where a variety of engines—diesel, gasoline, and steam—were used to demonstrate the fundamental principles of thermodynamics. The steam engines had engine-indicator diagrams that could trace out the pressure-volume relationships occurring inside the steam cylinders. The steam engine indicator was a mechanical device that could not be used with an internal combustion engine, because of the high speeds of the pistons and the extreme pressures generated within each cylinder. From his experience as the squadron's communications officer in Hawaii, Davis knew that an oscilloscope could be used to plot the physical parameters of what took place during such high-speed phenomena as occurred within the cylinder of an internal combustion engine.[4] After talking it over with the other instructors, Davis decided to visit the Dumont Laboratories, one of the leading companies in the oscilloscope business.[5] Dumont, which was located in Passaic, New Jersey, was only a few hours' drive from West Point.

Because Davis was on the staff at the U.S. Military Academy, he was able to get into Dumont and talk to their chief engineer. The engineer was not able to help with Davis' problem, but he knew about the work being done on a crystal pickup by the Radio Corporation of America (RCA). He suggested that Davis talk to them. So Davis drove to Camden, New Jersey, where the RCA research and development laboratory was. Once again, he managed to wangle his way in and talk to the engineers. They did indeed have a pickup, but it was too sensitive, more a vibration pickup than a pressure sensor. The RCA engineers suggested that Davis contact the Sperry Gyroscope Company, where a researcher had developed an electromagnetic pickup.[6]

The RCA people were referring to Joseph Lancor, an electrical engineer who had helped develop MIT's pressure sensor. Davis, during an interview conducted in 1973, described their first encounter:

> We sat down and talked about the problem [of plotting the cylinder parameters in real time]. He pulled out of the desk a little metal thing about the size of a spark plug, which was the pickup that you screw into the cylinder head if you have an extra spark plug hole like in aircraft engines where you have two spark plugs. Or otherwise you make an adapter; and then drill a hole and fit it in the place to get the pressure inside the cylinder head. He says, "Well, this puts out a signal which is rate of change or pressure. You've got to

integrate it, make an integrating circuit, in order to get what you want, which is pressure-time. Then you have to have a representation of volume, of course. Why don't you see what you can do with it[?]" He handed the thing to me, and so I took it back up to West Point and talked with the electrical engineering instructors trying to figure out what an integrating circuit was.[7]

When he got back to West Point, Davis also showed a sample of the sensor Lancor had given him to the instructors in the Electrical Engineering Department and was able to borrow an amplifier. He began fooling around with the equipment in the thermodynamics laboratory until he got it to work. Davis made photographs of the oscilloscope traces and sent them to Joe Lancor. Lancor forwarded them to Draper, who received them enthusiastically. In June, Draper came down from Boston to look at Davis' setup. He spent a week with Davis, as his house guest, while they made changes to the circuits, took pictures, and worked on a paper that, with Joe Lancor as a third coauthor, was presented at the Second Annual Summer Meeting of the Instituted of Aeronautical Science in Pasadena, California, between June 24 and 26, 1940.[8]

Draper's interest in Davis' work led the latter to transfer to MIT for a year of graduate study in aeronautical engineering during the 1940–41 academic year. There Davis learned about the gyro gunsight, which was in the early stages of development. As the reader will recall from chapter 5, it was Davis who arranged for Draper to use the range at the Watertown Arsenal to test the laboratory prototype against a moving target. Davis must have been present during the first day of the testing, for as he recalled some twenty years later, "Ed Bentley and Doc were working on the tracking problem by controlling the viscosity of the damping fluid by temperature adjustment. I couldn't quite understand what they were driving at, but Doc was very patient in explaining the dynamics of the problem of keeping the cross hairs on a moving target when the computing process had as its major input the motion of the gun which carried the gyros."[9]

After receiving his master of science degree in the spring of 1941, Davis returned to West Point as a mathematics instructor. A year later he was made the director of the West Point's ground school and was kept on for another year. At the beginning of 1943, Lieutenant Colonel Davis received orders to command a fighter-bomber group, flying North American A-36s, that was about to deploy overseas.[10] Before his transfer he flew to Alabama, where the squadron was in training, to meet his new boss and look over the A-36—the ground-attack version of the P-51 Mustang. "I was amazed to see that they still had ring and bead sights in the middle of the damn thing," Leighton exclaimed when he recalled the event in later years.[11] It was the same sight used in the P-12s that he had flown in 1936.

It appears that it was after returning to West Point that Davis journeyed to Norfolk for the meeting with Draper mentioned at the beginning of this chapter. Draper was involved in testing at the Dam Neck antiaircraft range, and Davis wanted to discuss the possibility of adapting the Navy's Mark 14 gyroscopically controlled gunsight to the dive-bombing problem.[12] Based on Davis' input, Draper sketched out a preliminary design, along with specifications, on a piece of grocery wrapping paper.[13] With Draper's assistance, Davis prepared a technical report on how a gyro computing gunsight could be used as a dive-bombing sight, which he submitted for endorsement to his commanding officer, Col. John Weikert, whom Davis had replaced at the Military Academy in 1936.[14] Colonel Weikert sent the report to Wright Field, Ohio, where it was analyzed by Dr. John E. Clemens, a civilian engineer employed by the Armament Laboratory (one of the activities later combined as the Air Force Research Laboratory) as its chief scientist. Davis' report must have been favorably endorsed by Wright Field, because his orders were changed by Gen. Henry H. "Hap" Arnold, the chief of the Army Air Forces, who just happened to be a friend of Weikert's. Instead of Alabama, Davis was posted to the Armament Laboratory to work on a lead-computing gunsight for dive-bombers.

When Davis reported to Wright Field in June 1943, he was assigned as the project officer for the new sight, to be developed with Draper's help.[15] That summer, a Mark 14 gunsight was modified and mounted on a Douglas A-24 dive-bomber—the Army's version of the Navy's SBD Dauntless.[16] Good results were obtained, and the Armament Laboratory initiated "Project MX-402" for the design of a gun-bomb-rocket sight for fighter aircraft.[17] MIT was awarded a contract to build and test three gunsights within six months at a cost of around a hundred thousand dollars.[18] The plan was to develop, prove out, and provide documentation by which sixteen preproduction models could be manufactured to be taken overseas by Colonel Davis for evaluation in combat.

Bob Seamans was one of the laboratory personnel assigned to the project. He started out by analyzing the ballistics with reams of graphs and calculations. Seamans spent a good deal of time looking through this information with Draper and using their findings to create the mechanical

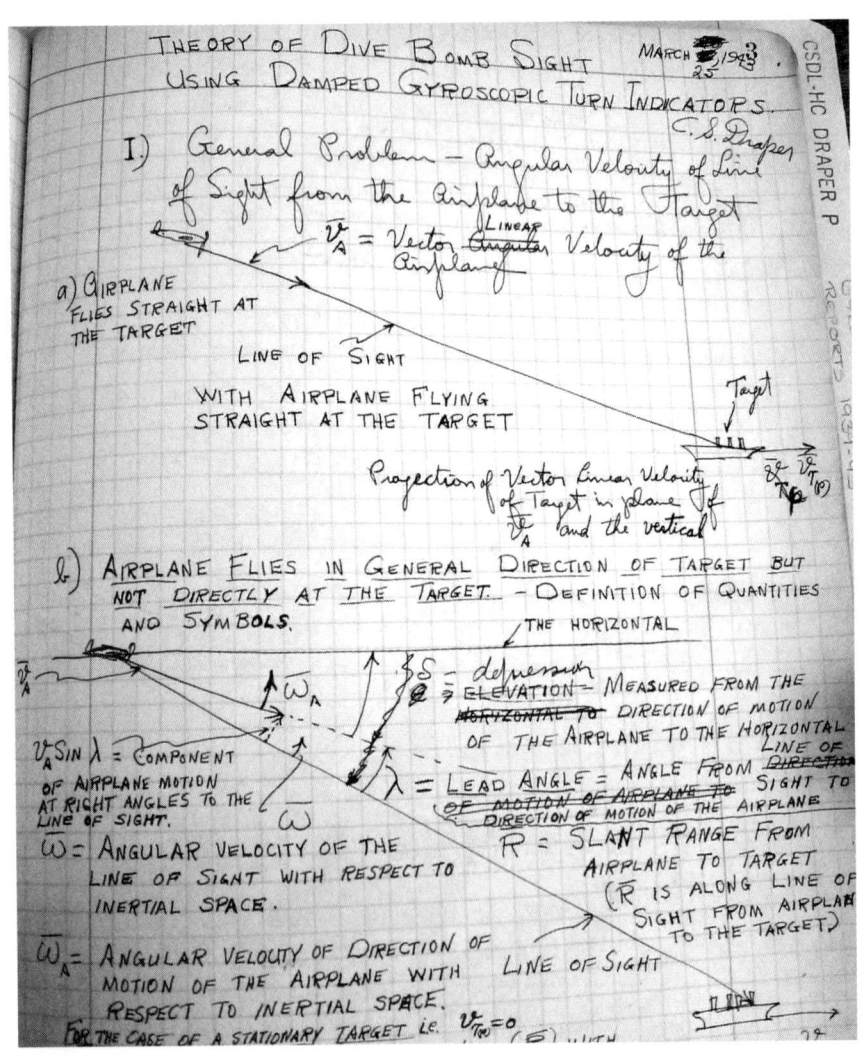

Fig. 8-1. A page taken from Draper's notebook dated March 23, 1943, showing his note on the theory of a dive-bomb sight using a damped gyroscopic turn indicator. *CSDL Collection, MIT Museum*

analogs for computing the proper lead. He worked out a design for a sight and was quite proud of it. Lee Davis and Draper looked at it. "That's fine on paper," they said, "but it's too complicated, it won't stand the rigors of airplane flight. We've got to build something that's more rugged than that if it's going to do the job."[19] So Seamans went back to the drawing board, came up with another design, and put it on a test stand.

Years later, during an oral history interview conducted under the auspices of National Air and Space Museum in 1987, Seamans left a pretty good description of how the design came together:

> Lee Davis was a man who wanted to move his program very fast, so in a matter of literally four or five months, we put together the instrumentation—these were gyroscopes with dampers and springs and so on. But instead of using mechanical constraints, for the first time we started using electromagnetic devices, and then instead of using direct connection between the gyros and the mirrors that moved the reticle around, there were servo-drives so you didn't have to have the computer right up in front of the pilot. It could be down somewhere else in the airplane, so only the optical equipment needed to clutter up the cockpit.[20]

The first experimental gunsight completed by the Instrumentation Laboratory computed only the dive-bombing solution.[21] Designated as the "A-1," it was installed by Seamans in a Lockheed P-38 and tested by Davis on the bombing range at Grenier Field, New Hampshire, during the early summer months of 1944.[22] Davis dropped dozens of bombs, struggling to overcome problems of tracking stability and solution time. Davis discovered that the experimental sight was unable to provide a proper solution to the ballistic problem in steep dives, developing a condition that Draper had previously termed "tracking instability." This phenomenon occurs whenever the rate of change of the lead angle changes faster than the motion of the sight line. At the steep dive angles that Davis was attempting the P-38 had minimal directional stability, making it very difficult for the pilot to keep the sight's reticle—frequently referred to as the "pipper"—on the target long enough to establish an accurate bombing solution. The "fix" was to decrease the dive angle.

Flight testing of the second experimental sight, which could be used for both bombing and gunnery, began at Elgin Field, Florida, also during summer 1944. The tests clearly demonstrated that the A-1 was superior to the fighter sights then in use.[23] The A-1 could compute leads accurately at ranges up to two thousand yards and solve bombing problems up to an altitude of ten thousand feet at dive angles of between fifteen and sixty degrees.

World War II ended before the new sight could be placed in production. After the end of hostilities in August 1945, the Army Air Forces authorized the installation of A-1 sights in the Republic P-84 jet fighter, then under development.[24] Additional funds for the A-1 project were approved for test flights in the Lockheed P-80A aircraft. The purpose of the tests was to determine what modifications to the sight would be necessary and how well it would operate with a ranging radar. By then, Davis was no longer directly involved with the project, having been promoted to technical executive of the Armament Laboratory. Nevertheless, he kept in close contact with Draper and the progress being made.

In March 1948, the U.S. Air Force (established in 1947) decided that both the F-84 and North American F-86 would be equipped with the newest version of the A-1 sight, the "B" model, which was considered to be more accurate than the K-18 sight that had entered service in World War II. When its radar was operational, the A-1 could lock onto a target out to a range of 5,400 feet. The pilot would then check visually to ensure that he was locked onto the proper target (if not, he pressed a target rejection switch and maneuvered until the radar locked onto the target desired) and "caged" the sight's gyros by pressing a button on the control stick grip. After placing the reticle dot on the target and releasing the caging button, the pilot had to keep the reticle on target for one-half to one second (the time needed by the sight to compute the ballistic solution) before opening fire in one-second bursts.[25]

Range was set manually whenever the radar was inoperative or the aircraft was below five thousand feet (when ground clutter would interfere with the radar return). Manual ranging was stadiametrically set by entering the target's wingspan on the sight head target span wheel and rotating the range control on the throttle grip until the reticle contracted to its minimum diameter. As in radar ranging, the pilot pressed the electrical

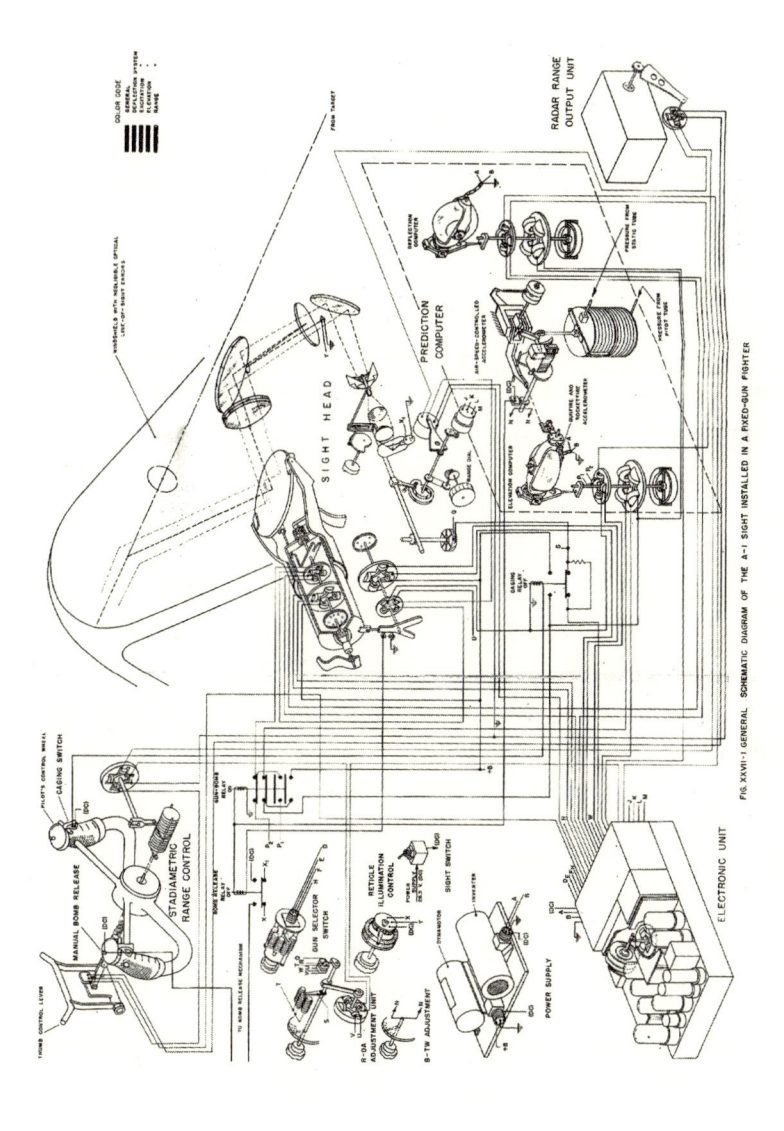

Fig. 8-2. Schematic diagram of the A-1C lead-computing gunsight. The sight computed kinematic lead based on the input from two internal gyroscopes based on range information supplied automatically by external radar, or manually via stadiametric control. *U.S. Air Force*

caging button and maneuvered the aircraft so that the reticle dot was on target. He established a smooth track and, when the target's wingspan filled the ranging circle, uncaged the gyros, waited a split second (i.e., one solution time), and began firing. If the computer or radar circuits of the sight were inoperative, the pilot could cage the sight mechanically and use the fixed reticle for rule-of-thumb gunnery.

During the fall of 1948, preliminary firing tests with F-84 and F-86 aircraft revealed a problem that was to persist throughout the life of the A-1 sight program—"reticle jitter." Whenever the pilot pressed the firing button, the vibration of the guns either drove the sight reticle entirely from view or caused it to oscillate so rapidly that it became an orange blur.[26] Flight tests to determine the best method of reducing reticle vibration to an acceptable level began at Muroc (later Edwards) Air Force Base, in California, in January 1949. The Sperry Gyroscope Company, which had been contracted to manufacture the first ninety-four A-1B sights, came up with a "fix" consisting of stainless steel stiffeners for the sight head mounting brackets. This reduced vibration considerably but did not entirely correct the problem.

That April, Sperry and the AC Spark Plug Company, the second source for the A-1B sight, both agreed to produce 551 A-1C sights, with improved computing features, for the F-86A, F-86D, and Lockheed F-94 aircraft. Full-scale production was scheduled to begin in August 1950, but the Air Force suspended deliveries of all A-1C sights until some method was found to make the sights more usable in the field. After a short period of review, the Air Force's Air Materiel Command, at Wright Field, authorized Sperry to modify thirty-five A-1C sights with a more efficient sight head and computer heating system, a brighter reticle, and special stiffeners to reduce the reticle vibrations. The result was designated as the A-1C(M).

The A-1C(M) was packaged in the nose of the F-86 Sabre, and its output was displayed in front of the pilot, who flew the plane so that the target and the reticle lined up. Once the target was properly aligned, the pilot released the caging switch, which held the prediction angle of the gyro gimbals to zero, and the sight began to generate a prediction angle. A period equal to approximately half the time of the projectile flight from gun to target was required to reduce the error of the prediction angle to 5 percent of the total angle. The sight corrected for velocity of both the target and the attack plane, rotation of the attack plane, gravity drop of the projectile, air density, and range.

Most aviation historians of the air war in Korea agree that the A-1C sights were far superior to those on the MiG-15 and gave the American pilots a great advantage—when it worked. The A-1C(M) and its radar differed radically from the gunsights they superseded. Spare parts, components, and test equipment for the new sight did not reach the Far East until mid-1951.[27] Both gunsight and radar were beset with a multitude of maintenance problems, owing in part to rough runways, which jarred delicate electronic components, but even more to a dearth of trained personnel and the test equipment needed to keep the system operational. Keeping the radar itself operational proved particularly daunting. It did not take long for maintenance personnel to discover that once an electronic component failed, the entire radar had to be replaced—not just the

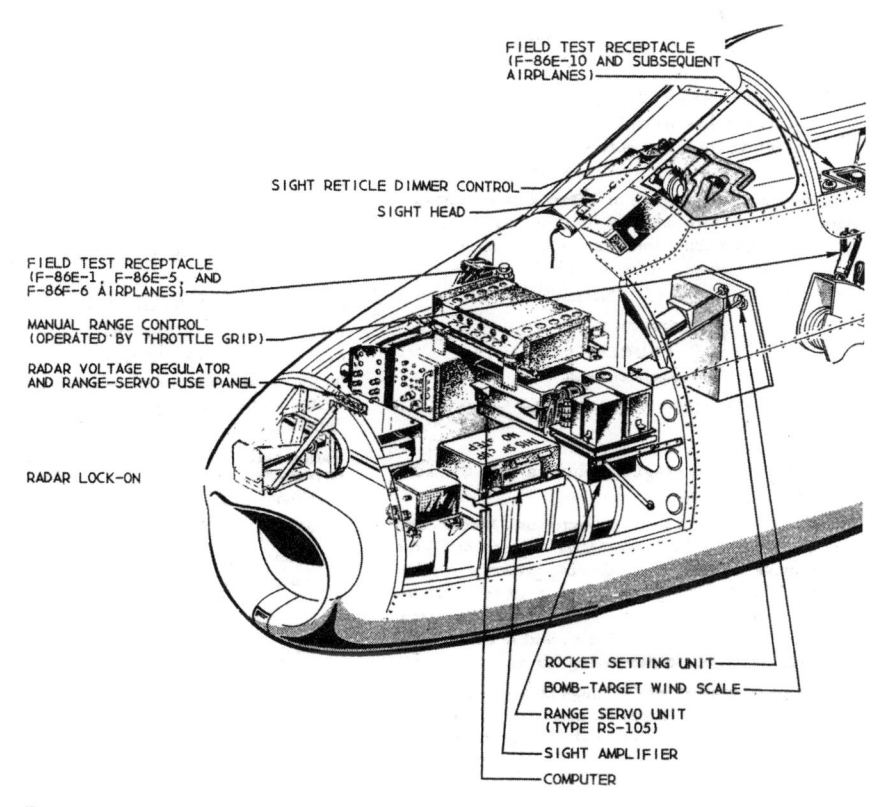

Fig. 8-3. A1-C and its associated support electronics and controls installed in an F-86 fighter. *U.S. Air Force*

faulty module.[28] At best, the radar performed erratically in clouds (due to water vapor), would sometimes break radar lock, and would not work at low altitude due to ground clutter.[29] All of the above, combined with a general lack of pilot training in the proper use of the A-1C(M) sight, soon led many of the Sabre pilots to become disenchanted with it. These pilots favored the clearly inferior, but dependable, K-18 sight, which continued in general use until the arrival of the F-86E, beginning in June 1951.[30] The Air Materiel Command later concluded that the decision to introduce the A-1C(M) had been somewhat premature; nevertheless, its use in Korea provided the service with invaluable experience in the practical problems of radar-based fire-control systems for air-to-air combat. (An improved version of the A1-C[M] gunsight, designated the "A-4," was to be produced by the Sperry Gyroscope Company and installed in the "F" model of the F-86, which arrived in Korea in the summer of 1952.)

As Draper made clear in remarks on the Instrumentation Laboratory in 1969, "The Airborne Fire Control system work afforded the laboratory many opportunities to begin significant activities in the work with flying systems. In particular, it became apparent that the limiting factor in tracking targets with fixed gunfighters was in the ability of the human pilot to accurately control the motion of his machine."[31] This led to an Air Force contract to study the dynamic coupling between an airborne gunsight and airplane dynamics.[32] The Instrumentation Lab, which dubbed this program the "Tracking Control Project," started work on it less than a year after the war ended.[33] Lee Davis was probably responsible for initiating the contract. He and Draper had become close friends and were continually brainstorming ways of improving air-to-air gunnery. Draper selected Bob Seamans as director of the project; he, of course, had worked on the A-1 sight and built the target acquisition system for the Mark 63 GFCS.

The project involved some thirty-five staff members and numerous graduate students during the four years that it ran under Seamans' direction. Seamans knew Davis from the A-1 project.[34] The two men discussed how to study the airborne dynamics of fire control before the contract was let. Seamans also remembers going with Draper to see Nat Sage about the arrangements that were to be made with the Air Force.[35] They sat around the table while Seamans described what he wanted to do. Sage's response was to "put that in a letter to me, to Nat Sage."[36] Seamans did just that: the next day he showed Draper the two-page letter he had written, then took it over to Sage. Sage liked to keep things as simple as possible, including contracts. he readily accepted Seamans' proposal—and so a two-page letter became the basis of the contract for the Tracking Control Project.

The Tracking Control Project began in 1946 as a program to measure the three-dimensional dynamics of the aircraft, its fire-control computer, its servos, and its radar and to use these measurements to work out analytically the dynamic performance of an aircraft as it flew.[37] Using a Douglas A-26 lent by the Air Force, Seamans and his project team intended to automate the control of an aircraft as it tracked a target, and to do so they measured the total response of the airplane and its instruments under various conditions and flight regimes. They eventually produced a big demonstration at Wright Field showing all the hardware they had built, including an airplane that could automatically lock on and track a target.

While the Tracking Control Project was under way, Davis, who had been moved to several successive positions in the Air Material Command, had eventually been placed in charge of the Armaments Laboratory. He became a frequent visitor to Cambridge and the Instrumentation Laboratory's Flight Test Facility at Hanscom Field in Bedford, Massachusetts, which Draper had established to support the laboratory's Air Force contracts. Davis would not hesitate, whenever he could, to take up one of the test aircraft that had been loaned to the Instrumentation Lab. Draper, of course, loved to fly and would accompany Davis whenever *he* could. At some point during the program they took off in the project's A-26 heading for a conference being held at the Navy's China Lake Ordnance Station near Inyokern, California.[38] Seamans flew out at the same time in a North American B-25. There was no copilot's seat in the A-26, just the pilot's seat and an engineer's seat. When they climbed into the aircraft Davis would take his rightful place in the pilot's seat on the left side of the cockpit while Draper settled into the engineer's on the right. Once in the air, they would swap places so Draper would actually fly the airplane.

At some point during the cross-country flight the A-26 started making pursuit passes (i.e., simulated interceptions) at Seamans' B-25. According to Chip Collins, who was head of the Flight Test Facility, Davis was in the left seat when they started making passes on Seamans' B-25 in an effort to demonstrate the problems they were having with the sight they were testing.[39] While Davis was making passes on the target plane, he and Draper had a running debate on the proper "stiffness," or restraint that needed to be incorporated in the sight for it to be effective in a pursuit-curve approach. Draper demanded to see for himself, and Davis relinquished his seat. Draper jumped in and proceeded to make several high-angle, high-G combat passes. As Chip Collins explained, Draper "was pressing the issue": he was making very tight intercepts to prove that stiffness in the sight was acceptable.

When Seamans landed, he approached the crew chief who had flown in Draper's plane and found the man visibly shaken.[40] "I'm so glad when I finally get out here," he said. "When that little fat man climbed into the controls, I never thought we were going to make it." Seamans told this to Draper, who thought it was a fine joke, even though he was somewhat overweight at the time.

Collins' version of the story is a little different. He claims to have gotten it from the crew chief after they returned to Hanscom Field.[41] As Collins tells it, Davis, Draper, and the crew chief were not in an A-26 but in a B-25, en route not to China Lake but to the gunnery range at Eglin Air Force Base, Florida, accompanied by an A-26. This crew chief was scared to death: he knew that Colonel Davis could fly, but what was this little fat guy with the cauliflower ears and a pugilist nose doing?[42] When they landed to refuel, he rather angrily asked, "Who the hell is this short, fat civilian who was flying all over the sky?"[43] When told that "it was the famous Charles Stark Draper," the crew chief elected to switch airplanes. He flew the remainder of the trip in the A-26.[44]

9

The "Immaculate Interception" and Other Air-Defense Activities

By mid-December 1950, Draper's extensive experience with radar-directed antiaircraft fire and air-to-air gunnery led to his appointment to the Air Defense Systems Engineering Committee (ADSEC), established by the Chief of Staff of the Air Force, Gen. Hoyt S. Vandenberg, to determine "the operational development of equipment and techniques—on an air defense basis—which would produce maximum effective air defense for a minimum dollar investment."[1] The committee, informally known as the "Valley Committee" after its chairman, George E. Valley, met for the first time on December 27, 1949, and on a weekly basis beginning on January 20, 1950.[2]

Valley, an associate professor in the MIT Department of Physics, had been serving on the Basis Research Panel of the Air Force Scientific Advisory Board (SAB) when in August 1949 Air Force intelligence officers advised the board that the Soviet Union had exploded a nuclear weapon and had the long-range capability necessary for delivering it to the United States.[3] Valley, gravely concerned about the status of U.S. air defenses, arranged to visit a radar station operated by the Air Force Continental Air Command to investigate the country's defenses against possible nuclear attack. What he saw appalled him.

After collecting more information on U.S. air defenses, Valley wrote to SAB chairman Theodore von Karmon suggesting that the board request permission to set up an Air Defense Committee composed of members from several of its panels.[4] Von Karman converted Valley's recommendations into a SAB proposal and forwarded it to General Vandenberg, who immediately authorized the formation of an ad hoc committee to study air defense under Valley's leadership. All of the ADSEC (Valley Committee) meetings took place at the Air Force Cambridge Research Laboratories (AFCRL), at 224 Albany Street, which was a convenient location for the majority of the board members, one of them Draper, who were either members of the MIT faculty or worked in the area.

For several months the committee (including Draper) studied air defense in general and the current status of U.S. radars. The ADSEC determined that the weakest link in the nation's air defenses was its long-range ground radars. Because the range of these radars was limited by the horizon, the radars were unable to detect low-flying aircraft beyond the horizon.[5] The solution proposed by ADSEC was to use a large number of small radars that could operate unattended for thirty days at a stretch. This approach, however, would impose a still greater burden on the already overloaded information exchange between the radar operators, the area commander, and his field units. That problem, declared the ADSEC, could be solved using a digital computer that would perform the "prime function of each Data Analyzer—that is, to compute the position of all the aircraft in range of the associated observation posts."[6]

The digital computer, or "data analyzer," as they were calling it, that was needed was already on hand. It was being developed for the U.S. Navy; the project's engineer was Jay W. Forrester. The computer, called Whirlwind, was located in the Servomechanisms Laboratory, administered by MIT's Department of Electrical Engineering. Valley learned about the Whirlwind computer project through a casual encounter with Jerome Wiesner, director of MIT's Research Laboratory for Electronics. Wiesner, Valley, and Forrester met for lunch at MIT's faculty club on Friday, January 27, 1950, to discuss the committee's work on air defense and the possible use of Whirlwind.[7] On the following Monday, Valley, in company with three other members of the committee—Draper, H. "Horton" Guyford Stever, and John W. Marchetti—visited Forrester's laboratory to inspect the work being done on his digital-computer project. After the tour, they discussed how Whirlwind could be integrated into the air-defense system proposed by ADSEC. In his notes on the meeting, Forrester recorded that "all of the men seemed to be very enthusiastic about the project."[8]

From then on, Forrester was a regular participant in ADSEC's efforts to develop a computer-based air-defense system. Among his tasks was the necessity of preparing Whirlwind to receive and process digitized radar signals. To test the feasibility of his approach, AFCRL established a digital data link between an early-warning radar setup at Hanscom Field and the Whirlwind computer, in the Barta Building in Cambridge. This arrangement was successfully tested in September: "While military observers watched closely, an aircraft flew past the radar, the digital radar relay transmitted the signal from the radar to the Whirlwind via telephone line, the results appeared on the computer's monitor."[9]

While hardware tests of the transmission system were taking place, the software engineers in Forrester's lab were working on a methodology for automatically tracking all targets within a selected range. The programmers also came up with a simple guidance procedure to direct defending aircraft to an interception. On April 6, 1950, after several months of trial and error involving flight tests, the system successfully directed a B-25 from the Instrumentation Lab's Flight Test Facility to within three to five miles of the target point, despite persistent range-coding errors.[10] By the second week of April the engineers had solved the range-coding problem and were ready to try another Whirlwind-guided interception. Three aircraft were at the team's disposal, according to the detailed history of Whirlwind written by Kent Redmond and Thomas Smith: "the Instrumentation Lab's B-26, a [North American] T-6 from the Servomechanisms Lab, and a [Beechcraft] C-45 from the National Guard. The B-26 was the interceptor; the other two planes were the targets."[11] Since the radar could not supply altitude data, aircraft altitudes were preassigned and computer control was exercised in only two dimensions. The B-26 made three successful interceptions in a row.[12] "For the first time, radar data on an approaching 'enemy' aircraft had been sent over telephone lines and entered into an electronic digital computer, which had almost instantaneously calculated and posted instructions directing the pilot of a 'defending' aircraft to his target."[13]

Details of the interceptions as described by the pilot (probably Chip Collins) were recorded by Robert Wieser for posterity and transmitted by letter to Forrester:

> [The pilot] estimated that each interception he was guided to within 500 to 1000 yards of the target. He had had experience testing the Instrumentation Lab's airborne intercept radar and auto pilot system and he expressed the opinion that our midcourse guidance would have permitted completion of the interception by the airborne equipment.
>
> The interception technique is, briefly, as follows: "The target ships are given predetermined courses (we have no direct communication with them) in the northeast area 30 to 60 nautical miles from Bedford. Targets and interceptors are detected by the Bedford MEW [Microwave Early Warning radar], their coordinates are sent to WWI [Whirlwind] over a telephone line. They are identified on PPI [Plan Position Indicator] scopes at WWI, and tracking of the interceptor and one target is initiated by means of a 'light gun' on the PPI. After initiation, tracking while scanning of both ships is automatically done by WWI, and the proper magnetic headings are computed to guide the interceptor on a collision course. Headings are computed once per antenna revolution (15 seconds) and displayed as binary-coded decimal numbers on a set of indicator lights. A man reads the headings off the lights and relays them by telephone to Bedford, where another man relays them by voice radio to the pilot of the interceptor.
>
> "After initiation of tracking, the interceptor was guided for about 10 minutes (about 35 miles) to the interception point. The interceptions took place at a range of about 40 nautical miles. The whole system worked well, and more experiments of this nature will follow."[14]

While feasibility tests of the Whirlwind computer and the Bedford radar were being conducted, steps were being taken within the Air Force leadership to approve the air-defense system proposed by ADSEC. On December 15, 1950, General Vandenberg wrote to James Killian, president of MIT, requesting that the institute contribute to the realization of the air-defense system proposed by ADSEC by setting up a laboratory devoted to the air-defense problem. "The Massachusetts Institute of Technology," he wrote, "is most uniquely qualified to serve as contractor to the Air Force for the establishment of the proposed laboratory."[15] Killian asked if MIT could first evaluate the need for and scope of the proposed laboratory. The resulting study, named Project Charles, was conducted by a group of twenty-eight scientists between February and August 1951. In its final report, Project Charles endorsed the concept of the digital computer–run, centralized air-defense system proposed by ADSEC and "came out unequivocally in support of the formation of a laboratory dedicated to air defense problems."[16] This led to an Air Force contract for Project Lincoln, formalized by a charter dated July 26, 1951.

Many of those involved in Project Charles joined Project Lincoln, which became the Lincoln Laboratory in 1952. Before they moved to their new quarters in Lexington, Massachusetts, the project's members frequently took over one of the MIT campus lounges, posting an armed guard at the door. This "elicited complaints from other MIT staff and students, a reaction that foreshadowed more substantial protests against MIT's military and intelligence ties in the 1960s."[17]

By 1952, MIT was the nation's leading recipient of defense-related subsidies from the federal government for academic science.[18] A special Division of Defense Laboratories was established within MIT to manage this work, much of which occurred in Draper's Instrumentation Lab and in Lincoln Laboratory. "Although Killian initially attributed some of the institute's participation in federally sponsored research to the national-security crisis of the early 1950s, the close relationship between M.I.T. and the government was maintained and even intensified in subsequent years."[19] Killian justified MIT's heavy involvement in federally supported research. "We know from extensive experience," he wrote in 1953,

> that sponsored research can enrich our educational program, and we do not share the extreme view sometimes expressed that sponsored research *ipso facto* is bad for education. . . .
>
> In addition, we have recognized an inescapable responsibility in this time of crisis to undertake research in support of our national

security which under normal conditions we would choose not to undertake. We propose to see this research through and to make our special competence available so long as national policy and need indicate that we should. When these conditions no longer hold, we shall withdraw from classified emergency research with enthusiasm and relief.[20]

Two years later Killian created a new organization called the Division of Defense to isolate these classified programs from the institute's budget. The Instrumentation Laboratory, the Lincoln Laboratory, and an Operations Evaluation Group were all placed in this division.[21]

With the assistance of Air Force units at Hanscom Field, Project Lincoln set up an experimental air-defense sector named the "Cape Cod System," for its location—the Cape was a good test site and convenient to Lincoln Laboratory. The Cape Cod System was intended to serve as a proving ground capable of being developed into "a complete model air defense system suitable for tactical evaluation."[22] It consisted of several L-band long-range radars with a nominal range of two hundred miles and a number of gap-filling, shorter-range, S-band radars to detect low-flying aircraft.[23] Data from the radars was transmitted over leased telephone lines to a command center in the Barta Building, where it was fed into the Whirlwind computer, which processed it in real time to track aircraft, assist operators to perform command functions, and guide fighter interceptors.

By January 1952, several aircraft were being outfitted with an eight-digit, compass-heading radio data link designed by the Air Force Cambridge Research Center that enabled them to receive data from Whirlwind.[24] One of these aircraft was a B-26 belonging to MIT with an autopilot that could take digital input.[25] Chip Collins suggested that this autopilot be set up to receive the radio signals from the data link so that vectoring instructions could be sent directly to it.[26]

Sometime that spring, Chip took off in the B-26 to attempt what some would later call the "Immaculate Interception." Ground personnel listening to the radio transmissions between the aircraft and the controllers heard Chip say, "Let George do it," referring to the autopilot.[27] (Autopilots in that era, as today, were commonly referred to as "George," a colloquial term that seems to have originated from the early autopilots, which used gyroscopes to provide stability. An autopilot would be called a "gyroscopic system," abbreviated as "G-system," and finally morphed into "George," which at that time was the phonetic-alphabet term for the letter *G*.)[28] The radar operators saw Chip's plane closing the target. "Tally-ho," he radioed upon sighting it.

Formal trials of the Cape Cod System began the following month, with flight tests conducted two afternoons a week.[29] The first incarnation of the system, which became fully operational in September 1953, was used to gather preliminary design data for SAGE (Semi-Automatic Ground Environment System), a large-scale air-defense system initiated by Lincoln Laboratory's Technical Memorandum 20.[30] More than five thousand sorties were eventually flown against the Cape Cod System, which was continually upgraded and expanded until the end of its operational life in 1956.[31]

The Instrumentation Laboratory's work on gunsights and interceptor control led to yet another Air Force contract for gunnery fire control, one that Draper dubbed the "Dummy Gun Project," because it created an artificial "gunline."[32] Begun in 1952, the project was designed to incorporate the technology of the Draper-designed inertial platform developed for naval fire control into the Air Force's planes. Instead of steering the plane on a course mandated by a lead-computing gunsight such as the A-1, the Dummy Gun System allowed the pilot to fly any course while the sight developed the lead angles needed to hit the target with guns or rockets. The pilot could then turn to the required orientation and fire. The system simplified the pilot's task, making it an ideal method for launching rockets against an incoming bomber "stream." (The fact that rockets had more hitting power than .50-caliber or 20-mm projectiles had caused the Air Force to favor unguided rocket salvos over gunfire to bring down enemy bombers.)

The Dummy Gun System used an inertial platform similar in design to that of the Navy's Gunar. The inertial platform determined the aircraft's orientation and established an artificial gunline along which guns (or rockets) needed to be aimed to strike the target. The gunline was stored in the system's computer so that the guns or rockets could be fired when the aircraft was aligned with the gunline. In July 1953, the name of the project was changed to "Warrior System." A year later, it was changed again, to "Black Warrior." In 1956, a production prototype was fitted to an F-94B all-weather inceptor and extensively tested at Eglin Air Force Base. A project summary states that the "Black Warrior system was used in interceptor aircraft," but if so, it must have entered service under a different designation.

A variation of the Black Warrior's three-axis, internally stabilized fire-control system was used to control the M61 six-barrel, 20-mm rotary tail gun that was installed on Convair B-58 supersonic bombers.[33] It was operated by the aircraft's Defensive Systems Operator (DSO) in the third cockpit. The system would automatically lock onto a target that appeared on his radar scope, compute the lead and windage, aim the gun, and notify the DSO when to fire, whether or not the target was in sight.[34] Few B-58s were produced, and the system did not get widespread use, although it may also have been installed in some B-52s.

In addition to these Air Force programs, the Instrumentation Laboratory conducted several air-to-air interception projects for the U.S. Navy. These projects included a ten-month air-to-air missile guidance study conducted in 1960 to investigate the problem of midcourse guidance for a long-range missile and a project to design the fire-control system for the Terrier and Tartar antiaircraft missiles.[35] The latter ran concurrently with the Black Warrior project and replaced the earlier beam-riding scheme with the lab's semiactive homing system.

10

Inertial Navigation

Here I am in an airplane in the fog and I can't see a thing. I don't know where the hell I am or where I'm going. I don't know whether I'm right side up or wrong side up. I ask myself, "If you can wear a wristwatch to tell you what time it is, why can't you have something that will tell you where you are with respect to space?"

CHARLES STARK DRAPER[1]

Draper liked to claim that he got the idea for an inertial navigation system "out of a bottle of whiskey."[2] It was a story that he told so often that the actual facts became muddled over the course of time. Inertial navigation was something that he had been thinking about for a long time when he broached the subject to Lee Davis on their way back to Wright Field from Flint, Michigan, in mid-August 1945. They had flown to Flint in a Beechcraft C-45 to look over the equipment that AC Spark Plug—the manufacturing contractor—was using to test the A-1 gunsights that were being prepared for an overseas tryout.[3] Halfway through the trip, they heard over the radio that Japan had surrendered, ending the war. As Draper was fond of telling his son James, Davis reached under the pilot's seat and pulled out a bottle of scotch.[4] As they celebrated the end of the war with Draper's "soothing syrup," Davis talked about the rearrangements in spending that would come as weapons that were no longer needed were canceled.[5] Draper saw, however, an unusual opportunity to gain support for a project that had been in the back of his mind ever since the early 1930s: the development of a self-contained inertial navigation system that could give the pilot an airplane's location in bad visibility without external inputs. Existing radio aids to navigation, for example, were notoriously unreliable and easily interrupted by bad weather.[6]

Davis had no trouble following the principles behind Draper's idea, because he had devoted a fair amount of his time to studying gyro principles as a graduate student at MIT. Both men realized that such a system—if it could be developed—would overcome the severe navigation problem that had been experienced by the crews of the Army Air Forces' long-range bombers during the war. The two discussed the details the rest of the way back to Wright Field.

When Draper returned to Cambridge, he discussed the idea of an inertial navigation system with key members of his staff. Although gyroscopes at the (then) state of the art could not be used, Draper and his colleagues "felt very strongly that self-contained [inertial navigation] systems were possible and that with existing motivation, useful results could be brought to realization in a few years."[7] After reviewing the theoretical and technological issues, Draper returned to Wright Field to discuss an inertial bombing system with Davis and a small group of Armament Laboratory engineers led by John Clemens. As Draper would write in 1969, "With the war just finished, the problems of accurate bomb and rocket deliveries after long flights over unfriendly ground environments were large in the minds of Colonel Davis and his scientists, Dr. John E. Clemens and Dr. Ben Johnson."[8]

Because gyros drifted over time, the Armament Laboratory insisted on adding a stellar-sighting system to provide the accuracy needed for the long-duration flights for which the system was intended. Once an aircraft equipped with the inertial guidance system reached cruising altitude, the stellar system would lock onto a celestial body and correct any errors in the flight path caused by gyro drift or other minor inaccuracies in the inertial system. Draper regarded the star tracker as a temporary necessity, given the limited accuracy of the gyros then available, but he felt that it was a "messy and inelegant" approach to the problem.[9]

On August 23, 1945, the Instrumentation Laboratory submitted a proposal to the authorities at Wright Field for a Stellar Bombing System, designed primarily for operation in jet-propelled aircraft as a bombsight but having in mind "the possibility of eventually robotizing the system for use with guided missiles."[10] Draper was well aware that stellar observations were subject to interference by weather, aurorae, meteors, and countermeasures. Although Draper would have preferred a closed "black box" solution, he was a pragmatic engineer who understood the severe limitations of the gyroscopes then available.

Less than a month later, on September 14, the laboratory received a letter contract to study the possibilities of the inertial navigation system that the Instrumentation Laboratory had proposed. To help conduct the study, Draper recruited Walter Wrigley, whose knowledge of determining "direction of the vertical" was essential to solving the inertial navigation problem. Wrigley, a former doctoral student of Draper's, had spent the war years working as an R&D project engineer for the Sperry Gyroscope Company.[11]

How to indicate accurately the direction of the vertical (the line running from an aircraft's center of gravity to the center of the Earth) from a rapidly moving airplane was one of the key problems that had to be solved in order to construct a workable inertial guidance system. To many in the scientific community this seemed an impossible task, given Einstein's general theory, which postulated that an observer inside a closed box could not distinguish the effects of linear acceleration from the effects of a gravitational field.[12] One physics textbook published in 1942 went so far as stating that it was impossible to construct a device "to indicate the true vertical unaffected by accelerations of the airplane when in curved flight."[13] The authors of that work were unaware, no doubt, of Walter Wrigley's dissertation. In that work, supervised by Draper and submitted on March 9, 1940, Wrigley provided a comprehensive mathematical analysis of the methods available for indicating the direction of the vertical from moving bases.[14] Wrigley's conclusion that a "damped gyroscopic servo-controlled by a pendulum" offered the most practical solution laid the groundwork for constructing an inertial navigator. It convinced Draper that such systems were now a possibility.[15]

After the Instrumentation Laboratory submitted its initial study of a Stellar Bombing System, the Air Force gave the green light to proceed with an experimental program designed to test the possibilities of actually constructing such a system.[16] Work on the project began on November 21, 1947, under the name of the Stellar Inertial Bombing System (SIBS).[17] The name was later changed to FEBE, a variation of the sun god of the ancient Greeks and Romans, Phoebus, in a reference to the use of the sun for stellar tracking purposes.

Obtaining precision sensors of the accuracy needed over the five-to-ten-hour-long flights for which the system was intended was a major problem that had to be overcome if the project was to be successful. In an attempt to satisfy the Air Force's desire to use existing technology, an Arma* stable element, commonly used in U.S. Navy fire-control systems to determine the vertical, was installed in an Air Force DC-2.[18] The equipment, which was large and heavy, proved unsuitable for the task. Draper and his staff at the Instrumentation Laboratory concluded that an inertial bombing system could not be constructed using existing technology and that new sensors would have to be developed. This conclusion led to a rigorous investigation of the use of inertial space references for navigation purposes. The analysis was based on the bombing mission, an application distinguished by long flight times and low accelerations in an environment in which the heading errors produced by gyroscopic drift constituted the primary inaccuracy. The study, which was completed in February 1947, pointed to the gyroscope rather than the accelerometer as the key sensor.[19]

As Walter Wrigley had suggested in his dissertation, a servo-controlled, single-degree-of-freedom gyroscope proved to be the critical component of the stable platform that was at the heart of any practical inertial navigation system. A stable platform requires three, gimbal-mounted, single-degree-of-freedom gyroscopes. Each is affixed to a flat platform so that its precession axis aligns with one of the three primary axes of motion (i.e., roll, pitch, and yaw). Collectively, the three gyros are able to detect the motion of the board in any direction.

Suppose this assembly is placed in an aircraft with the board aligned parallel to the horizon so that a perpendicular line through its center establishes the direction of gravity and thus of the vertical. Let us assume that the gyros are constructed so that their outputs, as they precess, are proportional to velocity of the change in direction experienced by the gyro

---

* Arma Engineering Company, which merged with Bosch in 1949, became the American Bosch Arma Corporation.

rotors (due to the forces of acceleration acting on the board) as it begins to move through space. Integrating these signals generates a measurement of how much the board has been displaced along each of the three axes. These signals can then be sent to small motors attached to the platform's gimbals. When the aircraft's attitude or heading changes, the integrated signals from the gyros will cause the servo motors to move the platform back to its original starting position, keeping the board in a horizontal position, thereby maintaining a true indication of the vertical. This is how a stable platform works.

If the gyro was the most important component of an inertial navigation system, highly accurate accelerometers were also needed to measure the distance traveled in a given direction. Getting back to the hypothetical system discussed above, let's place two accelerometers on board the platform perpendicular to one another so that they can measure the acceleration in the north–south and east–west directions. When the signals from these accelerometers are integrated twice

$$\text{first } \int a = v,$$

$$\text{then } \int v = d$$

the results are the distances traveled in each direction over an interval of time. A vector summation of these two measures accurately indicates the position of the platform at the end of that time interval.

To start the system, the stabilized platform is aligned to the horizontal and positioned so that the sensitive axis of the north–south accelerometer is pointed to the north. The latitude and longitude of the starting point and destination are then set into the system, and the integrators are trimmed to zero. As soon as the aircraft begins its takeoff run, the accelerometers will sense the resulting accelerations and measure how far the aircraft has moved. These distances are then converted into corresponding changes in latitude and longitude and applied to the starting-point coordinates to show the aircraft's new position.

Although building an inertial navigation system was now conceptually straightforward, several difficult problems had to be overcome before a working unit could be fabricated. The most difficult of these was to develop a set of gyroscopes, accelerometers, and the other components small enough to fit into an aircraft, yet accurate enough to provide the precision required over the long flight times specified. A separate issue was the need to take into account the effects of gravity as the stable platform moved over the Earth's surface so that the platform remained at right angles to the Earth's radius.

To compensate for the Earth's rotation, Wrigley applied Schuler's Principle, envisioning an "Earth-radius pendulum." Maximilian Schuler had been working to improve his cousin's gyrocompass in 1923 when he hypothesized that a solution to the vertical could be achieved if the vehicle traveling over the Earth were attached to a pendulum whose center of gravity was at the center of the Earth.[20] As the vehicle moved, the pendulum would continue to indicate the direction of the vertical. A pendulum of this size could never be built, but, Wrigley realized, the disturbing effects of gravity on the stable platform could be removed by designing a simple feedback loop that continuously caused the platform to remain horizontal. "Such a system could be seen as working as if it kept horizontal by an earth-radius pendulum."[21] In order to work properly, such a feedback system would have the same eighty-four-minute* natural frequency period of Schuler's pendulum. This concept, apparently coined by Wrigley, is named Schuler Tuning.[22]

As mentioned earlier, the instruments that had been applied to the fire-control problem in World War II were useless for inertial navigation purposes. The gyroscopic elements used in these devices had been elastically restrained and viscously damped to generate output angles rapidly in the one-to-two-degree range. As Draper wrote in his paper on the origins of inertial navigation:

> Navigation and guidance had to deal with flights of 5–10 h [hours], with circles of erratic inaccuracies of no more than a few hundred feet and 1 mile or less terminal inaccuracies. Taking 1 min. of arc between local verticals on Earth's surface as representing 1 n. mi. [nautical mile] and 900 min of arc approximating the 15 deg/h [degrees per hour] of Earth's rate of rotation on its axis, the buildup

---

* A satellite circling the Earth every eighty-four minutes would have to be at treetop height.

of one-mile error in 10 h would mean a drift rate of about 0.1 one-thousandth of the Earth's rate as an approximation required from gyroscopic units for providing satisfactory angular references within self-contained navigation systems. A circular inaccuracy of some 5 s [seconds] of arc would be needed for defining terminal points.[23]

This standard was approximately ten thousand times better than the gyroscopes then available. If the closed-box inertial navigation system envisioned by Draper was to be built, the Instrumentation Laboratory would have to develop new types of gyroscopes that were up to the task. Improving the accuracy of these instruments became a priority within the laboratory; for it Draper hired William G. Denhard, an electrical engineer who had graduated from MIT in 1942, and Albert P. Freeman, a recent graduate of Northeastern University with a bachelor's degree in mechanical engineering.[24]

Although several other firms were actively engaged in developing inertial navigation systems for various U.S. Air Force programs, some in the scientific community remained dubious about this unproven technology.[25] George Gamow, a prominent physicist and a member of the Air Force's Scientific Advisory Board, was highly skeptical that this approach was applicable to the guidance problem facing the Air Force in its attempt to develop "air breathing" intercontinental guided missiles capable of attacking the Soviet Union. In February 1948, Gamow, noted for his brilliant mind and ebullient sense of humor, was working on the long-range navigation problem as a consultant to the Johns Hopkins Applied Physics Laboratory.[26] On the thirteenth of that month, a day Gamow referred to as "Black Friday," he addressed a scathing memorandum to Ralph E. Gilbert, the newly appointed director of APL, attacking the concept of an inertial navigation system.[27] In his memorandum, titled "Vertical, Vertical, Who's Got the Vertical?," Gamow argued that the inertial navigation systems then under development by the Instrumentation Laboratory and other Air Force contractors were impractical, because they would have to work flawlessly. If such a system was to function as intended, it would have to be capable of indicating an aircraft's initial position and velocity with perfect accuracy. However, an aircraft's position "is completely undetermined, unless its initial position and velocity are known exactly and the integration [of velocity] is carried on faultlessly all the way through." At the time this was written, no instrument existed that could satisfy this criterion. But Gamow underestimated the capabilities of Draper's laboratory, which, as we shall see in the pages to come, was able to develop such a device.

This was not the only criticism leveled by Gamow. A greater concern, in his opinion, was correcting the errors that would inevitably be introduced by inaccuracies and outside inputs. Draper would address this problem later on, but for the time being Gamow's memo, which was laced with vivid, mocking caricatures of the current attempts at developing an inertial navigation system, caused quite a row within the inertial navigation community.[28] Because Gamow was a member of the SAB's panel on guidance and control, his ideas carried considerable weight and had the potential to derail catastrophically Draper's program, as well as those of his competitors. As Doel and Söderqvist note, "The multiplicity of groups working on the [inertial navigation] problem aggravated the task of responding to Gamow's criticism," which became essential if Draper and the other contractors expected the Air Force to continue funding the development of this yet to be proven technology.[29]

What transpired during the next twelve months does not show up in the historical record. Although Gamow's memorandum did not mention Draper by name, its content was undoubtedly of great concern to him, especially since the Armament Laboratory was the only source of research monies for the Instrumentation Laboratory's guidance work. The situation facing Draper and Leighton Davis, his patron, was clearly put forth by Michael Dennis: "Few organizations were capable of supporting Doc's research; if others on the funding food chain perceived Draper's research as a technological 'dead end,' then Davis and Draper were in jeopardy."[30] Where, when, or if Gamow's memo was circulated or discussed is not known. But Draper, who was a member of the SAB's Guided Missile Panel (though not the Guidance and Control Panel) had the connections and political clout to do something about it. Using his contacts he arranged a classified conference on guidance at MIT in February 1949. The meeting, which was held under the auspices of the SAB, was titled a "Seminar on Automatic Celestial and Inertial Long Range Guidance Systems."[31] Although the stated purpose of the meeting was "a means of promoting a wider dissemination of information on the basic theory involved" in the guidance problem, Draper cleverly used it as a means of refuting Gamow's contemptuous opposition to inertial navigation.[32]

To participate in the seminar, Draper invited every major firm and component manufacturer working in the field to demonstrate the progress that had been made in the previous few years. Gamow was also invited, but, probably recognizing that the meeting was "stacked" against him, decided not to attend.[33] The instrument errors that Gamow claimed would make inertial navigation unusable could be corrected—according to Draper—by a process he termed "smoothing." As laid out in Draper's opening statement on February 1, "the amount of smoothing that can be used is limited by the fact that any increase in smoothing always brings with it an increase in the time required for a system to solve its guidance problem."[34] Draper went on to explain the importance of solving the conflict between smoothing and solution time, a challenge that would have a prominent place in the papers to be presented.

This was Draper's hidden agenda: he was in fact presenting the details of FEBE, the experimental inertial navigation system being assembled by the MIT Instrumentation Laboratory under his supervision. Although FEBE had yet to be flown, it was nearing completion and would soon be tested. Draper's staff was responsible for presenting eight of the twenty-five sessions during the three-day meeting. This was twice as many as the lab's nearest competitor, the North American Aviation Company, which had an Air Force contract to develop an inertial navigation system for the Navaho intercontinental missile.

FEBE was a demonstration system engineered to validate the design assumptions needed to create a true inertial navigation system—one with no external inputs. It was designed to investigate the dynamics of a closed-loop, automatic navigation system, study the various instruments and their organization, and establish a correlation between the results of flight tests and theory.[35] Although FEBE could operate at night using navigational stars, the sun was selected as the celestial reference, so that records of the actual ground track could be more easily made to ascertain the system's accuracy. Its sensors were based on marine gyrocompasses and the gyroscopic elements used in the World War II antiaircraft fire-control systems designed by the Instrumentation Lab.[36]

FEBE, which weighed four thousand pounds when fully assembled, was installed in a B-29 to be systematically tested in flight. It was flown for the first time on May 5, 1949.[37] This "shakedown" flight was followed by nine more experimental flights designed to test the system's accuracy and see how it behaved over long distances. The longest of these flights was a 1,737-nautical-mile trip between Bedford, Massachusetts, and Alamogordo, New Mexico.[38] Because of equipment malfunctions and abnormally erratic readings, only six of the flights produced acceptable results. When averaged together, the results of these tests yielded a mean error of five nautical miles. This was not accurate enough for the bombing mission for which FEBE was designed. Nevertheless, the results encouraged the Air Force to issue a follow-on contract to design, build, and test a navigation and guidance system that would depend only upon inertial and gravitational inputs. This contract was responsible for the development of SPIRE,* the first inertial navigation system to guide an aircraft across the country.

The B-29 guided by SPIRE during that revolutionary flight was flown by Chip Collins. Collins had met Draper in 1947 as an officer assigned to the Armament Laboratory.[39] Draper, then on one of his many visits to Wright Field, needed to get to the McGuire Air Force Base in New Jersey, and Leighton Davis asked Collins to fly him there. Collins was told to have the airplane in front of the laboratory at 1300.[40] Draper showed up at the appointed hour and was introduced. According to Collins, the relationship between the civilian "ground pounders" at Wright Field and the aviators was not very cordial at the time, and like most pilots, he was suspicious of all engineers. "Now Chip," Colonel Davis assured him, "if you have any problems with the instruments just ask Dr. Draper about it. He could probably fix it since he designed most of it." Nevertheless, Collins remained skeptical of his civilian passenger until they were well into the flight.

Collins, after completing his combat tour in the Pacific in B-29s, had gone through test-pilot school and so was allowed to fly multiplace airplanes, such as the C-45 used that day, without a copilot. As Collins explained years later, "We took off with no other crew and I set my course.

---

* Space Inertial Reference Equipment; and see list of abbreviations.

An hour out of Dayton, as we neared Pittsburgh, I looked back and saw this little man with rolled up sleeves held up by elastic garters sprawled in the aisle on his hands and knees . . . busily scribbling on a 6-foot roll of paper."[41] As the flight continued, Collins "was streaming along and the weather otherwise was good and I looked back and here was this little man with a pugilistic nose and cauliflower ears on his hands and knees on the floor of the airplane with a role of paper and he's doing second order equations. Man I said to myself, I've got one on this trip."[42] At about that time Draper got up and walked forward. He had the biggest wristwatch Collins had ever seen.

"We ought to be abeam of Pittsburgh at 13:25:30," he told Collins.

"Yes, I think so," Collins replied and quickly proceeded to do the calculation on his E6B computer—a circular slide rule–like device used to solve aerial navigation problems. Draper was right on the money; now, he estimated when they would be over Harrisburg, Pennsylvania, and then their time of arrival in New Jersey. Draper was right again on both. Collins' impression of the "little man" changed radically. He quickly realized that Dr. Draper "was not the typical civil service type engineer we met at places like Wright Field."[43]

About a year after ferrying Draper to McGuire, Collins was asked to evaluate the condition of the B-29 bomber that the Air Force intended to loan to MIT for the FEBE test program. Chip, having flown thirty-five bombing missions over Japan in the B-29, knew everything there was to know about the big, four-engine bomber, but he had no idea how much work would be needed to put this particular B-29 in order. Once airborne, Collins quickly determined that it was so "war weary" and worn that MIT would need the equivalent of a depot maintenance shop to keep it flyable. In particular, he knew that if any of the cylinder heads on its R-3350 engines exceeded 170 degrees Fahrenheit, it would swallow a sodium-filled exhaust valve putting a hole in the piston through which would leak out all eighty-five gallons of oil that supplied the engine and the five gallons in the hopper tank that supplied the hydromatic pitch control on the propellers. Collins recommended that before transferring the aircraft to MIT, the engines be changed from normally aspirated to fuel injection, that the props be changed from hydromatic to Curtiss electric with full reversing, and the bomb-bay doors, which were electrically operated and were slow to retract, be changed to pneumatic so they would open and close instantly.

The next morning Collins got a call from Gen. Albert Boyd, chief of the Flight Test Section, to which the B-29 was assigned. "Yes, sir, General, how may I help you?"

"Collins," said the general, "I'm the one who is going to fly that airplane. I'm the one who is going to determine fuel injection, a Curtiss prop instead of hydrostatic prop and bomb-bay doors changed to pneumatic."[44]

The next day, General Boyd flew the aircraft and experienced the same problems. Chip was exonerated, and MIT got a B-29 with fuel-injected engines, full reversing propellers, and pneumatically controlled bomb bay doors. The B-29 was being loaned to Draper because he had realized that the lab needed to control all aspects of the testing if the inertial navigation systems were to work properly.[45] Testing on an ad hoc basis was just not working. To fly the B-29, Draper needed a pilot he could trust. Colonel Davis recommended Chip.

In late October or early November 1947, Collins, who was expecting to be promoted to major shortly, flew to Boston to interview for a flying position with MIT.[46] Al Coleman, an administrator who worked for Draper, offered Collins seven hundred dollars a month to work as a test pilot and to build and manage a flight facility for the MIT Division of Industrial Cooperation. Collins was making more than that in the Air Force, and his promotion would hike his salary even more. He also received thirty days' sick leave and medical insurance. He was flattered by the offer but told Coleman he could not accept it in light of the salary and benefits that he was receiving as an Air Force pilot. Chip flew back to Dayton that same day in the Douglas C-47 supplied by the Armament Laboratory under the guise of a training flight.

That night, after Collins had returned to Dayton, Draper called him to ask what had happened. Collins explained, "They're treating me like a sack of lard from college. They aren't paying enough money. I can't do that and I said I have a capacity that nobody you have on your staff has and it's got to be worth more money for me to give up a career."[47]

"Goddamit, you get your butt here. I'll take care of the money."

Collins took Draper at his word, got himself discharged from the Air Force, and reported for work day before Thanksgiving in November 1947.

(His supervisor was a "young and eager engineer" named Robert C. Seamans Jr.[48] Draper later revealed that he was grooming Seamans to take over the lab when he retired. But Seamans, whom we've met already, was too ambitious to wait that long. In 1955, Seamans was to join the Radio Corporation of America to become manager of its Airborne Systems Laboratory. He would become chief engineer of RCA's Missile Electronics and Control Division in 1958, be appointed deputy administrator for the National Aeronautics and Space Administration in 1968, and serve as Secretary of the Air Force from 1969 to 1973.)

One of Collins' first assignments was to determine the location for the flight facility that Draper wanted for investigations and testing of aircraft and airborne equipment. Colonel Davis suggested an abandoned site at Hanscom Field; formerly a barracks, it looked usable, even though the land was unprepared and would require extensive work.[49] MIT signed a thirty-year lease for the site with the Commonwealth of Massachusetts in May 1948. Later that year, a two-bay steel hangar built in the 1930s at the Hunter Army Airfield in Savannah, Georgia, was dismantled, barged up the Intracoastal Waterway, and installed in Bedford.[50] Draper had planned to celebrate the official opening of the Flight Test Facility by "christening" it with a bottle of scotch. It was a cold day, however, and on the eighteen-mile ride from Cambridge the scotch, as William Denhard recalled, "disappeared from the bottle."[51] In any case, by that time five aircraft—one B-25, one C-47, two A-26s, and one B-29—were already operating there.[52]

11

Floated Gyros and SPIRE

> Although many people worked on the design of the single-degree-of-freedom gyro, there is without question no single person who put in the totality of the effort that Draper himself gave to the development of the instrument.
>
> ROBERT A. DUFFY[1]

While FEBE was still in the design stage, experiments with and studies of the various gyroscopic instruments revealed that friction between ball bearings and gimbal axes was the major cause of drift. Further investigation indicated that below a certain level of ball-bearing friction the accuracy needed for long-range flights could not be achieved. This led to the conclusion that gyro gimbal supports were impractical for the inertial navigation system that Draper envisioned.[2] After the laboratory tried and abandoned various schemes for overcoming this problem, Draper and the staff settled on the most practical means of providing frictionless support for the gyro rotor: flotation, in a hollow housing. A useful by-product of this arrangement was the application of viscous drag to integrate the output.

Some consider the development of the floated integrating gyro to be the Instrumentation Laboratory's greatest contribution to inertial navigation.[3] Floating the gyro wheel allows it to precess with almost zero elastic restraint while minimizing sources of parasitic torque. According to the experts, "The flotation fluid provides the additional benefit of integrating the applied rate signal through its viscous damping reaction torque exerted on the float. That is, the floatation angle measured by the [signal generator] is proportional, not to the applied angular rate, but to the total angle the gyro is rotated through. This rotation angle is the information needed to calculate the navigation solution; consequentially, the direct integrating feature of the floated gyro is [a] highly beneficial design feature."[4]

As one authority argues, "a large portion of the credit for the original design and development of the Floated Integrating Gyro belongs to Charles Stark Draper."[5] Without doubt, Draper's "ability to amplify and implement his own ideas and to inspire new and original ideas in others" was largely responsible for this achievement.[6] A good example can be found in the work of Sidney Lees, one of Draper's students, who was responsible for investigating the performance of viscous shear damping. Understanding this phenomenon became a key factor in the design of the floated gyro. (Lee would use the initial studies as the basis for his thesis in 1948 for a master's degree in aeronautical engineering.[7] Lees was to extend his research on floated gyros for another two years as part of a doctoral program, which, under Draper's direct supervision, he completed in 1950.)[8]

The laboratory's first floated gyro, the 30-X-1, was developed in 1947 using a special floatation fluid developed by Merrill R. Fenske at the Petroleum Refining Laboratory at Pennsylvania State College.[9] The 30-X-1 was designed with an externally pressurized hydrodynamic bearing to assist the flotation fluid. A thermostatically controlled heater was required to keep the fluid at the proper temperature. Two microsynchros, developed by the Instrumentation Laboratory in 1942, were used as signal and torque generators. The float was sealed with O-rings—which leaked, creating more problems than they solved. The hydrodynamic-bearing approach was discontinued when it was discovered that the torque from the bearing not only supported the float but tended to rotate it, as a function of the precession angle.

Research into the properties of various damping fluids revealed that a class of compounds known as "halogenated hydrocarbons"* possessed the

---

* Organic polymers that contain one of the halogens: a group of elements that includes chlorine, fluorine, bromine, and iodine.

ideal density and viscosity characteristics for a floated gyro. In 1948 or thereabouts, Fluorolube®, manufactured by the Hooker Chemical Company, was selected for use in the lab's gyros.[10] As in previous designs, heaters and temperature controls were used to control viscosity. This led to the development of the "hermetically sealed integrating gyro" (HIG), which was developed only after the Instrumentation Laboratory had overcome design problems with filling the gyro, O-ring sealing, and contamination of the damping fluid.

The commercially manufactured HIG-5, which was used on the A1-C gunsight, and the other early gyros like it were designed for fire-control applications. The drift rates for these devices, which was about two hundred degrees per hour for operating periods of twenty seconds or less, was greatly in excess of that allowable for inertial navigation.[11] For inertial navigation, gyros having drift rates of less than a fraction of a degree per hour operating for ten hours or more were needed. The angular momentum of the rotating wheels in these gyros was one to two orders of magnitude greater than in those designed for fire control. This created additional design problems; among these was the change from Invar, an alloyed steel that was strong but heavy, to aluminum. To eliminate the O-rings and the problems associated with them, the Instrumentation Laboratory went to soldered joints, which required electrocoated plating before assembly. This approach had its own problems, which led to the introduction of adhesive seals; that in turn brought on new problems of its own, which were solved only after the Instrumentation Laboratory developed its own family of special adhesives. As Duffy remarks in his biographical essay, "Doc relied on the MIT material scientists during this process."[12]

The laboratory's design team faced similar challenges in engineering the specific force receivers, commonly called "accelerometers," which were also integral parts of an inertial navigation system. While design work on the laboratory's first floated gyro was still under way, Draper had the opportunity to inspect the inertial guidance system developed by Fritz Mueller during World War II for use in the V-2 ballistic missile.[13] Although this system was too crude to be used for the extended time needed for inertial guidance, it had an integrating gyro accelerometer that Draper would have found particularly interesting. Known as the "Mueller Mechanical Integrating Accelerometer," it consisted of a pendulous gyro mounted along the missile's major axis and coupled to a ball-and-disk integrator (see figure 11-1). When the gyro was displaced due to changes in acceleration, it activated an on-off contact switch connected to a torque motor that rotated about the gyro's precession axis. The gyro's precession axis was mechanically linked to the analog integrator connected to a cam switch that was used to initiate engine throttle-down and shutoff.

In 1982, Donald Mackenzie asked Draper during an interview about his thoughts upon inspecting Mueller's accelerometer. "Well, hell," Draper answered, "I could fix that"—and so he did, by floating it.[14] But that did not happen overnight. The task of creating an accelerometer suitable for inertial navigation, as opposed to the much less demanding application for inertial guidance, was attacked with the same means the laboratory's staff had used to improve the floated gyroscope.[15] The spinning rotor in Mueller's design was replaced by an unbalanced structure contained within a floated cylinder with internal signal and torque generators. As in the development of the floated gyro, integration, in these first examples of the floated-pendulum (FP) accelerometer, was done by electronic circuits external to the sensor. Integration was later incorporated within the instrument, leading to the development of the "pendulous integrating gyro accelerometer," or PIGA.[16]

Project SPIRE, a continuation, funded by the Air Force, of the effort to develop an inertial navigation system begun under FEBE, was to make use of the lab's floating gyros and pendulous accelerometers. Although the design for SPIRE was begun in 1949 (before testing of FEBE was concluded), physical work on the system did not get under way until 1950.[17] Unlike its predecessor, which had been constructed with a stellar tracker, SPIRE would depend only on inertial and gravitational inputs. At the heart of the system was a gimbaled inertial platform containing three single-degree-of-freedom 45 FG gyros so mounted that their input axes were at right angles to each other (see figure 11-2).[18] Each of these gyros, the precession of the gyro wheel of which was constrained to a single rotational output axis, was connected to a servodrive that kept the inertial platform in alignment. One of the gyros was aligned with the Earth's axis and was connected to an electronically driven torque motor that caused the gimbaled system to rotate with the Earth. The other two gyros, mounted perpendicular to the Earth's axis, were connected to torque motors that kept the platform stable with respect to what Draper termed "inertial space."

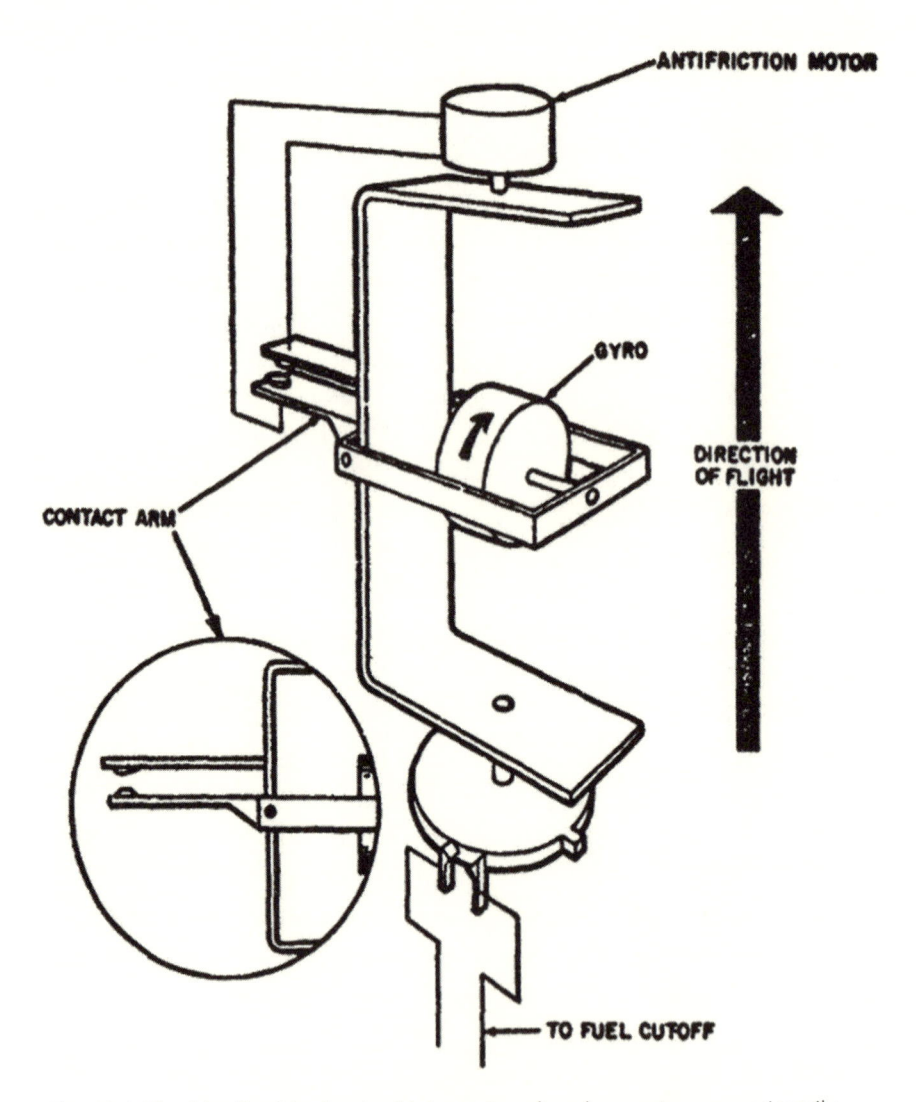

Fig. 11-1. The Mueller Mechanical Integrating Accelerometer was primarily a range device to cut off propellant when the rocket attained the velocity required to reach the target. Because the gyroscope is supported in an unbalanced position, it processes at a rate determined by the acceleration. The angle through which the gyroscope precesses is a measure of the integrated accelerations, hence the missile's velocity. *Mueller,* A History of Inertial Guidance

SPIRE was designed to navigate along a Great Circle route programmed before the flight began. This was achieved by turning the range isolation gimbal (see figure 11-2). about the line of nodes and locking it in. Two model 125 FP accelerometers were used to keep the aircraft on course and indicate its distance along the programmed track. One of the accelerometers was mounted with its output axis aligned with the Great Circle course. It generated a null signal when the airplane was on course and a deviation signal when the airplane strayed to one side or the other.[19] The other was mounted at right angles to the programmed course so that its output signal represented progress along the desired track.

In addition to the large gimbaled unit that contained the inertial platform, SPIRE had a large electronic console, a navigation panel, and a pilot's panel. All but the latter were installed in the bomb bay. Because practical results had to be obtained as soon as possible, no attempt was made to miniaturize any of the system's relatively large and cumbersome components.[20] The entire system weighed 2,800 pounds and (except for the pilot's console) took up the entire bomb bay.[21]

On its first test flight, to Los Angeles in February 1953, SPIRE's error rate had been five miles per thousand miles of travel. During the next two years, the Instrumentation Laboratory worked to improve gyros and accelerometers while conducting test flights between Holloman Air Force Base, New Mexico, and the Los Angeles airport from the flight test facility in Bedford. By 1955, the Instrumentation Lab had reduced the error rate to between one and two miles per thousand miles of travel.[22]

The success of SPIRE led to a follow-on contract from the Air Force to develop a lighter, more accurate inertial navigation system, to be called SPIRE Jr. Less money was to be spent on its design. The primary objective of the project was to improve in the system's performance in terms of a smaller margin of error.

Based on its experience with SPIRE, the Instrumentation Laboratory, under Draper's direction, concluded that better performance could be achieved if the sensors (the gyros and accelerometers) were redesigned to provide greater accuracies.[23] The Air Force contract funding SPIRE Jr. provided the resources necessary to improve the floated gyro, which was achieved by concentrating on the instrument's ability to accept and process command signals. Similar efforts were applied to the specific force

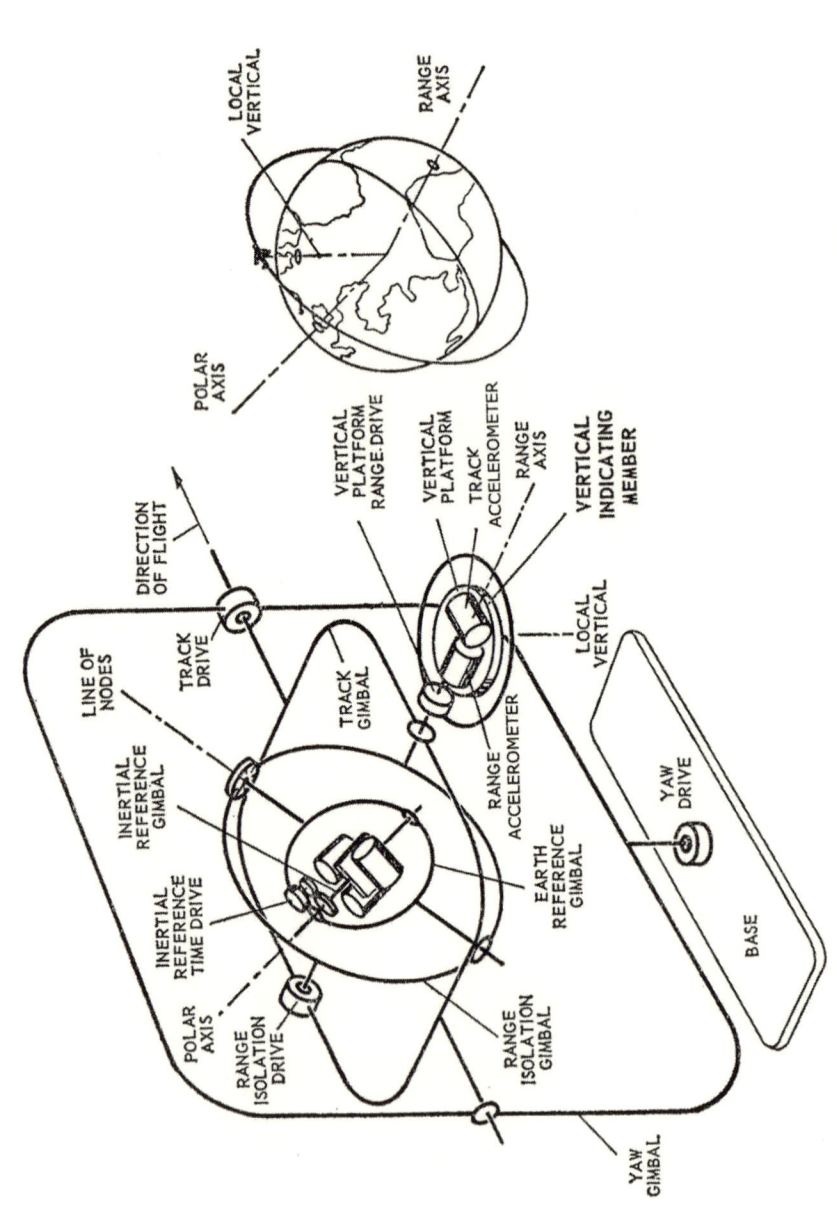

Fig. 11-2. SPIRE system functional diagram. *Instrumentation Laboratory, MIT*

receivers (accelerometers), with particular attention given to incorporating integration within the instruments themselves.

The design for SPIRE Jr. incorporated the second generation of gyros and accelerometers developed by the Instrumentation Laboratory. These were the 10 FG–series integrating gyro and the 25 PIG–series pendulous integrating gyro, the first gyros constructed with self-contained integrators (see figure 11-3).[24] SPIRE Jr., at 1,500 pounds (half the weight of SPIRE), made its first flight, in a Boeing C-97 cargo plane, on November 7, 1957.[25] Although the Instrumentation Laboratory had made great strides in reducing the size and weight of SPIRE Jr., its improvement in performance was not of the magnitude that Draper had hoped to obtain.[26] While the new sensors, which had been under design for some years, showed promise, they had not reached their full potential and were not ready for use in operational systems.[27]

Nevertheless, SPIRE Jr. brought unexpected recognition and acclaim for Draper and the Instrumentation Laboratory when it was brought to the public eye by the CBS Television Network. It was nationally televised as part of the *Conquest* science series on April 13, 1958.[28] By then, the development of inertial navigation for long-range bombing had become overshadowed by that of intermediate-range ballistic missiles (IRBMs) and intercontinental ballistic missiles (ICBMs), which, because of their relatively short flight times, required inertial sensors of different design characteristics. The Instrumentation Laboratory was already involved with this form of guidance (see the discussion of Thor in chapter 14). This work, like all of the lab's research under military contract involving inertial guidance and sensors, remained classified until the SPIRE Jr. program aired. Now the Air Force, which was competing with the other services for funds to develop a family of ballistic missiles, saw fit to publicize SPIRE Jr. as a means of winning the public's confidence in the technology of its ballistic-missile programs.[29] Sputnik (the first artificial Earth satellite), put into space by the Russians six months earlier, had severely shaken the nation's confidence in the military's technical prowess. To many it appeared that the United States had fallen behind the Soviet Union in the effort to develop ICBMs. Reassuring the public was undoubtedly in the minds of the senior leadership of the Air Force when it authorized the publicizing of SPIRE Jr.

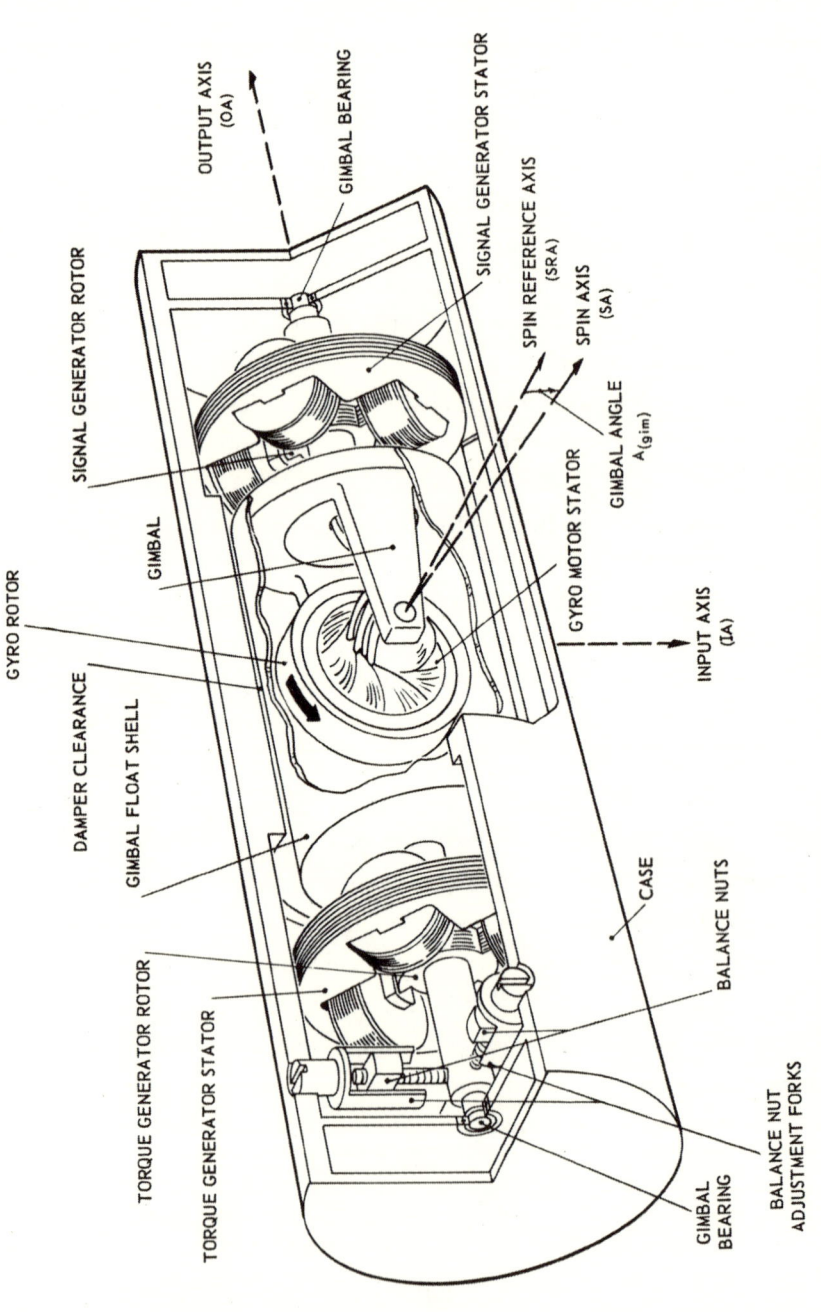

Fig. 11-3. Pictorial diagram for single-axis integrating gyro unit. *Instrumentation Laboratory, MIT*

On March 7, 1958, CBS News reporter Eric Sevareid and a camera crew accompanied Draper and a coterie of the lab's design engineers on a cross-country flight in an Air Force C-97 equipped with the SPIRE Jr. inertial navigation system.[30] The flight from Boston to Los Angeles and back was filmed. The press, in its account of the resulting CBS program, described SPIRE as a "remarkable device—the most precise and most advanced yet devised—can aim, fly and steer an object from one point to another without human control."[31] It was, news reports declared, a completely self-contained instrument that needed no outside references or information other than the electronic data fed into it before its use.

The *Conquest* show was unusual in that it was one of the few public efforts to promote Draper's highly classified work. "Much of what [Draper] did was more routine. On trips to visit his sponsors in the armed services, he would bring along 'Doc's dollar bills,' wallet-sized graphs showing the performance of his latest instruments (without a scale, to avoid broaching secrecy), to help reassure supporters and convince skeptics that the problems in producing 'inertial quality' devices were being overcome."[32]

As Donald Mackenzie pointed out in a workshop paper presented in 1984,

> Sponsors were not the only people who had to be kept "on board" if the new technology was to succeed. Those who worked with the system had to be persuaded often to act in new ways. The laboratory janitor, for example, had to be convinced that if he knocked a test table with his broom, he should report it, or else an unexplained jag might show up in the gyroscope's output. Those who assembled gyroscopes and accelerometers had to be shown the dangers inherent in facial hair and holidays—debris from moustaches and sunburnt skin could play havoc with delicate instruments.[33]

Draper's passion, or perhaps obsession, with the single-degree-of-freedom gyroscope was familiar to those who knew him well. Improving and upgrading the performance of the single-degree-of-freedom gyro was the driving force behind the Instrumentation Laboratory's ongoing efforts to develop the most accurate inertial navigation and guidance systems possible.

When inertial navigation systems began to appear in the 1950s, there were two schools of thought with regard to which gyro technology would be best suited for this application. The "East Coast school," led by Draper, opted to concentrate on the single-degree-of-freedom type. The "West Coast school," led by the Autonetics Division of North American Aviation, opted to develop the two-degree-of-freedom type. Both groups had to contend with the two major sources of gyroscopic error: rotor mass unbalance and bearing torque.

The Autonetics group benefited from its early work on the guidance system for the Navaho missile. Autonetics also had the good fortune to have the services of John M. Slater, whom one historian tags a "gyro genius."[34] Before joining Autonetics, Slater had been employed by the Sperry Gyroscope Company, where he worked closely with Walter Wrigley on a gyroscopically controlled autopilot.[35] What he learned from Wrigley pertaining to the theory of inertial navigation is not known, but it seems improbable that he did not gain some insights into the problem from his coworker.

For Draper's group, the basic engineering paradigm was to subject each successive design to a goal-oriented process calculated to produce instruments of ever-increasing accuracy.[36] As Mackenzie remarks, "no detail escaped attention in the search for the sources of inaccuracy and the means to remove them."[37] Under Draper's oversight, the staff in charge of gyroscope design focused on improving materials, methods of construction, and manufacturing techniques. Although the Instrumentation Laboratory was central to the solution of these problems, other parts of MIT (e.g., the Department of Metallurgy) were drawn in, as were outside companies with special expertise.

Autonetics took a broader-based approach in which no single device dominated.[38] This meant a wider range of research that led to the development of air-lubricated gas spin bearings, the use of two symmetrically opposed motors to correct for drift, and the application of two-degree-of-freedom gyros.[39] As Anthony Lawrence notes in his 1998 treatise on inertial technology, "The choice between one-axis and two-axis gyro was not based on very powerful cost or performance criteria: each performed well and system costs did not seem to differ much."[40]

When the SPIRE segment aired on national television, Draper presided over an Instrumentation Laboratory staff of more than eight hundred people.[41] Despite its growth in size, Draper knew exactly what was going on within each of the laboratory's many government-funded programs under way at any one time. Bob Seamans, who worked in the lab for many years, later told his interviewers that Draper loved "to come in the morning, take care of the minutiae, he was not strong on details of administration, and then as soon as he could, get out in the lab, you know, find out what kind of problem a machinist was having on a lathe, and he knew where the sensitive points were in the lab, the things that were going to make the difference between success and failure."[42] Draper, according to Seamans, had a seemingly endless supply of energy. He would take people out to dinner at a place called the "Fox and Hounds," have a few martinis with dinner, then come back to the office and continue to work until two or three in the morning. The next day he would already be in his office when everybody else came in to work. He even worked on Saturdays and Sundays.

12

SINS

The Submarine Inertial Navigation System

While Draper and the Instrumentation Lab staff were putting together the FEBE system, the laboratory received a contract from the Navy's Bureau of Ordnance to develop an advanced stable element similar to that being developed for FEBE. A stable element measures the level and cross-level angles of a deck-mounted weapon as it is subjected to the roll and pitch of the ship. Its output is fed to the fire-control computer, providing a stable horizontal reference from which to compute a fire-control solution. As the Navy moved into the guided-missile era, it needed a system that was capable of generating the explicit geometrical relationships between ship coordinates, target coordinates, and horizontal coordinates. This was the objective of the project established under the Navy contract known as "Project MAST" (Marine Stable System) issued on September 3, 1948.[1] The development contract called for a system capable of establishing the horizontal plane to within half a minute of arc and the geographic meridian to within five minutes of arc under shipboard conditions in north and south latitudes of up to eighty degrees.

A year earlier, the Navy had begun development of the turbojet-powered Regulus cruise missile, which could be launched by solid-rocket boosters from surface ships or surfaced submarines. The missile was remotely controlled throughout its flight by a radio-command guidance system that could be operated from ground stations, aircraft, or ships anywhere along its flight path. The Navy wanted the missile to carry a three-thousand-pound warhead to a maximum range of five hundred nautical miles at Mach 0.85 with a CEP* of 0.5 percent of the range. In order to hit the target within specified limits of accuracy, the launching ship would have to have accurate knowledge of both launching and target positions as well as a stable vertical to keep the missile on track while it was launched. Though none of the documents viewed by the author explicitly state this, it appears that hardware developed for Project MAST, which combined a stable vertical with a marine gyro, performed the latter function.[2]

The studies and preliminary tests of the hardware for Project MAST led Draper to believe that a complete inertial navigation system for naval vessels could be built.[3] This idea was presented to the Office of Naval Research, which entered into a contract with MIT in June 1950 to study the feasibility of an inertial system for the Navy's submarines.[4] Although Draper was listed as the principal investigator, the task of putting together the documentation that would emerge from the laboratory's work on the project was assigned to John Hovorka and Forrest E. Houston. They were aided by two students in the Navy's graduate program, Lt. Robert H. Cook and Lt. Boyd E. Gustafson, who were supervised by Walter Wrigley.[5]

The team's first effort, a report titled "Theoretical Background of Inertial Navigation for Submarines," was published in March 1951. The report began with the physical factors involved in inertial navigation, went on to a detailed discussion of Schuler tuning, and finished with an analysis of several idealized systems for inertial indication of position. Hovorka did the actual writing, which was then put in final form by Houston, who was responsible for the large number of illustrations and block diagrams. All mathematical work, as Wrigley noted in the report's preface, was derived using Draper notation, a convenient system of self-defining symbols (see chapter 3) invented by Draper to simplify the description of the geometrical and physical concepts involved in his work and that of the Instrumentation Laboratory.[6] The notation only used symbols that appeared

---

* Circular error of probability, the distance from the target within which, on the average, half of the warheads fall.

on standard typewriter, so that carbon copies could be easily made to facilitate dissemination of the material.[7] The notation also reflected the interests of the Instrumentation Laboratory, with special symbols to represent the performance conditions associated with certain self-referential devices it had developed.

Five months after issuing the first part of its study on the feasibility of inertial navigation of submarines, the Instrumentation Laboratory released the second. This final report examined the potential use of an inertial navigation system from a practical standpoint. It reiterated the conclusion of the first part, that "a servo-driven platform on which were mounted two accelerometers (or pendulums) and two gyroscopes could be used to indicate the vertical, and, with the addition of a third gyroscope, to indicate heading"; in a submarine it could show the direction of the Earth's gravity field and indicate the projection of the Earth's polar axis in the horizontal plane.[8] This three-axis-stabilized platform became the basis for the various methods of position indication discussed in part two of the study's report. Houston and Hovorka had determined that a self-erecting latitude indicator, operating in conjunction with an internal gyrocompass having a vertical indicator mounted concentrically with a controlled member to track the polar axis, represented the best and most practical approach to achieve the objectives established by the Navy. This concept involved a five-gimbaled arrangement in which the inertial reference package of gyros was permitted to tumble in the Earth's field.[9]

After reviewing the second part of the Instrumentation Laboratory's feasibility study, the Navy decided to proceed with the development of a prototype inertial navigation system for its submarines. On November 6, 1951, the Inspector of Naval Material, based in Cambridge, forwarded the specifications prepared by the Bureau of Ships (BuShips). The document required equipment that would "permit accurate navigation of submarines over a period of ten hours without requiring the vessel to disclose its presence to hostile forces."[10] This material was sent to the attention of Forrest Houston.

Houston, a Naval Academy graduate who had served in the Pacific during World War II as a gunnery expert, was the ideal candidate to manage the Submarine Inertial Navigation System (SINS) project.[11] He received his MS degree in electrical engineering from MIT in 1948 while still in the Navy. His thesis "Stabilization and Tracking Control of a Proposed On-Mount Antiaircraft Fire Control System," covered material that would be useful background for Project MAST. Now a civilian, Houston had been hired by the Instrumentation Laboratory in 1950 and assigned to the initial study of submarine navigation.[12]

When the BuShips specification was received, he was promoted to assistant director of the Instrumentation Laboratory.[13] At first glance, this would place him in the top leadership of the laboratory, but there were a number of assistant directors in the laboratory at that time. Draper had organized it in a matrix structure in which people with similar skills and knowledge bases were pooled to work on particular projects, managed by three tiers of senior staff. Instead of using the traditional titles that would be used in business organizations, such as "project engineer," "project leader," or "division manager," Draper adopted the university system used to rank professors. At the highest level were the deputy directors, who were equivalent to full professors. Next in line were the associate directors, who were equivalent to associate professors. Next came the assistant directors, like Houston, equivalent to assistant professors. It is not known whether Draper chose these titles to enhance acceptance in the outside world (as most people would be impressed by a "director") or because it was a simple, convenient arrangement and familiar to him.

The first SINS prototype was complete by the end of 1953. It was tested on shore in January 1954, in a van driven from Boston to Newbury. Then followed a number of test runs conducted on roads all over the eastern United States.[14] The first shipboard tests were conducted on board the fleet oiler *Canisteo* (AO 99) during a fifteen-day cruise between Norfolk and Long Island. In February 1955, the system was installed in the Navy cargo ship *Alcore* (AK 256) and tested on a round-trip voyage from Norfolk to Naples.[15] Additional tests runs were carried out in the Mediterranean and Caribbean. "The results, in terms of latitude errors and longitude errors, were erratic and tended to have mean values in the vicinity of 5 n. mi."[16] Although good performance was obtained for a period of as long as 108 hours, otherwise satisfactory operation was intermittent. As later noted by W. F. Raborn and

John P. Craven—two key figures in the development of the what would become the Polaris missile,

> The need for improved gyros and accelerometers was evident, resulting in an effort to reduce drift rates and uncertainties in these components. Among the problems to be tackled were air and outgassing in the flotation fluid; improvement of bearings and bearing materials, anisoelasticity, resulting in mass unbalance under acceleration; uncertainties, resulting from the use of the Microsyn for signal creep[;] . . . temperature variation, resulting in mass shifts; torques, resulting from the power leads to the gyro; mass shifts in the rotor caused by deflections of the ball bearings, creep in the materials, etc.[17]

Nevertheless, the prototype proved the feasibility of an inertial navigation system for submarines.

As a result of these trials, BuShips revised its design requirements and requested bids for an experimental model. Two firms responded, and both were awarded contracts: one went to the Autonetics Division of North American Aviation, the other to the Marine Division of Sperry.[18] The Sperry system, delivered by September 1956, was based on an MIT design in which the gyroscopes remain in fixed orientation in inertial space with respect to the stars.[19] This approach created unforeseen problems for Sperry, whose capabilities in precision manufacturing were not up to the standards of MIT's Instrumentation Laboratory. As Graham Spinardi writes in his history of the Fleet Ballistic Missile program, "this led to a system that rapidly lost accuracy. As the earth rotated and the submarine's position changed the gyroscopes were subjected to a varying gravity field. The slightest mass imbalance of their rotors would lead to significant errors. But achieving perfect or near perfect mass balance was an exceedingly difficult task, especially as one moved outside the laboratory."[20]

The North American system, the "N6A Autonavigator," had been originally developed for the Navaho missile. Unlike the Instrumentation Laboratory's design, it was a local-level system kept horizontal at all times so the gyros were not subject to change in direction of gravity. It also incorporated an innovative approach to the drift problem, one developed by John Slater of Autonetics.[21] Slater's solution relied on three sets of G2K reversable gyros that were mounted in pairs along each of the three axes. The sets were operated in alternation, one set spinning in one direction for a time, then passing control to the second set. The original trio now spun down and then spun up in the opposite direction, averaging out their drift. The reversing spin-up/spin-down pattern was called a "NAVAN cycle."

While work on all this was going on, the Navy established the Special Projects Office (SPO), to embark on an urgent program to develop a submarine-launched ballistic missile (SLBM), to be known as the UGM-27 Polaris. The development of a submarine inertial navigation system was deemed crucial for the launching vessel's navigation and its primary role, that of initializing the missile's guidance system. Solving the navigation problem was a critical aspect of the SPO program, therefore, because of the effect that it had on overall system accuracy. "Minimizing the time between exposures for radio navigation and satellite fixes for 'resetting' the SINS was a critical requirement to ensure stealth for survivability."[22]

To meet these challenging requirements, the SPO conducted sea trials on the *Compass Island* (EAG 153) to compare the performance of Sperry's SINS Mark 1 with a version of the Autonetics N6A-1 Autonavigator modified for naval use. The *Compass Island* had been converted to a navigational research test vessel under the Polaris program budget.[23] The ship operated along the eastern seaboard testing equipment and training personnel until March 13, 1958, when she sailed from New York for experiments in the Mediterranean, returning to New York on April 17.[24] The Autonetics system performed much better than the Sperry system did and was installed on the *Nautilus* (SSN 571), where it played an essential role in guiding the "boat" under the ice during the submarine's record-breaking voyage to the North Pole in August 1958.

Despite the performance advantage exhibited by the Autonetics system, the Navy continued to go forward with both variants. They were modified and improved with the help of the Instrumentation Laboratory, which provided technical support to the two companies under separate Navy contracts.[25] It is likely that Sperry, which had been selected to integrate all of the navigation equipment for Polaris submarines (three SINS units, a stabilized periscope for celestial navigation, a radio navigation system, and two Navigation Data Computers, or NAVDACs) lobbied hard for the continued development of its production version of SINS, the Mark 3. This new version of Sperry's Gyronavigator had interchangeable gyroscopes and an improved and integrated polar mode of operation for better operation under the ice.

According to Vice Adm. Levering Smith, once head of the Navy's Fleet Ballistic Missile (FBM) program, the Navy initially intended to install Sperry's SINS Mark 3 in the first five *George Washington*–class Polaris submarines and the Autonetics version, SINS Mark 2, in the follow-on *Ethan Allen* class.[26] Apparently this arrangement was reversed when Autonetics was able to produce the SINS Mark 2 ahead of the Sperry equipment. After a few years all of the Sperry systems were replaced, and the Autonetics Mark 2 (and its various modifications) became standard on FBM submarines. A total of fifty-six Mark 2 Mod 0 SINS were produced, three (for redundancy) installed in each of the submarines that carried the 1,200-mile Polaris A1 missile.[27] (As will be noted in chapter 16 [see p. 167], Draper was not pleased with Sperry's handling of the Instrumentation Lab's design for SINS.)

In April 1958, two years before the first Polaris A1 was deployed, the SPO began efforts to extend the missile's range to 1,500 nautical miles.[28] The longer-range missile would become known as the Polaris "A2" and was scheduled to become operational in 1962. An even longer-range (2,500 nautical miles) and more accurate missile (to become the A3) entered the planning stages a year later. To meet the accuracy needs of this missile, which was to have a CEP of 0.5 nautical miles, the shipboard performance of SINS would have to be improved. In 1959, Draper was asked by the Polaris Steering Task Group to review the status of submarine inertial navigation with an eye to forecasting future developments that would improve the accurate determination of the submarine's firing position. Draper presented to the committee a written report on October 22, 1959.[29] "The allowable uncertainty in firing-point position," it concluded, "must be less than one quarter mile for any fleet-ballistic-missile system that will be satisfactory during the 1964–1970 period. . . . [T]his level of uncertainty should exist during a time period of approximately two days after the last fix that is based on accurate information from sources outside the missile carrying vessel." Draper argued that such performance could be achieved by inertial systems taking advantage of the nulling action of untorqued gyro units.

Draper believed that the Instrumentation Laboratory could deliver by 1963 an improved SINS having the following characteristics and performance:

- Accuracy of a quarter-mile over a fifty-hour operating period.
- System contained in a cylinder approximately twenty-five inches in diameter and twenty-six inches high, weighing less than 225 pounds.
- Reliance on monitors needed only once every fifty hours, for the purpose of alignment.
- Maneuvering restriction to exist for one hour only during realignment.
- Component life of 20,000 hours.

The new system would incorporate improved floated gyros geometrically positioned with the respect to the floatation fluid and a magnetic suspension system, "Ducosyn." On the basis of laboratory tests then being conducted by the Instrumentation Laboratory, he expected that the new Ducosyn-fitted gyros would show "an improvement of over an order of magnitude" over the performance obtainable with conventional gyros.[30] On March 22, 1960, the Instrumentation Laboratory received a Navy contract (NObs-78456 [FBM]) to design, construct, and test a new inertial navigation system for FBM submarines, to be called "SINS Mark 4."[31] Tests of the SINS Mark 4 Mod 2 on the *Compass Island* were conducted dockside from October through December 1964.[32]

Unfortunately for Draper and the Instrumentation Laboratory, the Navy chose not to develop the SINS Mark 4 into a production unit. It decided that the improved accuracy needed for the Polaris A3 missile could be accomplished by upgrading the SINS Mark 2 Mod 0, produced by Autonetics, into what would be the SINS Mark 2 Mod 2. This was easily accomplished by replacing the original gyros with the Autonetics G7B, a single-degree-of-freedom floated gyro in which the conventional ball-bearing supports were replaced by gas-spin bearings.[33] Autonetics found that gyro performance could also be significantly improved by certain changes in design and materials. By redesigning the induction motor and the rotor, its engineers were able to increase the gyro's angular momentum by a factor of 2.7. This reduced the effect of external torque acting on the float, lowering the random drift rate of the gyro. All Polaris A3 submarines were subsequently equipped with the Autonetics SINS Mark 2 Mod 2.

13

Professor, Prodigious Worker, Family Man

In 1939, MIT's aeronautic program, which had been part of the Department of Mechanical Engineering curriculum, was made a distinct department under Professor Jerome Hunsaker. Hunsaker, who had previously been head of the mechanical engineering department, became chairman of the Department of Aeronautics and remained its head until 1951, when he decided to relinquish this position so that he could concentrate on his work for NACA and the special committees on which he served.[1] Hunsaker, who had been suffering from gallbladder trouble for some time, had been trying to find a suitable replacement, but as Draper later remarked, "the academic types were not . . . used to struggling with the problems of organizations, people, and money and the management kind of thing."[2] Hunsaker found several people, but the job "either made nervous wrecks out of them or else they didn't do it."[3] As a favor to Hunsaker, Draper took on the job of deputy head of the department until a suitable department head could be found. He knew all of the department's people well, and by this time had quite a lot of experience managing the Instrumentation Laboratory. Draper volunteered to do this without giving any thought to whether or not he would be made head of the department. He was mildly surprised, then, when he was made head of the Department of Aeronautics on July 1, 1951. He did not think it was any particular honor. He was just going to continue to do what he had been doing for the past couple of years.

As head of the department, Draper was expected to maintain a proper balance throughout the department between scholarly achievement, creativity, collegiality, professional competence, leadership, and the desire of the faculty to teach.[4] He was also responsible—in accordance with MIT policies—for cooperating with other departments, promoting the welfare of the institute as a whole, and suitably recognizing individual achievement and service. Further, he was expected to "nourish" teaching and research (the primary functions of MIT), by "efficient and imaginative administration." None of this was particularly new to Draper. He had been teaching and innovating within the instrumentation area for a number of years, had consistently worked with other departments and outside vendors, and was an exceptionally strong mentor and supporter of his students.

The five-year Aero program he established, for instance, provided continuing guidance to the students in the honors course, which began in a student's senior year and continued through a second year of master's work. Students in Draper's honors course had considerable latitude in making substitutions for regularly prescribed senior subjects.[5] Upon graduation at the end of the second year, he or she received a bachelor's degree and a master's degree in aeronautical engineering. One of the "perks" of belonging to the Draper's honors course was an invitation to the monthly dinners hosted by Draper. Robert G. Chilton, another Draper protégé and an honor student (whose master's thesis discussed "Distance Measurement by Inertial Means"), thought going to these dinners "was great fun."[6] He described Draper as a "real culinary artist," who would take his guests to fancy restaurants with various ethnic backgrounds.

"Once you were one of his graduate students, you were part of the club"—this according to Vice Adm. Thomas Weschler, a former MIT student, who later recalled that Draper "would invite you to Locke-Ober's, which is a wonderful restaurant in Boston [the admiral spoke before 2012, when Locke-Ober closed]. About every quarter, Doc would have a mid-afternoon dinner for those who were members of his class. We would go over there, and we had clams casino and lobster. It was really quite a feast, and we were his guests for this wonderful spread. He would talk about what he was doing, what his goals would be, what the possibilities were."[7] It was inspiring for his students, who got motivated, even inflamed, by him. He liked and was interested in his students, and his students loved him in return.[8]

Draper believed that one of the primary goals of higher education was to identify and develop individuals who showed sparks of creativity. He felt that such individuals had the ability to "deal directly with situations in nature by identifying basic problems and contributing effective and original solution for these problems."[9] He considered research in an atmosphere of educational advancement an excellent means of developing the full creative potential of such individuals. In his opinion, the best path for nurturing this talent depended on the effectiveness of the opportunities presented to these students.

Draper taught engineering by *doing* engineering: learning and research were interchangeable.[10] Under his artful teaching, the student's experience became stimulating and exhilarating. Draper would come to class wearing a green eyeshade and accompanied by assistants to help handle his teaching props. "Today," he might announce to his class, "we are going to discuss the genesis of an invention we made overnight."[11] Using Draper notation, which took his students some time to learn, he would proceed to illustrate the creative brand of engineering he was using to advance the state of the art for instrumentation.

He never taught any of his instrumentation courses out of a book.[12] Every one of the two dozen or so courses Draper developed was based on his experience in the laboratory. That experience was the source of his inspiration, and the fact that he had to reduce it to a format that could be used for teaching was good for the laboratory too: it tied down his ideas and validated his theories. The courses were all based solidly on mathematics. As Bob Seamans later explained,

> There was an attempt by Doc throughout everything that he did to try to simplify the mathematics so you could get useful generalized ideas about how things would perform. Instead of trying to squeeze the last tenth of 1 percent out of the mathematics, he felt it was better to make some assumptions that would permit a better understanding of the physics and what had to be done to improve the performance of the equipment. . . . He was a great believer in following two simultaneous paths, one a theoretical path and the

other an experimental path. He kept matching them up as he went along, and that way, determined from the experiment what the important parameters were, so he could weed out the unimportant mathematical terms.[13]

Through the judicious use of mathematics, Draper was able to predict how equipment would operate without having to make the hardware to test it.

As head of the department, Draper was responsible for overseeing at least 15 professors and the curriculum for over 160 undergraduate and 70 graduate students. He also had administrative authority over four other Aero Department laboratories: the Wright Brothers Wind Tunnel, the Naval Supersonic Laboratory, the Unclassified Instrumentation Laboratory, and the Gas Turbine Laboratory. The latter was directed by Professor Edward Taylor, with whom Draper had worked as a research assistant years before. Although Taylor was in the Department of Mechanical Engineering, the laboratory was jointly administered with the Aero Department. The research programs conducted by Aero Department, funded mostly by outside sources, were intended to provide a creative atmosphere considered essential to a good engineering education.[14]

When Draper became head of the Department of Aeronautics he continued as director of the MIT Instrumentation Laboratory, which he characterized as "concerned with no-fooling technology, to actually produce something that worked."[15] In that way it differed from an academic department, which he felt should incorporate the most recent advances in science, technology, and engineering in its courses, which should in turn be integrated into an overall picture. The courses given by the Aero Department, he strongly believed, needed to be in the forefront of whatever industry and research organizations were doing.

In 1947, all of the Instrumentation Laboratory's classified work in Building 33 was removed and the space equipped as a general instrumentation laboratory, to be supervised by Professor Walter McKay.[16] The secret work, funded primarily by Navy and Air Force contracts, was moved to two floors in the Hood Building and three floors of the Whittemore Building, constituting the IL classified section, supervised by Walter Wrigley. By 1951 the classified section of the Instrumentation Laboratory had over five hundred employees and an annual budget of almost four million dollars (equivalent to $38.8 million today).[17] Many of the people who worked there were continuing their educations. Ernst Steinhoff, a contemporary of Draper's who was also an expert on inertial guidance, tells us that these engineers were disciples of Draper's work, people "who by their motivation and drive, carried the technological torch he [Draper] inspired towards their goal."[18] Many of these students were Navy and Air Force officers who became steeped in Draper's ideas and were awed by his personality. In later years they were to become known as the "Guidance Mafia."[19] They became Doc's external field representatives and spokesmen. In the months ahead, these Draper-trained military men would become highly valuable resources on whom he could count to support his technically sophisticated proposals for new projects.

Much of the Instrumentation Laboratory's success was due to Draper's leadership. As Robert Duffy writes in his biography of its founder and longtime leader, Draper "understood human beings and he understood how to challenge them. . . . [He] knew how to lead and how to get people to follow towards a common goal."[20] He was available to anybody who took the trouble to go and see him.[21] He knew the names of nearly all of the technicians who worked in the lab; the secretaries called him "Doc," and if he forgot their name, "darling" sufficed.[22]

Draper was a prodigious worker who was never content to sit at a desk. He loved to run the laboratory and was a leader who radiated energy and self-confidence.[23] Draper loved the fun of making something work, which he described as "defeating Mother Nature."[24] He listened to everybody, then made his own judgements after evaluating their inputs. Duffy tells us, "He tried to have at least two proponents in contention on every issue. He claimed over the years that his lab was really an Athenian democracy where talent ruled. The better solution would evolve from the contest—but if it wasn't Doc's preferred solution, it didn't always survive."[25] Draper could be ruthless, and he remained deeply convinced that the single-degree-of-freedom, floated, integrating gyro reigned supreme.

He was dogmatic at times and could be difficult to work with.[26] This was especially true when he was trying to come up with a new idea.[27] During these spells, he would often become very impatient with routine administration. Nathan Sage, who managed all of the lab's contracts, told

Bob Seamans that he could always tell when Draper was "going to hatch a new one, because he became so difficult." Sage encouraged Draper's work, and they were very close friends.

When in Cambridge and not on one of his many business trips, Draper presided over a daily ritual in his office at the end of each day. As the clock approached four, key people in the laboratory would begin to congregate at his door for an evening report. Before long, half a dozen people would be arguing around the conference table that abutted his large, wooden desk.[28] When they became too contentious, Draper would reach under his desk and push a button activating a relay that automatically advanced the hands of a large clock on the wall to 5:00 p.m. This alerted his secretary to announce "cocktail time."[29] Draper would break out the contents of the John B. Nugent Medicinal Aid Foundation; quiet would momentarily reign, and then the meeting would continue, less stridently.[30]

Draper's fondness for a good drink was notorious. Legend has it that he was once asked to explain the meaning of a dyne-centimeter, which is a very small unit of gyroscopic torque, and that he replied, "Well—that's about the force required to twist my arm to get me to take a drink."[31] The consumption of alcoholic beverages was a very normal aspect of social life during the 1940s and '50s and was a prominent fixture of the numerous parties that Draper hosted for his friends, associates, and laboratory staffers. His son James tells of an amusing challenge that was sometimes included in these festivities, when he was nine or ten years old.

The family lived in a three-story, Victorian-style home on Bellevue Street in Newton, Massachusetts. Facing the front door as one entered the house was a staircase leading to the upper floors. James' room was at the top of the stairs on the third floor. The room had a safety device for exiting the building in case the stairway was ever blocked by fire: a heavy rope that could be thrown out the window and climbed down. The rope played a key role in entertaining the revelers during Draper's parties, and "there were a lot of big parties."[32] They were held once every couple of months. The merriment went on well into the night. Most parties continued until two or three in the morning, at which point Ivy would prepare a breakfast of bacon, eggs, and toast for those who remained.[33]

Walter McKay, whom James always called "Professor," remembered one incident with the rope. "You see," he said, "you'd bring your date. In

One of the many jobs that Draper filled during World War I was that of town linesman. He is shown here climbing a utility pole, ca. 1917. *Library of Congress*

Draper (*right*) attended Stanford University between 1919 and 1922. This photograph was taken at the Stanford Carnival, ca. 1919. *Library of Congress*

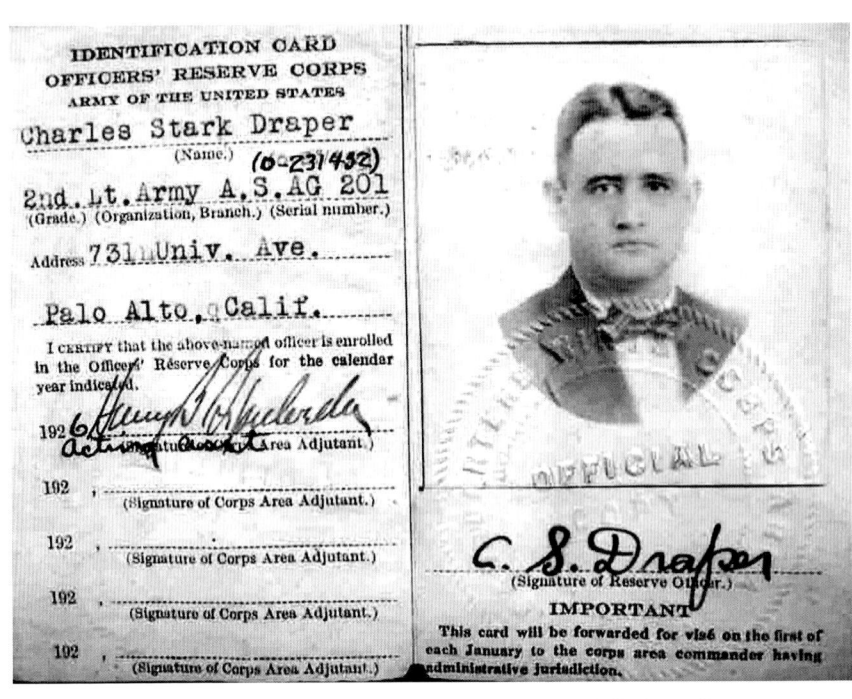

Draper was commissioned a second lieutenant in the U.S. Army Reserve Corps when he graduated from MIT in 1926. This is the identification card he was issued. *Library of Congress*

Draper undergoing flight training in the Ruggles Orientator at Brooks Field, Texas, in 1926. *Library of Congress*

Draper (*right*) and an unknown assistant working on the engine indicator in the MIT Engine Laboratory, ca. 1930. *CSD Collection, MIT Museum*

This picture was taken during Prohibition in the early 1930s, when Draper worked as a bartender in a local speakeasy to help pay his way through MIT. *CSD Collection, MIT Museum*

The prototype of the Mark 14 gunsight pictured above was produced by Doc's Confidential Instrument Development Laboratory at MIT in order to prove the concept to the skeptical leadership of the Navy. *Courtesy the Charles Stark Draper Laboratory*

The "Gyro Group" at the Confidential Instrument Development Laboratory, responsible for developing the Mark 14 and A-1 gunsights, ca. 1941–45. Front row (*left to right*): Joe Picardi, Walter McKay, Walter Loeb. Back row (*left to right*): Bob Seamans, Paul Shaffer, Bill Gedritis, Clarence Haskell, Santo Liquido. *Library of Congress*

Draper tests one of his lead-computing gunsights mounted on a 40-mm gun displayed on the MIT campus during World War II. *Courtesy Charles Stark Draper Laboratory*

Cockpit of a North American F-86A Sabre photographed on July 26, 1950, showing its instruments and A-1C(M) gunsight. The A-1C(M) was added to the second production run of F-86s. *NARA*

The Instrumentation Laboratory's SINS-1. *CSD Collection, MIT Museum*

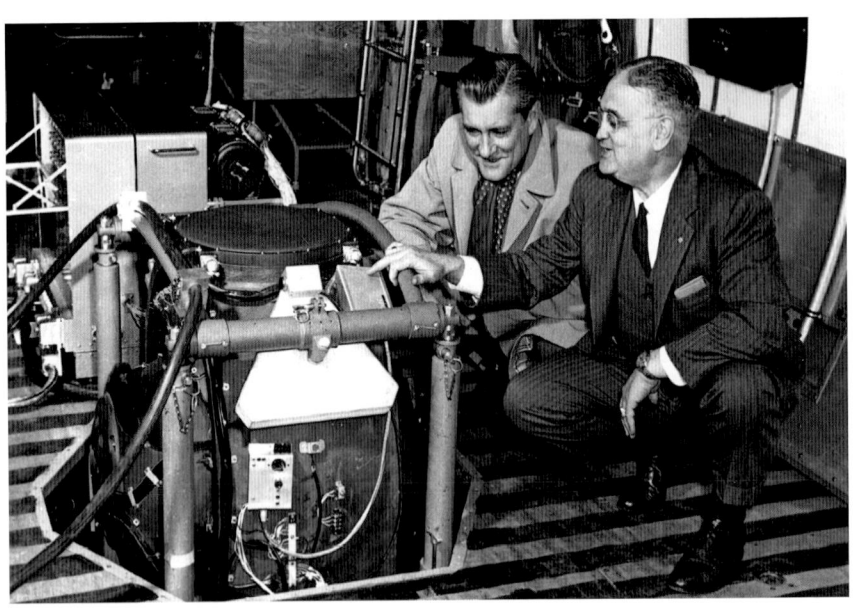

CBS news reporter Eric Sevareid (*left*) inspects SPIRE Jr. while Draper describes some of its features, ca. 1944–45. *CSD Collection, MIT Museum*

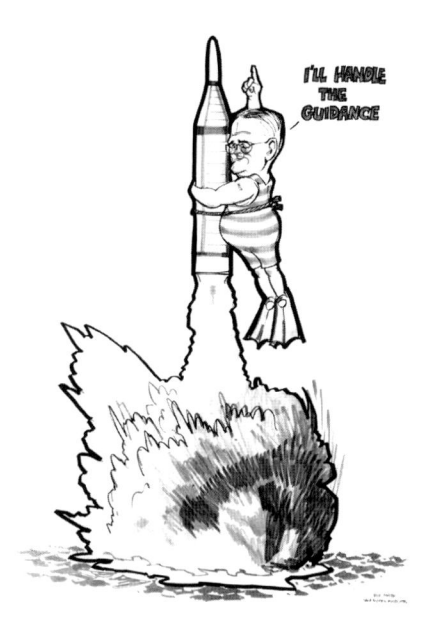

One of many cartoons that characterized Draper's love of flying, gyros, and inertial guidance and navigation. *CSD Collection, MIT Museum*

Draper standing next to a mockup of the Apollo guidance system, ca. 1955. *CSD Collection, MIT Museum*

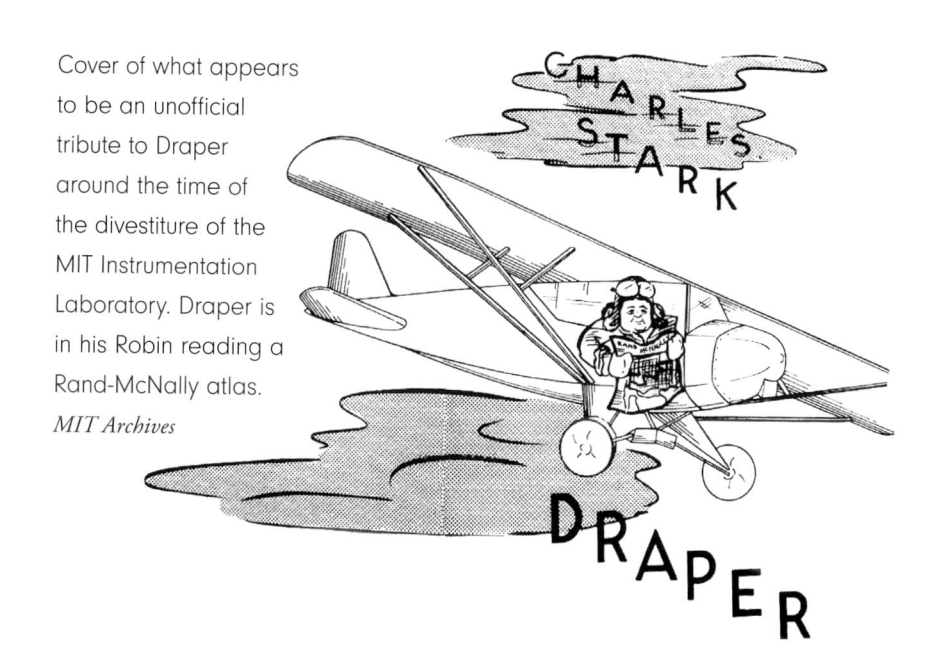

Cover of what appears to be an unofficial tribute to Draper around the time of the divestiture of the MIT Instrumentation Laboratory. Draper is in his Robin reading a Rand-McNally atlas. *MIT Archives*

A page from the unofficial tribute to Draper.
MIT Archives

FIG. 6 FUNCTIONAL DIAGRAM SHOWING ELEVATION OF TWO-DEGREE-OF-FREEDOM PEDAL EXTREMITIES, DEPRESSION OF LINE OF SIGHT, ROTATION OF TRAVERSE AXIS

Charles Stark Draper holding a small gyroscope, the hallmark of his research, ca. 1950.
CSD Collection, MIT Museum

those days, if you had a date you might date four or five women. It wasn't serious."[34] When the dates entered the Draper home, they would place bottles of whisky on a side table next to the stairs. By the time everyone arrived, there might be five or six bottles and a corresponding number of shot glasses on the table. When everyone was having a good time, according to a story that James tells, someone would say, "OK, let's go," and someone would open the front door (it was "a summer thing"). Next, they would open the window in the third-floor bedroom and throw out the rope. Then the men would line up and the first one would walk over to the table, down a shot glass of liquor, and going roaring up the stairs as fast as he could, go into James' room, climb down the rope, go around the house to the front door, and return for another shot. By the time the first player got to the second flight of stairs, the next in line would be on the way up, and so on and so forth until five or six inebriated young men were going up the stairs and down the rope. This went on until they were too drunk to continue, the winner being the one who by then had done it the most times.[35]

One day, Walter McKay lost his grip on the rope, crashed to the ground, and broke his leg. Everyone got into their cars and rushed Walter to the Newton Hospital, where his leg was set. Then, as if nothing had happened, they went back to Draper's house for dinner.

Draper's use of liquor, especially scotch, as a social lubricant is aptly illustrated in another anecdote told by Chip Collins. Collins, in addition to being Draper's chief pilot, was also director of the Flight Test Center. At one point he was having a personnel problem with one of Draper's honor students, who had been assigned as Chip's administrative assistant. Instead of sticking with the paperwork aspects of the job, this fellow would climb up on the work stand and start poking around the aircraft, which did not endear him to the mechanics in charge of keeping them airworthy. The mechanics went to Chip en masse to say that if he did not get this youngster off their backs they "were going to stab him with a screwdriver."[36] The student's behavior could not continue. It was affecting the laboratory's programs and the progress being made on projects.

So Chip asked Draper's secretary to set up an appointment with him. She came back in a few moments and said that he wanted to talk with him.

Chip said, "OK," and went into the office.

"Chipper, what are you doing Saturday?" asked Draper.
"Nothing unusual."
"How about coming over and we'll drink some scotch."
"OK."

So it was that Chip Collins arranged to meet with Draper to discuss the problem he was having. When Chip arrived at Draper's house, he rang the little doorbell and was asked to come in. Chip entered the house and was shown into one of the two living rooms that were typical of Victorian houses. Draper had a pitcher of water, a bottle of scotch, and glasses ready. "Sit down, let's have a drink," he said, and started talking. After an hour and perhaps three or four drinks, Draper started telling Chip all about his own life. "It was very interesting," Chip explained. "You know he was offered the presidency of Sperry Gyro because they needed somebody like Doc. But again, his heart was in academia. He really loved teaching and he was all for the students." They never did get around to discussing Collins' problem. (There was to be no chance to ask Collins to fill in the details before he passed away in 2015.)

Draper's proclivity for alcoholic beverages was in fact legendary, and it is enshrined in an astonishing piece of memorabilia hidden away in the annex, known as the "barn," on the family's Newton property. It is a mannequin, the spitting image of Doc—that short little guy with the pug nose, dressed in his traditional single-breasted suit and wire-rimmed glasses. The real surprise is that the chest cavity opens, revealing a built-in bar. The figure was fabricated by the Charles Stark Draper Instrumentation Lab's art department for one of its Christmas parties.[37]

The house that James and the rest of the Draper family called home had been purchased in 1939. According to a family legend, Doc made a long-distance call to Ivy from his parents' house in Palo Alto.[38] A war is coming, he said, and I want you to buy a house and two cars—and so she did, in her own name. Because they were both "farm kids," they had a very rural "take" on what a house should be. They treated it as a shelter from the wind, rain, and cold, not as an aesthetic extravagance. When the house was re-sided and painted in 1958, much of the Victorian architectural decoration was torn out.

After a few drinks at the office on a typical workday, Draper would return home at around seven o'clock and say hello to the two dogs—his daughter Martha tells us that he was "a great dog person."[39] The dogs were part of the family, and every night he gave them liverwurst, which was one of Draper's favorite foods. After dinner he would go to sleep, soon get up again, and work until two or so in the morning, then go to sleep again. Ivy was not a "morning person," so Draper had the job of getting the kids (James, Martha, Michael, and John) up, fed, and off to school. When it was time to go himself, he would say something like "climb on the outside of your cathead," one of Draper's bizarre sayings that has mystified Martha throughout her life. Or maybe he would say, "Let's get out of this firetrap." It was his way of signaling that it was time for him to get into the car and go.

Draper loved Chinese food, a taste that he seems to have acquired as an undergraduate at Stanford. He took the family out for Chinese food at least once a week, usually on Sunday. Sometimes, they went to a Chinese restaurant in Newton Center, but more often they went to Boston's Chinatown. Everyone had to eat with chopsticks, even the children; if not, they did not eat. Even John, the youngest, only five years old, had to use them.[40]

Holiday dinners were another story altogether. Draper would invite whatever students were around and anybody else who was in town. After everyone had settled in and was seated, he would cut the roast. Then he would start reciting poems. His favorite was Robert W. Service's "The Cremation of Sam McGee." He also knew a German poem, whose origins no one seemed to know. It was "Greek" to his kids, which made it very funny. Draper had a retentive mind and would just reel the thing off, along with other bits "that he could recite off the top of his head."[41]

The single-degree-of-freedom gyro and its application to inertial navigation and guidance were not Draper's only interests. When he had time to pursue these activities is unclear. He liked reading about history, and he loved photography: he took pictures all the time—lots and lots of pictures of the family—which he developed and printed in a fully equipped darkroom in the basement. He was also fond of working with machinery. The basement was full of machine tools: a milling machine, a grinder, a drill press, jeweler's lathes, etc. He also had one of the earliest commercial oscilloscopes, which was a very fitting instrument for an individual whose professional career involved the measurement of minute physical phenomena. The machinery was important to him emotionally, but he spent most of his basement time in the darkroom.[42]

In addition to the workshop and photo lab, in the basement were also a squash court and a boxing bag. His career as a boxer in college, during which he had broken his nose several times, is what had led Chip Collins to refer to him, affectionately, as "a little man with a pugilistic nose and cauliflower ears."[43] There are several family legends relating to Draper's boxing. One episode supposedly took place outside a nightclub one night after he got into an argument with a Harvard man.[44] It must have happened during the winter, for Draper, as the story goes, looked around for a patch of sidewalk that was free of ice so he could get good footing. Then this guy comes at him, Draper would relate, and *pow!*, Draper gives him a roundhouse punch and knocks him down. For James this and other Draper tales were "like reading the gospels. Who knows what really happened?"[45]

On Saturdays during the spring and summer, Draper would listen to Boston Red Sox baseball games in the yard behind the house, setting up a card table with a portable Emerson radio under the apple trees. On these occasions he always wore a gray tee shirt, a gray sweatshirt, and khaki shorts from Brooks Brothers. He got all his clothes there: suits, pants, tee shirts, underwear, etc. He would sit at the table, listen to the game, and do his work—always on yellow legal pads. He had a big console radio in a room upstairs with large tables laid out against the walls with big lined pads of paper. Even when he was home, he would toil away at some project or other.

As for getting along with people . . . Martha, his only daughter, maintains that he was as much of a psychologist (he got his first degree in that) as he was an engineer, if not more.[46] He got along with people. "He could be pig headed as hell," she said, "but he did not come on with superior airs and stuff like that."

14

Inertial Guidance for Atlas and Thor

In December 1952 Draper, who had been a member of the Air Force Scientific Advisory Board since 1946 and was now assigned to the Guidance Committee, was appointed to an ad hoc committee chaired by Dr. Clark B. Millikan, to study the requirements for the Atlas ICBM.[1] U.S. intelligence reports indicated that the Soviet Union, which was now considered a major threat to U.S. security, was building a large variety of missiles and had developed a rocket engine generating 235,000 pounds of thrust. This was twice the size of any American rocket engine.[2] The Air Force Air Research and Development Command (ARDC), which had awarded the Consolidated Aircraft Corporation (Convair) a study contract for an ICBM, argued for full-scale development of Convair's Project MX-1593 missile, soon to be dubbed the "Atlas."

After conducting several meetings in the second week of December, Millikan's ad hoc committee unanimously recommended to:

1. Retain the program for the Atlas missile.
2. Relax the warhead requirement to three thousand pounds.
3. Reduce the accuracy requirement of the 5,500 nautical mile missile from a 1,500-foot CEP to one mile.
4. Retain Convair as contractor.
5. Adopt a stepwise development approach.[3]

The Air Research and Development Command did not agree with the committee's recommendation for lowering the accuracy requirement nor the committee's proposal to test components of the Atlas on board Viking and Navaho missiles. To resolve the differences between the Millikan Committee report and the Air Force ARDC, Maj. Gen. Donald L. Putt invited Dr. Marvin J. Kelly, director of the Bell Telephone Laboratories and a member of the SAB executive committee, to review the program along with other scientists. Among the proposals made by these scientists was the recommendation—initiated by Dr. Kelly—to consider the use of inertial navigation in place of the Azusa guidance system being developed by Convair.

The Azusa system used an inertial autopilot transponder-receiver and a pair of ground-based stations. Signals transmitted from the missile in flight were received at two ground stations. Phase variation, due to differences between the stations' radar receivers and the missile's transponder, were fed into a digital computer enabling the missiles flight path to be compared to an ideal trajectory. Corrective signals were then relayed to the missile guidance system until the point of nose-cone separation, at which point the nose cone would follow a ballistic course to the target. From then on, the principles that determined the warhead's flight to the target were the same as those that guided artillery projectiles.

The laws of physics determine how an object behaves under the influence of gravity. These laws can be used to determine the impact point of a projectile (in this case the warhead) provided that its position, velocity, and direction of travel are known, along with its height above the surface of the Earth. By working backward, one can calculate the velocity needed to reach a given target if the other parameters are known. Accuracy depends upon the precision knowledge of the point of separation, the distance to the target, the direction to the target, and the warhead's velocity. These principles, according to one writer, relied upon the traditional rules of navigation: "If you know exactly where you started, how long you've been traveling, the direction you've been heading, the speed you've been going the whole time, then you can calculate exactly where you are—and how to reach your destination."[4]

Convair's engineers envisioned that an Atlas missile firing would take place in accordance with the following sequence:

The rocket would be elevated to a vertical position on a concrete stand. After takeoff—with all 5 engines operating—the rocket would execute a programmed turn at approximately 15,000 feet altitude, placing it on a ballistic trajectory. Some 200 nautical miles downrange, the ground station would begin tracking the rocket. At 120 seconds after launch, the first-stage engines would be jettisoned. Next, the ground station would assume control of the stabilization system while the computer analyzed the flight data and applied corrections to keep the rocket on target. At 266 seconds the sustainer engine would shut down and a pair of small thrust rocket motors (verniers) would "trim" the final velocity for the remaining 30 seconds of powered flight.

Approximately 296 seconds after launch the verniers would cut off, the nose cone would separate from the airframe, and, by a final command from the ground, the warhead would be armed.[5]

Millikan's ad hoc committee was not the only group investigating the state of U.S. missile developments. In June 1953, the Department of Defense Armed Forces Policy Council ordered the establishment of a study group composed of the nation's leading scientists to evaluate the Air Force's strategic missile program.[6] To perform this task, Trevor Gardner, special assistant to the Secretary of the Air Force for research and development, assembled a group of scientists under Professor John von Neumann. It would become known as the Air Force Strategic Missile Evaluation Committee. Under von Neumann's leadership, the committee began to examine the impact of the recent breakthroughs in the development of the hydrogen bomb (which enabled them to be much lighter) on the development of strategic missiles and the possibility that the Soviet Union might be ahead of the United States in developing ballistic missiles. During its deliberations, the committee maintained very close working relationships with key individuals in the various fields of technology related to missile development. Especially important among these was "Stark Draper, the director of the Instrumentation Laboratory at M.I.T.," who knew more about the science and technology of inertial guidance "than anyone else in the Western World."[7]

After conducting an intensive series of meetings in close coordination with the Air Force officials, the Von Neumann Committee concluded that: "The state of the art in the relevant branches of technology had reached the point where a practical rocket-powered ballistic missile capable of carrying a nuclear warhead intercontinental distances and delivering it with sufficient accuracy could be built."[8] After the Eisenhower administration made the decision to go ahead with the development of ICBMs, the Von Neuman Committee focused its attention on the technical details of the system and established a rough outline of the design perimeters for an ICBM.

The specifications they established for Atlas, the first U.S. ICBM to be started, called for a missile capable of delivering a one-megaton warhead over a 5,500-nautical-mile range with a CEP of five miles or better. Draper, always the technical optimist, "said he could foresee much better CEPs than five miles."[9]

In 1953, Convair approached Draper to discuss the development of the inertial guidance system that Dr. Kelly had recommended to the Air Force.[10] They were impressed with SPIRE's performance, but not its hefty size.[11] Draper convinced the decision makers at Convair that he could do the job using the down-sized components for SPIRE that were under development. He persuaded them to provide the Instrumentation Laboratory with a study contract to develop an inertial guidance and control system for the Atlas ICBM that could potentially replace the ground-based Azusa system.

The contract funding allowed Draper to establish an ICBM guidance group, under the direction of Roger Woodbury, to work out the basic principles of an inertial guidance system for an ICBM.[12] Responsibility for the guidance effort was shifted from Convair to Air Force sponsorship in the early months of 1955, after the Western Development Division (WDD) was established under Brig. Gen. Bernard A. Schriever within the ARDC to coordinate the development of the Atlas. Schriever's decision to sponsor the IL was undoubtedly influenced by his guidance expert, Lt. Col. Benjamin P. "Paul" Blasingame. Blasingame, another of Draper's young protégés, had spent three years studying at MIT in the Weapons Systems Section of the Department of Aeronautical Engineering. Draper had established this program so that the Navy and Air Force officers sent to MIT could study instrumentation and the principles of fire control. Blasingame had graduated in 1950 with a doctor of science degree in aeronautics, having completed a dissertation on the "Optimum Parameters for Automatic Airborne Navigation." After leaving MIT, Blasingame had gone to the ARDC, where then-Colonel Schriever placed him in charge of development planning. When the WDD was formed, on August 2, 1954, Blasingame moved to it, along with Schriever, who was promoted to brigadier general and placed in command of the division the same day.[13]

The contract's scope of work consisted of a feasibility study, a theoretical analysis of the equations of motion of a ballistic missile and of potential means to control suitably the placement of the warhead in a free-fall ballistic path to the target.[14] The work was primarily concerned with determining what equations an airborne computer needed to solve in order to generate guidance commands to a ballistic missile that would ultimately bring the warhead to the target with sufficient accuracy. The Air Force contract also charged the Instrumentation Laboratory with investigating the feasibility of producing an inertial measurement system of the required size and accuracy of the Atlas ICBM and evaluating the inertial components needed to produce it.

Similar work was being supported by the Wright Air Development Command of the ARDC, albeit with different goals in mind. One contract issued to the Instrumentation Laboratory around this time called for a general study of inertial navigation and guidance techniques for both missiles and space vehicles. Another contract covered the development of inertial components, specifically gyroscopes and accelerometers.

When the Instrumentation Laboratory's work on the Atlas guidance system was initiated, Dick Battin and Hal Laning Jr. were assigned the task of analyzing the mathematical basis for guiding an ICBM.[15] Battin had joined the Instrumentation Laboratory in 1951 after receiving his doctorate in mathematics from MIT, becoming a research mathematician in the areas of fire control and inertial navigation.[16] Laning, who had been with the laboratory since 1945 and had obtained his doctorate in 1947, was another MIT-trained mathematician.[17] A year earlier, Laning, who had been working with the Whirlwind computer, had come up with a way of simplifying the arduous task of programming its mathematical functions by creating the first algebraic compiler,* affectionately called "George."

---

* A computer program that translates an entire set of instructions written in a programming language having a binary form known as object code that can be executed by the computer.

As Battin relates in a paper on space guidance published in 1982, there was "no vast literature to search for 'standard' methods of guiding ballistic missiles, so we 'invented' one."[18] The first method they developed was called "Delta Guidance." It relied on a preflight reference trajectory stored in the missile's guidance system. Battin used a Taylor series to create a mathematical model of a missile's preplanned trajectory. As the missile deviated from its intended flight path during the powered phase, the guidance system would bring it back by zeroing the velocity vectors expressed as the linear terms of a Taylor series. For this reason, the approach is sometimes referred to as the "fly-by-wire" system, where the "wire" is an imaginary reference trajectory. Using George, Battin was able to create computer programs that calculated the trajectories for the Atlas missile.[19]

Though simple in concept, Delta Guidance was not easy to implement in the analog hardware then available. The digital computer industry was in its infancy, and the Atlas onboard guidance system had to rely on an analog computer for control. Several steps were needed to create the guidance system envisioned by Battin. First, as he explained, considerable reference data had to be fixed; next, a complete navigation system was required; then, time-of-flight errors had to be compensated for (requiring additional hardware), or accuracy would suffer. Nevertheless, he was determined to make Delta Guidance work—until he visited Convair in San Diego, California, during the summer of 1955.[20]

Charlie Bossart, Convair's chief engineer, and Walter Schweidetzky, head of Convair's guidance group, introduced Battin to the concepts of the "correlated flight path" and the "velocity-to-be-gained" vector. The former was a predetermined, free-fall reference trajectory that would intercept the target. The missile would intersect the correlated flight path at the engine-cutoff point needed for maximum velocity to the target. The velocity-to-be gained vector at any point along the missiles flight path was the difference between the correlated-velocity vector and the missile-velocity vector. In mathematical terms,

$$V_g = V_c - V_m$$

In contrast to the Delta method, which relies on following a predetermined missile trajectory, the velocity-to-be-gained method asks, "Never mind where we are supposed to be—given where we are, what should we do?" The answer to this question is to apply acceleration (i.e., via engine thrust) in the direction of the velocity-to-be-gained so that the actual velocity becomes close to the correlated-velocity vector. When these values become equal (and the velocity-to-be-gained vector becomes zero), it is time to shut off the engine,

The problem facing Hal Laning and Dick Battin was how to solve the differential equations that described the three-dimensional vector matrix that in turn described the dynamics of the velocity-to-be-gained model—what they dubbed the "Q Matrix." As Battin later wrote, "To say that calculating the Q Matrix was no simple exercise would be a gross understatement."[21] The ultimate solution, while elegant in its simplicity, required fourteen pages of mathematical equations to express.

A report on the resulting guidance scheme, "Q-guidance," was presented on June 22, 1956, during the first Technical Symposium on Ballistic Missiles, held at the Ramo-Wooldridge Corporation in Los Angeles, California. The velocity-to-be-gained vector was only one element of Q-guidance. Of equal importance was a method of controlling pitch and yaw that became known as "cross-product steering" (an approach that also relied on driving vectors to zero). The big advantage of Q-guidance was that it could be easily realized using analog instruments. As it turned out, Q-guidance was never implemented on Atlas, as intended, but it did prove crucial for Thor.

The Instrumentation Laboratory was also issued contracts to design gyros that were smaller, lighter, and more accurate than those currently available. To meet the inertial guidance requirements of the Air Force's missiles the Instrumentation Laboratory developed the 25-series IRIGs (inertial reference integrating gyros) and 25 series PIGAs.[22] The "25" designation indicated the approximate size of the instrument case to the tenth of an inch (the 25-series cases, for example, were 2.5 inches in diameter). Both instruments were designed to be readily manufactured and yet meet the stringent performance requirements. The development of the PIGA, which was essential for precise engine cutoff and thus missile accuracy, preceded that of the reference gyro. The experience gained during its development in the use of new materials, bearing techniques, and magnetic suspension subsequently proved useful in designing the 25 IRIG.

With the design studies for the 25-series gyro, the laboratory set out to produce a gyro that would have a drift rate in the vicinity of 0.1 degrees per hour under a constant acceleration of at least ten times the force of gravity (10 *g*s) and vibrations up to two thousand cycles per second. The gyro had to be as small as practical (less than three by four inches) and weigh as little as possible. Although the gyro's minimum life expectancy was a thousand hours, five thousand was set as a goal, to broaden its potential applications. Last, but not least, was a desire to create a modular design that would break down the manufacturing functions into independent subassemblies each having a minimum number of parts.

The design effort was organized as a three-pronged attack, aimed at minimizing mass shifts, friction uncertainties, and the dirt and misalignment defects introduced in manufacture and assembly. To reduce mass shifts, the lab's engineers:

1. Designed a spherical float structure for greater rigidity.
2. Added an elastically preloaded, adjustable ball-bearing spin wheel.
3. Pressed the spin-motor stator outside the wheel into the gimbal structure to provide more rigidity.
4. Used beryllium as a construction material to provide a greater stiffness-to-weight ratio, more angular momentum in a given volume, better matching of thermal expansion, good corrosion resistance, and high thermal conductivity, coupled with good dimensional stability.
5. Employed symmetrical construction throughout the device.

To minimize frictional uncertainties, they designed a magnetically supported bearing.

Their biggest achievement however, was a result of the thought given to the means of assembly. The extremely small manufacturing tolerances (which often dipped into the micro-inch range) for what was then a state-of-the-art gyroscope required large investments in time, money, and care. Lester R. Grohe, lead engineer on the project, and his assistants concluded that although it was possible to manufacture the required parts through "tedious and rigorous devotion to detail . . . the expense would be extraordinary, the rejection rate overwhelming, and the output meager."[23] Instead, they incorporated adjustment mechanisms to make precision corrections wherever a critical operation was undertaken, such as on the ball bearings in the spin wheel, where differential screws permitted a preloading adjustment. To reduce final assembly time, they used a modular building-block approach that provided for numerous subassemblies, such as the plug-in Microsyn. These were fabricated from printed circuits coupled to magnetic shields attached in turn to a "potted" Microsyn having hermetic lead-in terminals. Although the overall design contained more external adjustments than did previous gyroscopes, it had fewer, simpler, and more rugged parts.

In March 1956, after the Atlas program was under way, the Air Force was given the additional responsibility of developing an intermediate-range ballistic missile, Thor, with a specified range of only 1,500 miles. It was an interim missile that could be deployed one or two years sooner than the Atlas. Thor was expected to be a much simpler, less expensive missile that could be built using many of the components being developed for the Atlas. The initial schedule called for ten of the missiles to be operational by October 1958.[24]

Initially, Thor was to have a dual guidance system that used both inertial and radio guidance. But prior to the missile's first test launch, the Air Force's increasing confidence in the Instrumentation Laboratory's inertial guidance system led it to drop the requirement for radio guidance.[25] By then, the Air Force had selected the AC Spark Plug Division of General Motors to manufacture the guidance system, on the basis of a design developed by the Instrumentation Laboratory, which was now under contract to provide technical and engineering support for its implementation.[26]

AC, which had built a six-story plant on Milwaukee's northeast side in 1948 to manufacture precision inertial guidance systems, was selected because it was one of the few firms that had the capability to tool up and produce the precision gyros and accelerometers needed.[27] It had worked with the Instrumentation Laboratory before, on the A-1C gunsight, and it was now acting as a second source for the Air Force's K-Series navigation and bombing systems.[28] AC completed the design in just eighteen months, during which time it successfully put the guidance system through laboratory tests and then rocket-sled tests at the Naval Ordnance Test Station in Inyokern, California.[29]

When full-scale production of the guidance system began in May 1957, the AC plant in Milwaukee had 1,400 employees, most of them engineers, electronic specialists, and inspectors.[30] The latter examined every part under thirty-power microscopes and applied dental tools, steel wool, and other equipment to eliminate imperfections. Many of the components in the inertial measurement unit (IMU) had to be machined to extreme tolerances, down to three millionths of an inch.[31] This is about a thousand times smaller than the diameter of a human hair. The IMU's three 10 FG gyros, used to correct the pitch, roll, and yaw, were housed in a container about the about the size of a five-inch-diameter juice can.[32] The IMU also contained the three 25 PIGA accelerometers that generated the three-dimensional velocity data needed to calculate the velocity to be gained.

It was intended that prior to launch, a computer on the ground would produce a mathematical model of the correlation trajectory and send it to the missile's guidance system. These equations were simple enough to be solved by analog computation. Also, when the AC system was first designed, a stellar check was built in. But the AC engineers found that whenever there was an error it was usually in the stellar check, so they dropped it.[33] Because transistors were not fully developed or reliable at the time, the designers had to rely on vacuum tubes and magnetic amplifiers.[34] When completed, Thor's guidance system weighed 650 pounds. The system was flown for the first time on December 7, 1957. It failed to provide a main-engine cutoff signal on that mission, but a follow-on test two days later went off without a hitch, and the missile's warhead was successfully guided to the target.[35]

15

Titan, FLIMBAL, AIRS, and the MX/Peacekeeper

A few months after the Instrumentation Laboratory received its contract to conduct a feasibility study on ICBM guidance, the Secretary of the Air Force approved the development of a second ICBM, in case Atlas did not meet expectations. On May 6, 1955, the Air Material Command (AMC) invited contractors to bid for a two-stage alternative to the Atlas ICBM, a weapon that would become known as the "Titan." Like Atlas, Titan was initially to be guided by a ground-based system. When the Bell Telephone Laboratories declined to bid on Titan's guidance system, General Schriever wrote to the firm, which the Air Force considered to have unusual skills in radio and radar tracking and guidance, asking it to reconsider.[1] Officers from the Western Development Division hand-carried the letter to New York City on October 5 for a meeting with Bell engineers. Schriever's letter, which declared that the attainment of a ballistic-missile capability was of greatest importance to the security of the United States, changed the Bell engineers' minds, and they agreed to undertake the development of Titan's ground-based guidance system.

In addition to Bell Labs, the Air Force committee in charge of selecting the guidance contractor for Titan had listed possible contractors capable of developing an inertial guidance system. Knowing that the Instrumentation Laboratory was already working in this area and seeking to obtain a second source for guidance (as was standard policy in the WDD), the committee suggested that the three top contenders—North American Aviation, Sperry, and the Arma Division of the American Bosch Arma Corporation—be invited to compete. Arma won and was awarded a contract to develop an inertial guidance system that could be used on either the Atlas or Titan missile. In May 1958, after several changes to the contract, the Air Force decided to install the Arma inertial guidance systems in all hardened Atlas missiles and Bell's radio guidance in the early Titans, as planned.

By then there was a growing concern within the defense community about the difficulties of handling cryogenically fueled missiles, such as the Titan, and their vulnerability to Soviet attack. Among these concerns were: the quarter-hour it took to raise the missile from its silo, load propellants, and launch it; the difficulty of handling extremely cold liquid oxygen inside a missile silo; the vulnerability of its launch-control centers and of, when they were deployed, the antennas needed for the radio-guidance system. The Air Force commissioned the Martin Company to study these issues and come up with a solution. Martin recommended the development of a new missile, subsequently named Titan II, which would use storable propellants, be launched from a silo, and use an all-inertial guidance system.[2]

By January 1959, it was apparent to the Titan program managers that they needed to find a second source for the Titan II's guidance system. One authority familiar with the history of inertial guidance suggests that the reason was organizational and managerial difficulties being experienced by Arma.[3] AC Spark Plug was the obvious candidate to replace Arma, but Lt. Col. F. M. Box, the WDD's director of guidance and control programs, dubious about authorizing AC as a sole-source supplier for the Titan II inertial guidance system, advised that a formal competition be held.[4] Schriever, after careful consideration, so ordered, and the WDD solicited proposals. WDD sent to a select group of contractors information on the inertial guidance techniques and components developed by the Instrumentation Laboratory under Air Force contract AF 04(6450)-9.[5] The selection of a contractor would depend in part on which could work best with the Instrumentation Laboratory and was best qualified to complete the project on a timely basis.[6]

AC Spark Plug scored high on both counts. It had the additional advantage of having recently strengthened its R&D capability. On April 14, 1959, AC Spark Plug was awarded a contract to build the inertial-guidance system for Titan II missiles on the basis of the development model constructed by the Instrumentation Laboratory.[7] Under its terms, AC and the lab would cooperate on the development and test of production prototypes.

Although the existing state of the art had been successfully built into the Thor IRBM guidance system, further improvements in gyro performance were needed to meet the more stringent requirements of the longer-range Titan II ICBM. Fortuitously, the Instrumentation Laboratory had already solved this problem with the newly developed 2 FGB gyro, which replaced the 10 FG developed for Thor.

Those in charge of gyro development in the laboratory had realized that the 10 FG gyro, which weighed about fifteen pounds, was heavier and larger than desired and not optimized for ICBM application.[8] Because of the short time of flight of a ballistic missile, the engineers in the lab's components group had realized that reducing the error-producing torques of the gyro (due to the forces of acceleration) was more important than reducing the error from drift. Consequently, they had proceeded to develop a smaller gyro with a more uniform elasticity. This effort culminated in the production of the 2 FGB floated beryllium gyro. The 2 FGB was a high-performance, floated inertial gyro of beryllium structure, low anisoelastic compliance,* and characteristics suited to the requirements of the high acceleration and short time of flight of the ICBM. As one author notes, the "intense effort to turn the laboratory version, created at the MIT Instrumentation Laboratory . . . into a device that could be mass produced finally came to fruition with the successful launch and flight of the first Titan II missile" on March 16, 1962.[9]

As mentioned earlier, the Instrumentation Laboratory had begun working on ballistic-missile guidance in 1954, when it was subcontracted by the Convair Division of the General Dynamics Corporation to investigate the feasibility of an inertial guidance system for the Atlas missile. In January 1955 the laboratory was directly contracted by the WDD to

---

* Approaching uniform elasticity.

continue this effort.[10] The contract funded the development of a new type of inertial measurement unit conceived by Philip N. Bowditch in 1957.[11] Bowditch, who had graduated from MIT in 1946 with a bachelor's degree in metallurgy, had joined the laboratory in 1950 and there was assigned to the mechanical design group, where he worked on SPIRE and SPIRE Jr.[12]

In a traditional inertial navigation system, gimbals are used to decouple the motion of the vehicle from the stable platform in which the inertial instruments are mounted. One of the problems associated with this technology is a phenomenon known as "gimbal lock," which occurs under certain extreme flight conditions when the axes of two of the three gimbals are driven into a parallel configuration, locking* the system into rotation in a degenerate two-dimensional space. This problem can be overcome by restricting the missile's trajectory or by adding a fourth gimbal. The former restricts the tactical performance of the missile, while the latter increases the cost, complexity, and weight of the IMU.

Bowditch came up with the idea of placing all the inertial instruments and their associated electronics inside a floating sphere so that they were hydrostatically centered in a close-fitting support structure that provided a "womb-like" environment. Floatation not only freed the inertial guidance system of gimbals but enabled the system to operate in the high-g environment of atmospheric reentry. (The latter would become important in improving accuracy in the third generation of U.S. ICBMs.) Bowditch's concept was eventually developed into a new type of IMU dubbed the "FLIMBAL," for floating inertial measurement ball.

Bowditch is given credit for the idea for FLIMBAL, but turning Bowditch's concept into reality became the responsibility of Kenneth Fertig. Fertig had joined the laboratory in 1950 after receiving his bachelor of science degree in electrical engineering from MIT and was assigned as a research assistant on the SPIRE system.[13] He obtained his bachelor of science in mechanical engineering from MIT in 1952 and was then made responsible for testing the SINS system. After that he conducted the flight-test program for the laboratory's first all-transistorized inertial platform for the F-94C Starfire interceptor.

---

* The use of the word "lock" is misleading, since the gimbals are not physically restrained.

While the FLIMBAL concept was attractive from a theoretical viewpoint, it took six years to solve all of the technical problems associated with its development.[14] In addition to his engineering responsibilities, Fertig continually had to "sell" the program to the Air Force for funding, in what he termed a "stay alive exercise."[15] He was a great showman, just like Draper. To show the benefits of the FLIMBAL-like "womb," Fertig would lift it up and drop it on the floor. In addition to its ruggedness, a free-floating, gimbal-less IMU based on the FLIMBAL concept had another advantage: it could operate during the reentry phase of the flight, thereby improving the missile's accuracy. The engineers in the Ballistic Missile Office of AMC were so impressed with FLIMBAL that they initiated a program to develop it into a highly accurate guidance system that would be capable of guiding an ICBM all the way to the target.[16]

In late 1963, the Instrumentation Laboratory began work on "Sabre," the Air Force program to develop a gimbal-less inertial navigation system based on the FLIMBAL concept.[17] The laboratory had overall responsibility for the design and development of the system, but the components were to be developed in conjunction with various suppliers, who were then expected to manufacture them. The manufacturers included the Eclipse-Pioneer Division of the Bendix Corporation, which was responsible for the accelerometers; Bendix and the Nortronics Division of Northrop, which dual-sourced the gyros; Univac, which developed the computer with the help of the Boeing Company; and Autonetics and AC Electronics, which were to build the prototypes.

Due to budgetary constraints, the Sabre program was abruptly terminated in 1969, when the Department of Defense opted to achieve greater missile accuracy by improving the Air Force's Minuteman II guidance system, built by Autonetics. Unlike the Draperian guidance systems, which relied on three single-degree-of-freedom gyros floated in Fluorolube®, the Autonetics designs relied on two two-degree-of-freedom G6B4 gyros that were suspended on gas bearings filled with hydrogen.[18] In order to meet the accuracy requirements of Minuteman II, however, Autonetics had to replace their VM4A accelerometers with the Instrumentation Lab's 16 PIGA-Gs.

The FLIMBAL concept, dubbed the "beryllium baby" by historian Donald Mackenzie because of the beryllium sphere that formed its outer shell, was resurrected a few months later as a way to increase the accuracy of ICBMs enough to constitute a counterforce capability—that is, effectiveness against hardened strategic targets. The continued development of the "beryllium baby" was funded under the Missile Position Measurement System (MPMS) program, intended to assist in the testing of missiles. In this guise, it was test-flown riding "piggyback" on a Minuteman III missile in 1976.[19]

By then, the Air Force had initiated an advanced development program for a new ICBM to replace Minuteman. Mackenzie maintains that "although counterforce was still largely alien to stated national strategic policy in the period, it was by then well entrenched in the Air Force, and there seems to have been little question but that the new missile, MX as it was christened, was going to be designed to have as great a counterforce capability as possible."[20] AIRS was the perfect choice for guidance.

In May 1975, the Instrumentation Laboratory's MPMS design was transferred to Northrop for further development as the Advanced Inertial Reference Sphere (AIRS). The heart of the 126-pound, basketball-sized AIRS IMU was a beryllium sphere containing three pendulous integrating gyro accelerometers and three gas-bearing floated gyros. The beryllium sphere was floated on fluorocarbon fluid within another shell and so could rotate in any direction, eliminating the potential for gimbal lock. The fluid's temperature was controlled by freon-cooled heat exchangers and the sphere's alignment by three hydraulic thrust valves directed by the inertial sensors in the sphere.

Northrop, which received contracts worth $1.6 billion for work on the MX (renamed Peacekeeper in 1982) IMU, discovered that it was much harder to build AIRS than it, or the Air Force, had anticipated. Despite extensive support from the Instrumentation Laboratory, Northrop found it extremely difficult to transfer the handcrafted laboratory design to a production environment. Although extremely accurate—the drift rate of its gyros was less than 1.5×10^{-5} degrees per hour—the AIRS IMU was extremely complex, requiring some 19,000 individual parts.[21] The penalty for its extreme accuracy and tremendous complexity was cost. In 1989 a single AIRS accelerometer, of which there were three per unit, cost $300,000 (over $500 million in today's dollars) and took six months to manufacture.

When the first guidance unit was delivered in May 1986, Northrop was 203 days behind schedule. Much of the blame, according to congressional investigators, was attributable to the Air Force, for having set too ambitious a pace for the program. The service had continually pressed Northrop to speed production of the guidance system, even though it was unclear whether the MX missile program would survive repeated assaults on Capitol Hill.[22]

Northrop's manufacturing problem was compounded by two factors: few of the parts needed to produce the IMU had ever been produced before, and that very few units were needed—only 239, to support the hundred or so MX missiles the Air Force was now authorized to deploy. As John Cushman Jr. of the *New York Times* reported, it was "hard to automate production in small numbers without wasting money. The traditional aerospace 'learning curve,' in which performance improves as the pace of production expands, is not steep."

For its part, Northrop's production and testing practices raised serious questions about the reliability and performance of the MX missile. Nevertheless, data from seventeen flight tests and dozens of simulated launchings confirmed the accuracy and reliability of the missile's guidance system. By then, the MX missile project had undergone several changes, including the decision to base the new missile, now called the Peacekeeper, in existing Minuteman missile silos.

The first Peacekeeper missile was placed on operational alert on October 10, 1986.[23] The Air Force had initially planned to install 114 of the missiles in modified Minuteman III silos at the F. E. Warren Air Force Base near Cheyenne, Wyoming, but deployment was capped by Congress at fifty in 1990. These were removed in 2004 under the START (Strategic Arms Reduction Treaty) II agreement.

The FLIMBAL concept, as implemented in the AIRS IMU, was the pinnacle in the development of mechanical, gyro-based inertial guidance. The gas-bearing floated gyroscopes used on the Peacekeeper were the ultimate achievement in gyroscopic technology. By the time of their operational debut in 1986, advances in other technologies, such as the ring-laser gyro and the interferometric fiber-optic gyroscopes, along with improvements in microelectronics and software, had superseded the mechanical gyro.

16

Polaris

In the fall of 1954, President Eisenhower, in response to growing alarm about the strategic threat from the Soviet Union, established the Technological Capabilities Panel of the Science Advisory Committee to conduct an in-depth study of the nation's current defense measures. James R. Killian Jr., president of MIT, was appointed the committee's chairman. Killian's report to the president of February 14, 1955, urged him to accelerate the development of Atlas as a national priority, recommended the development of an IRBM (which became Thor), and suggested that the National Security Council recommend that part of the IRBM force be sea based.[1]

As a result of the last-listed recommendation, the Navy, which had been locked out of the strategic nuclear weapons arena since the Key West Agreement of 1948, was soon authorized to proceed with the development of a sea-based IRBM. On September 9, 1955, the secretary of defense established a Joint Army-Navy Ballistic Missile Committee to oversee the adaptation of the liquid-fueled Jupiter missile for naval use. The Army was to develop the missile, the Navy the shipboard launching system. To handle the special problems associated with a ship-launched IRBM, the Secretary of the Navy created Special Projects Office (SPO). To ensure that the Navy requirements for the missile's characteristics were understood and were being met by the Army, a cadre of officers from the SPO was established at the Redstone Arsenal in Huntsville, Alabama, which was the design agent for Jupiter. Although the original plan called for the deployment of a surface-launched, liquid-fueled missile similar to Thor, the program quickly moved toward the development of a solid-fuel missile, which would be less hazardous for shipboard operation and could be launched more quickly. Within a year the Army's assignment for the missile's development had been shifted to the SPO, which now had full responsibility for the project, later named Polaris.[2]

Meanwhile, the Instrumentation Laboratory, which had already begun work on the guidance system for the Air Force's Thor, had been given a contract to investigate a shipboard stabilization system for the Navy's version of the Jupiter. This brought laboratory personnel working on Thor into contact with members of the SPO's Fire Control Branch, "SP-23." Ralph Ragan, a member of the Instrumentation Lab's team working on ballistic-missile guidance, discussed the possibility of the lab's providing a guidance system for the Navy's missile with Cdr. Samuel A. Forter.[3] Forter, a World War II veteran, had attended the Navy's postgraduate school at MIT after the war.[4] He graduated in 1947 with a master's degree in electrical engineering, with a thesis on a light-weight telemetering system. He was assigned to the SP-23 in 1955. There he worked with the German scientists designing the guidance system for Jupiter.

Forter arranged for Ragan to meet with Capt. Levering Smith, SPO's technical director. Smith, then considered the Navy's preeminent expert on rockets and solid propellants, had been drafted by the SPO's director, Rear Adm. William F. "Red" Raborn.[5] When Raborn established a Steering Task Group of senior representatives of the various organizations involved, Smith was charged with organizing the group's tasks. Dr. William F. Whitmore, the first chief scientist of the SPO, suggested, only half-jokingly, that the criterion for selection to the Steering Task Group should be, "The senior guy who will be fired if this thing doesn't work."[6] Among those selected was Charles Stark Draper. The steering group spent three months defining the total program, including schedules, costs, performance goals, and distribution of tasks among the members. Once the program was defined, the steering group met regularly to agree on changes, modify plans, and adjust resources.

Two of the steering committee's most important goals for the program concerned the missile's range and accuracy. To ensure that accuracy standards could be met, Levering Smith arranged for the Instrumentation

Laboratory to be contracted to investigate the guidance-system requirements for the as-yet-unnamed Navy IRBM.[7] Thus, Forter and his associates at SP-23—like David Gold, who had worked on Project MAST—were introduced to Q-guidance, which Hal Laning and Dick Battin were developing for the Air Force. At one point, according to Ralph Ragan, IL's David Hoag and Sam Forter tried to explain the principle of Q-guidance to Smith.[8] The two became confused themselves as to how it really worked, and it was left to Smith to clarify it for them.[9]

At the end of the summer of 1956, the 160,000-pound liquid-fueled Jupiter design had been replaced by a concept for an as yet unnamed missile based around a 30,000-pound solid-fuel launch vehicle. By then Levering Smith was convinced that Q-guidance was particularly suited for use with a solid-fuel missile, because it would not need to adjust the thrust program in flight: unlike liquid-fueled missiles, there was no practical means to do so. Q-guidance was attractive also because it shifted much of the computation (such as the components of the Q Matrix) outside the missile and even the submarine to a computer on shore. Even though launch conditions at sea could not be known in advance (as they could be for land-based missiles), only fairly simple tasks were left for the submarine's fire-control system and the missile's onboard computer.

The Instrumentation Laboratory was to be selected over the Redstone Arsenal to provide the guidance for the Navy missile in large measure because of its involvement with, and development of, Q-guidance. As Smith later recalled, "It did appear that we could work more closely with Draper than with Huntsville [i.e., the Redstone Arsenal], partly because I thought the [Draper] fluid floated gyro would adapt easier to the solid motor accelerations, but to my way of thinking the choice was driven more by Q-guidance than anything else."[10]

The selection of a guidance contractor warranted SP-23's immediate consideration, although the official go-ahead to develop Polaris would not be received for several months.[11] On October 10, 1956, Admiral Raborn flew up to Cambridge with several members of his staff to interest Draper in developing the inertial guidance system for the new missile.[12] Why the head of the SPO had to make the trip personally is subject to interpretation. But it appears that Draper was not pleased with the outcome of the contractual arrangements with Sperry that had been used to develop the Instrumentation Laboratory's version of SINS. If so, that had probably left a bad taste in his mouth. He must have been reluctant to turn over any more designs to outside contractors.

As we saw earlier, the Sperry system, which was delivered in September of that year, had not performed up to expectations, opening the door to Autonetics, which ultimately became the Navy's primary source for SINS. This was not a pleasant outcome for the Instrumentation Laboratory. Draper could not have been happy to see his work on SINS go for naught. This was not the first time he had experienced difficulties getting one of his designs into production, either. The Sperry Gyroscope Company had also had trouble manufacturing the first production models of the Mark 14 gunsight. On that occasion Draper himself had had to find an outside vendor capable of manufacturing the unit's gyros, which Sperry was unable to do. He had also been obliged to cajole, push, and practically force Sperry's engineers to make the design modifications needed to make the gunsight effective. It would be understandable if he was now unwilling to enter into any new contract that was likely to cause similar problems.

When Admiral Raborn and his entourage were ushered into Draper's office, Draper proceeded to lecture them about all the terrible things that had been done with the systems that he had designed for the military services in the past. The military, according to Draper, had messed them up by turning them over to industrial contractors to replicate: "They [the military] screwed up."[13] (As the episode illustrates, Draper could be stubbornly tenacious and "at times dogmatic.")[14] Rear Adm. Robert Wertheim, then a young lieutenant who was there to take notes, later recalled what happened next. "Admiral Raborn figuratively swore on a stack of Bibles that Draper would have full responsibility for the new guidance system they wanted him to build for the Navy. The transition from R&D practice to industrial practice would be under his oversight and control."[15]

The Instrumentation Laboratory was duly awarded a contract to design the guidance system for Polaris. The laboratory, however, had no production capability; the Navy would have to rely on a contractor to supply the guidance systems needed to equip the operational missiles. The

General Electric Company was selected for this purpose and contracted to manufacture the guidance system and provide technical support for installation, operation, and maintenance. This set the pattern for the development of inertial guidance systems in all of the fleet ballistic missiles deployed by the Navy over the next forty years.

The Mark 1 Guidance System

Working on a very tight schedule—which became even tighter after Sputnik was launched on December 4, 1957—the Instrumentation Lab designed and developed the Navy's Mark 1 Guidance System. It consisted of an IMU containing three 25 IRIG gyros, three 25 PIGA gyro accelerometers, and an electronics assembly.[16] Although it would be similar in theory and function to Thor's 650-pound guidance system, it needed to be reduced in size and made much lighter to fit into Polaris, which was to be smaller than Thor and one-third its launch weight. To save weight, the gimbal platform was made of beryllium. The mechanical and physical characteristics of this metal—which is extremely stiff, light (one-fourth the weight of steel), and has excellent thermal stability—made it ideal material for the Polaris IMU. Beryllium dust is toxic when inhaled, however, complicating the manufacturing process and making it very expensive to fabricate.

Another weight-saving approach that was considered early on was replacing Thor's analog instruments, which relied upon vacuum tubes and magnetic amplifiers, with a digital device.[17] Although the conversion to digital data would be complicated, a group at the Instrumentation Lab began to investigate the potential of using a digital computer for the onboard calculations. Whether the Polaris' guidance electronics should be analog or digital was widely studied and debated.[18] As Eldon Hall explains in his history of the Apollo computer, emulating "Thor's analog computation system was the less risky choice, but a digital system offered greater increased accuracy. The result was a compromise. The Navy decided to accept the laboratory's proposal for a digital computer in the Polaris guidance system but continued to fund the development of an analog system as a backup."[19]

Space constraints led to a computer architecture chosen to minimize size. Only functions that were required for the missile's guidance were included in the design, computations were serial, and the few words of memory were stored in a hard-wired magnetic core. Because of the intense time pressure, the computer's designers had to use only components and packaging technologies then available: germanium transistors for the logic circuits, magnetic shift registers for memory, and standard printed-circuit boards for packaging. The result was a computer that was difficult to assemble, test, and repair, which in turn created severe reliability problems. None of the electronic components had any "track record," the manufacture of printed-circuit boards was immature, solder joints were unreliable, and the germanium-junction transistors were sensitive to heat and could be damaged during soldering.

Despite these problems, the first version of this computer, designated the Mark 1, was used to guide a Polaris SLBM in flight on January 7, 1960, in the first use of a digital computer for inertial guidance in a ballistic missile.[20] This historic "first" occurred two months and a day before the first Atlas flight with an onboard digital guidance computer. The Mark 1 occupied about four-tenths of a cubic foot, weighed twenty-six pounds, and consumed eighty watts of power.[21]

Even before the first flight of the Mark 1, Eldon C. Hall, its chief designer, began to explore design approaches for a second-generation computer that would be smaller and have better interconnection reliability and maintainability.[22] The result was the Polaris Mark 2 computer. It was based on the same architecture as the Mark 1, but

Table 16-1. **Polaris Mark II Computer Characteristics**

Architecture	Wired program DDA
Memory	Magnetic core shift register
Words	12
Word length	17 bits
Logic	512 transistor NOR gates
I/O	Incremental data transfer
Clock rate	81.6 kHz
Size	0.1 cu ft
Weight	12 lb
Power	40 W

the germanium-junction transistors were replaced with silicon-planar transistors and the printed circuit boards with a modular construction employing welded-cordwood modules using wire-wrap interconnection. The new computer took up one-fourth the space of the Mark 1 and weighed half as much.

The Instrumentation Laboratory's interest in minimizing the size of the Polaris computer spurred commercial development of the integrated circuit. The germanium transistors employed in the Mark 1 computer represented one of the first large-scale uses of discrete transistor components in the military.[23] They were supplied by Texas Instrument Company, largely through Hall's actions: "During my 1959 visit to the Polaris program's captive [manufacturing] line at Texas Instruments, Jack Kilby, the inventor of integrated circuits, demonstrated his experimental circuit, which Texas Instruments announced that spring. Later in the year, Kilby came to MIT/IL, discussed his work in greater detail, and made a proposal to design a logic circuit that would replace the NOR gate in the Polaris computer. My interest was sparked, and the Navy was willing to fund an order for 54 logic gates at $1,000 a piece."[24]

After visiting Texas Instruments, Hall went to Fairchild Semiconductor to discuss semiconductor circuits with Robert Noyce, Fairchild's founder and coinventor of integrated circuits. Under Noyce's guidance, Fairchild developed production techniques to reduce the integrated-circuit idea to practice. The first commercially available hardware was to be produced in 1961.

The Mark 1 guidance system was complex, temperamental, and expensive. Each cost a quarter of a million dollars, according to one account.[25] It was built for the Navy at General Electric's Ordnance Systems plant in Pittsfield, Massachusetts. One knowledgeable engineer has called the plant "a technical wonderland of electro-mechanical engineering."[26]

The guidance system was so vital and secret that an engineer had to accompany every shipment—in a sealed, temperature-controlled container—from the Pittsfield plant to Charleston, South Carolina, for installation at the Navy's nearby Polaris Missile Facility. Before it could be fitted into a missile, however, each system had to be tested and calibrated. As Charles Wright, a former guidance system specialist for General Electric at the facility, later explained:

Each guidance system was not perfect—it had errors in the gyros and accelerometers that could make the missile miss the long-range target. Therefore, we had to determine exactly what these errors were (e.g., like the speedometer on our car being in error by a few miles per hour). We would carefully measure gyro and accelerometer errors so that the submarine could take them into account and offset them before the missile was launched.

As a vastly oversimplified example, if the guidance errors would cause the missile to hit 10 miles to the left of the target, the submarine would apply factors to aim it 10 miles to the right. Six to eight hours were required to calibrate and test each system, the intricacies involved were staggering.[27]

Once the guidance system was installed in the missile and the missile had been loaded on board a submarine and the boat sent to sea, the weapon could be prepared for launching, by the Mark 80 Fire Control System. The Mark 80, installed in each Polaris submarine, was responsible for:

- Aligning the inertial guidance system based on the direction of north and the vertical, received from the submarine's SINS.
- Providing the navigational and targeting data needed by the guidance system.
- Verifying the status of the other shipboard systems needed to launch the Polaris missile.
- Controlling the actual launching sequence.[28]

The fire-control system had to tell the guidance system what trajectory the missile needed to fly from the launch position to the target. Aiming data was provided in the form of the Q-matrix and velocity-to-be-gained needed for that trajectory. Calculating these initial conditions in real time, using data from the SINS navigational suite, would have required a digital computer that in those days was much too large to fit into the submarine. Instead, these calculations were made ashore at the Naval Ordnance Station in Dahlgren, Virginia. The resulting targeting data was provided to each submarine in the form of target cards, each usable for a mission in a given twenty-square-mile area of ocean to a selected target.[29] Data was read from the appropriate card and entered into the Mark 80 Fire Control System, using knobs and dials. The target card also contained the solution to a test problem that verified the manual settings. So many precomputed punch cards were needed for a sufficiently fine grid of all conceivable patrol positions that they had to be provided on microfilm. To produce the individual cards, a microfilm reader and keypunch were placed on each nuclear-powered ballistic-missile submarine (SSBN).[30]

The Mark 80 Fire Control System was installed on the first ten Polaris submarines. It was replaced by the Mark 84 FCS on the thirty-one *Lafayette*-class SSBNs. The newer system used a digital computer—a militarized version of the CDC 1604—to perform real-time targeting computations. With it the Mark 84 was able to handle the complex guidance functions that were previously solved at Dahlgren to produce target cards. This system, with the first digital fire-control software, became operational in 1963 with the Polaris A2 missile.

The A2 was an improved version of the Polaris A1 with an extended range of 1,500 nautical miles, as authorized by the Department of Defense in September 1960. The greater range was achieved by lengthening the A1's first-stage motor by thirty inches and employing a high-performance propellant and lighter-weight inert components in the second stage.[31] There was no change to the guidance system.

The third and final variant of Polaris, the A3, presented new challenges to the Instrumentation Lab's guidance team. When design studies for the new missile began in 1959, the initial objective was to increase the range of the Polaris while making it capable of carrying a larger warhead, in the one-megaton range (versus the six hundred kilotons of the A1 and A2).[32] A thousand miles was added to the Polaris' range by improving overall missile design and incorporating better propellants in a larger second stage.[33] Because of concern over the possible deployment of a Soviet anti–ballistic missile system, the A3 was designed to carry three two-hundred-kiloton warheads that would create a triangular impact pattern with about the same destructive power of a one-megaton warhead while greatly complicating the interception problem.

The Mark 2 Guidance System

To meet SPO's range goal for the Polaris A3, Lockheed, the missile's contractor, wanted a much smaller and lighter guidance system.[34] The engineers in the Fire Control Branch of the SPO decided to separate the IMU from its electronics. This would simplify the contractual arrangements for its manufacture and allow the guidance system to be shipped in two parts. Separating the IMU from the electronics would also make it easier to find space for the guidance system in the missile. As noted in previous pages, Eldon Hall had already begun work on a computer, the Mark 2, that would suit the needs of the A3. The A3, like its predecessors, was designed as a "counter-city" weapon that was supposed to deter nuclear war through the principle of "mutually assured destruction." Its limited accuracy, coupled with its relatively low-yield W-47 (six hundred–kiloton) thermonuclear warhead, made the A3 unsuitable for counterforce strikes. The SLBM force itself, however, because it could not effectively be taken out by the enemy, was in any case the ideal retaliation weapon. Thus, no improvement in accuracy was considered necessary, aside from that needed to meet the 2,500-mile range specification.

To satisfy these weight and size requirements of the A3's guidance system, the Instrumentation Lab replaced two of the three 25 PIGAs with the smaller 16 PIPA (that is, a 1.6-inch-diameter pulsed integration pendulum accelerometer). Unlike the PIGA, with its gyro wheel and internal gimbal, the PIPA consisted simply of a pendulum that is held in the null position by pulses sent from a signal generator to an electromagnetic torquer. Although not as sensitive as PIGAs, they were simpler, easier to build, and therefore cheaper to make. Since forward acceleration was a more critical measurement to ensure accuracy in range, a single PIGA was retained to measure this parameter.

Like the Mark 1, the Mark 2 guidance system was manufactured in Pittsfield by the General Electric Company. It was half as large and a third as heavy as the Mark 1 and was more accurate, yielding a CEP of 0.5 miles. As Graham Spinardi spells out in *From Polaris to Trident*, "the increase in accuracy was more a consequence of lessons learned in developing the original Polaris" than a major shift in the missile's strategic role.[35]

17

Poseidon and Trident

Draper's pioneering work on inertial guidance and its application to ballistic-missile guidance brought unexpected publicity. On January 2, 1961, he was honored to receive public acclaim when his photograph appeared on the cover of *Time* magazine in company with those of twelve other American scientists selected by the editors of *Time* as "Men of the Year."[1] Draper was the only engineer in the group of scientists, who included twelve present or future Nobel laureates. Although guided missiles and space flight had captured the public's interest, Draper and James Van Allen, the discoverer of the layer of charged particles surrounding the earth that would bear his name, were the only members of the group involved in these endeavors. Draper's newfound celebrity status had little impact on his work. He was already considered the nation's authority on inertial guidance, and his Instrumentation Laboratory continued to enjoy ample financial support from the Air Force and the Navy.

As a follow-on to the Mark 2 guidance system for the Polaris A3, the laboratory received a Navy contract to develop a "strapdown" guidance system to replace the Mark 2 that would be easier and cheaper to build.[2] In a strapdown system, the gyros and accelerometers are attached rigidly to the frame of the missile. The computer in this application functions as an analytic gimbal, eliminating the need for a stable platform. This greatly reduces the cost, size, and weight of an inertial navigation system. Strapdown systems, which require few moving parts, are also easier to maintain and more reliable over time. The Instrumentation Laboratory conducted tests that demonstrated the feasibility of the strapdown system as requested by the Navy, but the project never entered production.

Another system developed during this period, called MIGIT, was designed around a small gimbal system and an electronics assembly that was more densely packaged than the Mark 2. The size of the MIGIT IMU was reduced to a ten-inch diameter (as compared to the Mark 2's twelve-inch diameter) by replacing the 25-size IRIGs with 18-size IRIGs.[3] The MIGIT IMU project was particularly useful in developing advanced electronic packaging techniques and was responsible for introducing "flatpack" integrated circuits on plug-in cards.

In November 1960, the Navy's Special Project Office was given the go-ahead to develop a follow-on missile to the Polaris A3. Although larger than the A3, the new missile, initially designated as the "B3," was to be installed in existing Polaris submarines, by removing some of the shock protection, which had been overdesigned in the Polaris submarines. The diameter of the B3 was seventy-four inches, versus the A3's fifty-four.[4] The larger diameter provided the opportunity for designers to increase either the missile's range or allow a bigger payload. The latter opened the door to a larger warhead and a refined guidance system for greater effectiveness. Antisubmarine warfare projections and the availability of forward-based tender support did not suggest the necessity for a further increase in missile range, so the missile's designers focused on increased payload flexibility.

For a while, SPO and Lockheed based their warhead studies around the Mark 12 reentry vehicle, designed to carry a 150-kiloton warhead. Several of these could be carried by the larger B3. The possibility of carrying six Mark 12s was considered at first, but the simple ejection method used to disperse its warheads could not be employed for six multiple reentry vehicles (MRVs).[5] Thus the warhead delivery problem became a critical aspect of the B3's design.

In the early part of 1964, Lockheed began to investigate three different approaches to the warhead-dispersal problem code-named respectively Mailman, Blue Angels, and Carousel. Mailman, a technology adapted from the Minuteman missile, relied on a platform, or "bus," containing both a guidance and propulsion system, that would release the reentry vehicles one at a time. It required a change in approach from the "implicit" Q-guidance used on Polaris to one based on "explicit," direct knowledge of where the missile was. Blue Angels would retain Q-guidance, but each vehicle would require its own guidance and propulsion system. Carousel, which early was considered the method least likely to be selected, used centrifugal force (exerted by spinning the missile) to scatter the warheads.

The bus system, with its multiple independent reentry vehicles (MIRVs), was recommended by the Navy's senior leadership and approved by the Secretary of the Navy, because it would potentially result in an SLBM that could compete with the highly accurate Air Force land-based missiles. Thus, in November 1964, the SPO was officially instructed to proceed with a MIRVed Polaris B3 with guidance improvements for greater accuracy. Two months later, the missile's name was changed to the Poseidon C3 by President Lyndon B. Johnson, in order to rebuff criticism that the administration was not developing any new strategic missile systems. The Poseidon, as envisioned by its designers, was to have a 2,500-nautical-mile range and carry up to fourteen MIRVs, each of which could be individually programmed to strike separate targets.

Poseidon's flight time was up to five times longer than that of Polaris and had a "footprint" (the size of the area over which its MIRVs would strike) on the order of hundreds of miles.[6] In response to pressure for greater accuracy from Secretary of Defense Robert McNamara, Rear Admiral Smith, having taken over the direction of the SPO, agreed to make Poseidon 50 percent more accurate than Polaris. This figure was set as a goal, but not a requirement.[7] In addition to improved accuracy, Poseidon now had to be capable of surviving in a hostile environment, making radiation hardening essential. Both requirements placed significant new demands on the guidance system.

As before, the SPO commissioned the Instrumentation Laboratory to develop the guidance system for Poseidon, albeit there may have been a movement afoot to award the contract to a commercial firm. That prospect drew the intervention of MIT vice president James McCormack. McCormack spoke with both Admiral Smith and Adm. Harold Brown, the director of defense research and engineering, reportedly with regard to the Instrumentation Laboratory's involvement, although it is not clear what his objective actually was or why he had to contact them.[8]

The Navy quest for another guidance contractor, if indeed it was substantial, may or may not have been triggered by the Instrumentation Laboratory's resistance to moving away from Q-guidance. As Sam Forter, an important member of the Polaris guidance team, later explained to Graham Spinardi, the laboratory's guidance team felt they "had a horse that ran right in the proper direction."[9] So why change it? But the decision to use a maneuvering bus dictated that the missile constantly know its position and velocity in order to bring each warhead on target. This in turn required more onboard computational capability than was available in the Polaris system. But advances in component technology and the progress being made by the Instrumentation Laboratory in the Apollo manned-spacecraft guidance computer (see chapter 19) opened the door for an explicit system. Drawing on its experience with Apollo, the laboratory designed a general-purpose computer for the Poseidon guidance system using integrated circuits having five thousand NOR gates, as opposed to eight hundred for the Mark 2 computer. The new computer had a 100KB read-only memory (ROM) to store programs and guidance formulas and a 12KB wire-braid memory for stored variables. The latter was one of the techniques used to harden the electronics against radiation; others included radiation-resistant integrated circuits.[10]

The task of determining the proper guidance algorithm and identifying the associated guidance presettings was allocated to the Naval Surface Warfare Center (NSWC), in Dahlgren, Virginia, which had designed the fire-control system for Polaris and provided the missile's precomputed guidance settings.[11] Dahlgren developed a trajectory simulation for Poseidon, employing its large general-purpose computer to develop the targeting techniques and algorithms that would direct the missile's warheads to their targets.[12] The presetting data, including fuze settings, were designed to include the effects of forecasted reentry winds, atmospheric densities, gravity anomalies, and guidance perturbations. The number of presettings into the Poseidon guidance system was twenty times greater than that for Polaris, and the equations developed by Dahlgren were ten times as large. To handle this information, Dahlgren developed the Mark 88 FCS, which replaced the Mark 84 on the thirty-one *Lafayette*-class SSBNs retrofitted with the Poseidon C3.[13]

Poseidon's Mark 3 IMU was a traditional Draper design with three gyros and a like number of accelerometers. Poseidon's greater accuracy—

0.3 nautical miles (nm) versus the A3's 0.5 nm—was achieved in part through an improved 1.6-inch diameter PIPA with a permanent magnet torquer instead of an electromagnetic one.[14] Three of these units replaced the single 25 PIGA and older-model 16 PIPAs used in Polaris' Mark 2 IMU. The permanent-magnet PIPA was a continuation of the evolutionary refinements in component design that had been continuously funded by both the Air Force and the Navy. While the substitution of a permanent magnet for the electromagnetic torquer might seem a minor change, it was quite revolutionary for Draper, who was suspicious of permanent-magnet torquers. Other improvements in the Mark 3 IMU included a thirty-six-speed precision resolver and improved mechanical stability.

The Mark 4 Guidance System

When Lockheed began assessing the various possible delivery systems for Poseidon, the Air Force was developing a very large multimegaton reentry vehicle, the "Mark 17," that was designed to provide a hard-target-kill capability and thereby a counterforce weapon.[15] The SPO preferred to equip Poseidon with a large number of the smaller Mark 3 reentry vehicles as better fitted to the retaliation mission for which the Polaris was intended. In any case, the Navy, intent on preserving its turf, was reluctant to use a reentry vehicle developed by the Air Force.

SPO resisted attempts by advocates of a hard-kill missile who attempted to add to some Poseidons a capability to carry the heavier reentry vehicle, creating a mixed force of reentry vehicles. SPO argued that "the Mark 17 was going to be expensive, it was going to require a logistical nightmare . . . a specially configured missile assigned to special targets as opposed to submarines which could go on patrol with the flexibility to be targeted from one kind of a target to another without having to worry about what you had in the tubes."[16] Whether or not it would have to develop a guidance system for the Mark 17 was not the only issue facing the guidance division. Engineers from contractors outside the SPO/Instrumentation Laboratory circle began to push for the addition of stellar sightings to provide the greater accuracy needed for a hard kill. Their idea was to supplement the missile's inertial guidance system with information derived from star sightings taken while the missile was in flight. The Air Force had tested a stellar acquisition system using spare Polaris A1 missiles, but no production contract had yet been given to the Kearfott Division of the General Precision Corporation, which was the main proponent of stellar sighting. Kearfott offered the Navy a greatly simplified version of its stellar system called "Unistar." The company promised that Unistar would bring a considerable increase in missile accuracy at relatively little cost. Kearfott's proposal initially met considerable skepticism within the SPO. Admiral Smith jokingly referred to it as "Mexican arithmetic."[17]

The Instrumentation Laboratory too was opposed to the addition of stellar guidance. "Not Invented Here" played a part, but there was also concern about adding unnecessary complication to the guidance system as well as about the possibility that a high-altitude nuclear explosion from an antimissile system might make a stellar system inoperable.[18] Stellar guidance also threatened the laboratory's fundamental approach to its technological line of development, one that focused on improving the accuracy and reliability of its gyros and accelerometers. Draper, always an unshakeable proponent of the single-degree-of-freedom gyro, insisted that if you had a good enough gyro you didn't need a star sensor.[19] Draper and his people were sure their inertial guidance system did not need "these crazy things called stellar sensors."[20] Ben Olsen, the deputy associate director responsible for the technical design of the Polaris and Poseidon guidance systems, certainly believed that: "We don't need to improve it with a star-tracker," he told Forter.[21] Nevertheless, the laboratory eventually agreed to include a star tracker in Poseidon's Mark 4 guidance system. "If the guy with the money says he wants it," Forter told Graham Spinardi, "you convince yourself quite easily that you agree with him."[22]

Stellar guidance for Poseidon seemed to gain traction in March 1969, after the secretary of defense, now Melvin Laird, asked Congress to authorize $12.4 million for the development of significantly improved guidance for Poseidon. As Spinardi points out, however, once "its visibility increased, stellar-inertial guidance became controversial"—controversial because increased accuracy would make Poseidon a "first-strike," counterforce weapon, rather than a retaliatory system suitable to the idea of mutually assured destruction, which had become the public rationale for the possession of nuclear weapons.[23] The political fallout forced President Richard Nixon to declare, "There is no current U.S. program to develop a so-called 'hard target MIRV capability.'"[24] Although other hard-kill programs involving Air Force missiles continued to be funded, the Mark 4 had to be canceled because of its prominence. This was fine for the SPO, which considered the Mark 4 guidance system a nonessential program; its elimination meant "buffer" funds that could be redirected to more critical areas if needed.[25]

Trident and the Mark 5 Guidance System

While development of Poseidon was still under way, the Instrumentation Laboratory began on a fifth-generation SLBM guidance system. In late 1966, a study called "Strat-X," conducted by the Institute for Defense Analyses, examined ways to counter a Soviet anti–ballistic missile (ABM) threat.[26] The alternatives were judged by comparing the cost of a potential Soviet ABM system to that of the development, procurement, and operation of a ballistic-missile system capable of countering it. The Navy entry judged the most cost-effective was a very large, quiet, nuclear-powered submarine; it would be based in the continental United States, from where its sorties could be guarded by U.S. forces, and it would carry twenty-four large missiles with a range of 4,500 to 6,500 nm.[27] This concept became known as the Undersea Long-Range Missile System (ULMS). An advanced-development directive for ULMS was issued by the Chief of Naval Operations on February 1, 1968; Admiral Smith of SPO was named as project manager in March. In July, the Special Project Office was renamed the "Strategic Systems Project Office (SSPO)," and Admiral Smith's responsibility was broadened to include all Navy strategic systems.[28]

In September 1971 the ULMS program was directed toward the development of a new large, higher-speed submarine and an advanced-technology missile capable of a four thousand-nautical-mile range. The new missile was to fit in the circular cylinder designed for the Poseidon C3 so that it could be used in existing Polaris submarines.[29] The accuracy of the missile was to be the same as that of the two-thousand-nautical-mile Poseidon C3. The overall weapon system was to be known as "Trident," the submarines were eventually to be known as the *Ohio* class, and the missile to be carried, provisionally, the "C4," was named the Trident I.

The longer range of the Trident I was needed to increase the sea room the Trident submarines would have in which to counter the threat from

Soviet antisubmarine forces. The new SSBNs were designed to have much longer periods of submerged cruising, during which time they would be without access to external sources of navigational information.[30] To meet the accuracy requirements of the missile under these criteria would require improvements in navigation, fire control, and guidance.*

The Instrumentation Laboratory—which had now been divested from MIT and renamed the Charles Stark Draper Laboratory (see chapter 20)—was selected once again as the design agent responsible for developing the new missile's guidance system, the Mark 5. The laboratory had extended roles in several support areas as well: system testing and evaluation, materials evaluation and approval, reliability assessment, test equipment development, and flight test program support. As Draper historian Paul Dow Jr. has written, "The combination of greater range and increased time between external position updates for the submarine navigation system created significant challenges for the guidance design. The added time required to attain the higher velocity needed for longer range, because of time and acceleration-dependent error sources in the sensors, required design changes for sensors and gimbals."[31]

To meet the enhanced performance goals of the Mark 5 guidance system, the laboratory's engineers developed new types of gyroscopes, made changes to the gimbals, and added a star sensor mounted on the stable platform of the IMU. The latter was the work of John Brett, formerly president of Kearfott. Brett was a key proponent of stellar inertial guidance. When it was rejected on Poseidon, he had decided that the "only way to get this done was to become part of the power structure."[32] Brett, who was well connected in Washington, became an under secretary of defense with responsibility for turning the C4's specifications into reality.

In addition to adopting Kearfott's Unistar stellar-inertial concept, the Mark 5 guidance system used just two of Kearfott's dry, tuned-rotor,** two-degree-of-freedom gyros.[33] These were less expensive to produce and

---

* Details of the navigation and fire-control improvements needed to meet Trident I's accuracy goals are beyond the scope of this work. Readers interested in these details should consult Spinardi, *From Polaris to Trident*, and Gates, "Strategic Systems Fire Control."

** A tuned-rotor gyroscope is "dry" because it does not employ fluid or gas to suspend the rotor. Instead, the gyro wheel is mounted on a flexure specially designed so that the wheel's rotation cancels the spring effect of the support.

were smaller and lighter than three Draper instruments. The weight savings was important for stretching the range of the C4 to almost double that of the same-sized Poseidon. Elimination of the third gyro also "made room for the stellar sensor, which because of its ability to compensate for errors elsewhere in the system, helped undercut the argument that Draper gyros were more accurate."[34] Trade-offs between the two types of gyros were considered before the Draper Laboratory, as SSPO's design agent, had concluded that either approach would meet the system goals, although it had recognized that a Draper design would probably be more expensive, though less risky.[35]

The intense pressure on the design team to minimize system weight dictated the choice of accelerometer. Like Poseidon's Mark 4 guidance system, the Mark 5 relied upon an improved PIPA that was easier to manufacture.[36] Although a PIGA would have been more accurate, the SSPO considered the PIPA to be good enough to meet the accuracy goal and light enough to meet the range goal without the extra cost of developing a new accelerometer.

Incorporation of a stellar sensor was a major challenge to the laboratory's IMU design team. The sensor's purpose was to measure the line-of-sight bearing to a selected star relative to a set of axes defined by the accelerometers. It was essential that the stellar sensor and the accelerometers be in close proximity to minimize errors in this measurement, on which were based vertical and horizontal corrections to the guidance system, based upon a star map preloaded in the guidance computer.[37] In addition to correcting launch-point errors, these values were to correct for gyro drift. The need to make corrections throughout the postboost phase of the flight required much more complicated fire-control software.[38] Because stellar sightings could not correct for gravity anomalies, the SSPO identified "areas [on the Earth] where the gravity anomalies were known to vary and just stay[ed] away from them."[39] Another difficulty was arranging the gimbals of the IMU in such a way that the star tracker could be properly aimed prior to launch while keeping accelerometers and gyros in the desired orientation relative to gravity and the trajectory plane.[40]

In addition to the IMU and stellar tracker, the Draper Laboratory was responsible for designing the guidance computer and control electronics. These were packaged in a separate container known as the electronics assembly, or "EA." The EA had to be environmentally proofed and radiation hardened in the same way as prescribed for the IMU. Both were placed within the third-stage "bus" near the third-stage rocket nozzle.[41] Computational software for Trident and its fire-control system were furnished by the Dahlgren Division of the Naval Surface Warfare Center.

The first flight-qualified Mark 5 guidance system was delivered on schedule in early 1976.[42] A Mark 5 successfully guided the Trident C4 in its first flight test, conducted from Cape Canaveral on January 18, 1977.[43] It took another two years of development and testing before the first Trident missile was launched from a submerged submarine and successfully guided to its target by the Draper-designed guidance system.

The Improved Accuracy Program

Throughout the 1960s, as we have seen, SPO had resisted pressure to meet increased accuracy requirements for the Fleet Ballistic Missile force. Requests from the offices of the secretary of defense and the Chief of Naval Operations met with a standard response: "SPO would attempt to meet accuracy 'goals,' but measurement and understanding of FBM inaccuracy was not good enough to promise to meet 'requirements.'"[44]

Toward the end of 1973, a new secretary of defense, James Schlesinger, asked the Chief of Naval Operations for a presentation on possible improvements in accuracy for the sea-based strategic missile system. After discussing the issue with Admiral Smith on several occasions, Schlesinger announced a program to measure and improve the accuracy of the Navy's SLBMs. The Draper Laboratory was tasked to undertake this work in late 1974, under the Improved Accuracy Program (IAP).[45] The basic objectives of the program, were, first, to gain an understanding of SLBM error sources and their interrelationships and to assess the accuracy-improvement potential of improved components and advanced system concepts; and second, to conduct advanced development of components and concepts that proved promising.

The Improved Accuracy Program started while Draper's development of Trident I was at its peak. To prevent interference with this effort, a separate program office was established within the Draper Laboratory to manage the IAP tasks. To ensure that the results could be realistically applied to a future FBM program, FBM managers and senior engineering personnel were given supervisory roles in the IAP in addition to their FBM duties. During this program, which ran between 1974 and 1984 and reportedly cost on the order of six hundred million dollars, the Draper Laboratory investigated new technologies for precision guidance accelerometers and stellar sensors, tested new IMU platform configurations, and developed new methods for investigating and analyzing errors due to submarine velocity and position prior to and during launch. Because of the work done during the IAP, the Trident II guidance team was able to move ahead with the design of a next-generation weapon system that emphasized greater accuracy.

Trident II and the Mark 6 Guidance System

When the Navy moved ahead with the secretary of defense's "Advanced Development Program" for an "enhanced performance" SLBM in 1981, it was integrated into the IAP. The new missile, designated the Trident II (D4), was designed to utilize the entire volume available in the launch tubes of the Trident *Ohio*-class submarines, tubes that had been overdesigned specifically for this purpose. It would be almost one-third longer than the C4 and weigh almost twice as much. At 130,000 pounds, it was 4.6 times the weight of the Polaris A1. The Trident II (D5) was also the first Navy SLBM designed from the outset to have a hard-target-kill capability.[46] Primary among the new missile's performance objectives were improved accuracy and increased payload.

The Mark 6 guidance system that emerged from the Advanced Development Program was an improved, high-accuracy, stellar-inertial system that achieved accuracy through improvements in hardware and software. It had a new set of inertial instruments and an advanced "stable member" that provided a more stable mechanical and thermal environment. It also had a new, higher-resolution star tracker.[47] The IMU of the Mark 6 was similar to the Mark 5's in having four gimbals, two two-degree-of-freedom, dry-tuned gyros, three pendulous integrating gyro accelerometers, and a stellar sensor system consisting of telescope optics and a camera detector.[48]

The accelerometers were Draper size-10 (one-inch diameter) PIGAs originally developed under an R&D contract and refined under the

IAP program.[49] These accelerometers represented a major technological advance, characterized by improvements in accuracy, size reduction, and producibility. Although the first units were produced by Draper, after that the project was turned over to Honeywell for manufacture in quantity. During the IAP program, new gyros had been developed by Litton Industries and Singer-Kearfott. After careful evaluation of accuracy, system integration, size, and power, Draper selected the Singer-Kearfott MITA-5 gyro, which was based on the MITA-4 gyro in the Mark 5 system.

Another component studied by the Draper Laboratory and pursued during the Advanced Development Program was the stellar sensor. The stellar sensor in the Mark 5 guidance system had been built around a Vidicon tube, but new advances in solid-state technology had opened the door for the use of a charged-coupled device (CCD). The primary candidate was a ninety-by-ninety-pixel CCD made by the Hughes Aircraft Company.[50] The laboratory also considered an improved Vidicon tube that contained multiple markings in the field of view for more accurate calibration and sighting. Those investigating the stellar sensor determined that both Vidicon and CCDs could meet the accuracy goals established by the SSP but that the Vidicon represented obsolete technology, whereas the solid-state devices were being adopted universally in electronic design. A properly designed solid-state device also offered longer life, which would simplify logistics in the fleet.

Because of new design requirements, a significant quantity of electronics had to be incorporated within the Mark 6's IMU. The minimal space available meant that the electronics had to be densely packaged. Also, the difficulty in dissipating heat from the IMU made it essential that the electrical components operate at low power. Reliability, also of utmost importance, was achieved by very-large-scale integration (VLSI) in an all-digital architecture.

This same design philosophy was followed for the EA, which housed the guidance computer that performed the navigation, guidance, and steering computations. The EA also contained a digital platform controller that provided functions for gimbal control, gyro control, and accelerometer control. In the Mark 5, these functions had been done by analog control systems. The EA also contained several power supplies.

The Trident II (D5) entered operational service in March 29, 1990, when the *Tennessee* (SSBN 734) deployed on its first patrol. According to unclassified sources, the missile's warheads have a CEP of three hundred feet.[51] As described to Congress, this extreme accuracy involves the interaction of six main subsystems:

1. The ship's navigation system, which uses sonar, a global positioning system, and other equipment to determine the missile's launching position.
2. The ship's fire-control system, which constantly processes the submarine's location, true North, target location, and other data to compute the proper trajectory for each missile.
3. The launcher system, which ignites gases that expand and eject the missile from the launch tube, through the water, and to the surface.
4. The missile's three-stage rocket-motor propulsion system.
5. The missile guidance system, considered the most complicated and sensitive of the D5 missile's six main subsystems, which is responsible for directing the missile on a corrected trajectory, compensating for submarine position and in-flight effects such as high winds, and triggering the reentry bodies for release toward the target.
6. The reentry-bodies system, including separation of the warheads toward the precise target, which is totally dependent on the missile guidance system.[52]

Trident II Mark 6 Life Extension Program

When production of the Mark 6 guidance system, which was based on early 1980s technology, ended in fiscal year 2001, the Navy's Strategic Systems Project Office (successor to the SPO) faced a new problem. For the first time in over four decades there was no new SLBM under development, yet within fifteen years the Trident II D5, which was designed for a twenty-five-year service life, would no longer be operational. Accordingly, the SSPO began efforts to extend the service life of the Trident II D5. In 2002, the Navy announced plans to extend the life of the submarines and the D5 missiles through 2040 via the D5 Life Extension Program, which had just gotten under way. As part of this program, the Charles Stark Draper Laboratory, the historical design agent for SSPO guidance programs, was awarded a three-year contract to study the technological issues and develop a program for extending the life of the Mark 6 guidance system.

In January 2005, the laboratory received an initial cost-plus-incentive-fee contract for $62 million to develop the system software and algorithms, system sensors, gyroscopes, and accelerometers for the next Navy guidance system, the Mark 6 LE.[53] In December, three months after it had completed a preliminary design review, the Charles Stark Draper Laboratory received a $101.1 million modification to this contract to provide a final Mark 6 LE proof-of-concept model by 2007. For the first time in its history, the laboratory would become the prime contractor for a guidance system, not just the design agent. In this role, the Draper Laboratory would build all the test beds and integrate the subsystems produced by the subcontractors: General Dynamics (system design and analysis), Raytheon (radiation testing, module design, and "breadboarding"), Honeywell (Mark 6 memory design, development and delivery of interferometric fiber-optic gyroscope), and Dynamics Research Corporation (communication timing system and test equipment).[54] The laboratory also received a separate contract to provide engineering support and repair services for the existing Trident II (D5) guidance system.[55]

Because of technological obsolescence and the disappearance of the industrial base needed to support the original Mark 6 IMU design, the Charles Stark Draper Laboratory had to redesign the IMU using the latest advances made in solid-state sensors and electronics. The same was true for the electronics package, for which components were no longer available.

Mark 6 Mod 1 Guidance System

Like its predecessor, the Mark 6 Mod 1 guidance system that emerged from the Draper Laboratory was based on a four-gimballed, inertially stabilized platform and used as much of the existing system (i.e., its hardware) as possible.[56] In place of the mechanical, dynamically tuned gyros, which had been a hallmark of all previous guidance systems, Draper developed the highly precise solid-state IFOG, or interferometric fiber-optic gyro.

The accelerometers were upgraded versions of the highly successful PIGA used in the Mark 6 guidance system. The stellar-sensing CCD was also improved to increase resolution.

A key challenge was finding a radiation-hardened memory to replace the existing 132KB plated-wire memory, which had been developed in the 1970s and was proprietary. Designing a completely new replacement memory would have been cost-prohibitive, so the laboratory took advantage of existing commercial technology. After extensive analysis and testing, the Draper worked with Honeywell to select MRAM (magneto-resistive memory) which was being developed for commercial applications by the Everspin Technologies Company.[57]

The first prototype of the Mark 6 Mod 1 guidance system was completed in 2007 and successfully tested in the "hardware-in-the-loop" simulation facility that had been developed as part of the Mark 6 LE program.[58] The finalized design was flight-tested for the first time at sea during the twenty-third Demonstration and Shakedown Operation (DASO), which took place on the Trident submarine *Tennessee* on February 22, 2012. Twenty-two years before, the *Tennessee* had served as the platform for DASO-01, when the sub became the first *Ohio*-class SSBN to launch Trident II (D5).[59] LE (life extension) missiles with the full complement of modernized systems were scheduled to begin entering the fleet in 2017.

18

Spy Satellites and Space Planes

The code name "Corona" was used for a highly classified program conducted by the Central Intelligence Agency (CIA) and the U.S. Air Force that produced the first successful photo-reconnaissance satellite, using the Discover space-exploration program as a cover. Launched into polar orbit on Air Force Thor boosters, Corona satellites took pictures of the Soviet Union while in orbit a hundred miles above the Earth. The exposed film capsule was ejected from the spacecraft, slowed by parachute, and captured by specially equipped aircraft.

The Instrumentation Lab's role in Corona began in early 1955 after a visit by Capt. James S. Coolbaugh, U.S. Air Force, an aeronautical engineer by training in charge of the Air Force's first satellite program. During one of his visits to Draper's laboratory, Coolbaugh learned that three Air Force Reserve officers enrolled in the master's degree program at MIT were doing their theses on a horizon scanner program for one of his office's reconnaissance projects.[1] The students had signed up for Draper's two-year program for military officers in the Weapons Systems Section of the Aeronautical Engineering Department.[2] The Air Force sponsored the students in exchange for a three-year tour of duty after they graduated.

The Weapons System Program, previously known as the "Armament Engineering and Fire Control" sequence of study, had been initiated by Draper after World War II to prepare military officers with sufficient science, engineering, and practical experience to fulfill assignments as research-and-development project officers.[3] His aim was to educate these officers so that they could "hold their own" with professional scientists and engineers should they find themselves in such assignments. Participating in the internships available in the Instrumentation Laboratory and completing the thesis requirements gave these officers a practical understanding of engineering, research, and the personnel who conducted them. Draper was in charge of the program until he became head of the department in 1951, when he relinquished his teaching duties and coordination of the program to Walter Wrigley.

When Coolbaugh discovered that Draper's students could contribute to the Instrumentation Laboratory's research on the Air Force's satellite program, he suggested, and Draper agreed, that it would be to the Air Force's advantage to assign several graduate students to work on a reconnaissance satellite that had been tagged Weapons System 117L (WS-117L) by Lockheed aircraft, the prime contractor. The satellite being developed under the WS-117L program had evolved from the project feedback proposal submitted by the RAND Corporation on March 1, 1954.[4] Lockheed's reconnaissance satellite was designed to photograph the Earth's surface with a television camera and beam its images back to the surface. It would deliver images to a resolution of a hundred feet from three hundred miles above the Earth.

Three Air Force graduate students, who just happened to be roommates, were assigned to the WS-117L project. They were Lt. John C. "Jack" Herther, Lt. Malcolm R. Malcomson, and Lt. William O. Covington. All three were given copies of RAND's proposal as background material for use in formulating theses that could be submitted in fulfillment of their degree requirements.[5] Herther and Malcomson teamed up to study the transitional control system that would place the satellite in a three-hundred-mile orbit. They proposed an inertially referenced, computer-controlled second-stage thruster fired at apogee (the orbital point farthest from the Earth). They based their design on a three-gyro, single-accelerometer inertial measurement unit for a satellite-stage ascent to provide the attitude control for that satellite's ascent stage. The system relied on the use of either reaction wheels or gas jets for three-axis control (roll, pitch, and yaw) and reorientation while coasting, for accurate addition of vernier velocity to achieve orbit, and initiation of Earth-pointing

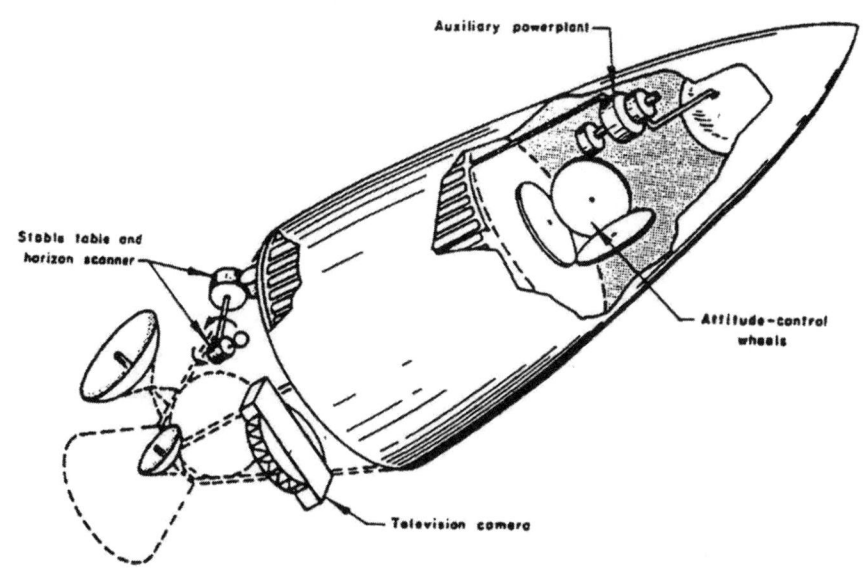

Fig. 18-1. Schematic drawing of reconnaissance satellite described in project feedback report. *Rand Corporation*

gravity gradient on-orbit stabilization. Roll jets would be used during thrusting phase, while the gimbaled engine controlled pitch and yaw.[6]

Covington produced a complimentary study on an on-orbit control system that used passive on-orbit gravity stabilization. This design relied on solar panels to recharge batteries that provided electrical power to run the rate gyros and flywheels that provided inertial stability. This method was practical only for a high-altitude reconnaissance mission that was free from the disturbing torques of atmospheric drag.[7]

The control concepts developed by Herther, Malcomson, and Covington during their internships at the Instrumentation Lab would eventually become collectively the foundation for the ascent guidance and three-axis passive gravity gradient on-orbit stabilization used on the Lockheed Agena spacecraft that carried Corona's photographic capsule. The WS-117L program, under which this work was originally funded, came under critical scrutiny after the Soviet Union surprised the world by launching Sputnik into orbit. This feat shocked the U.S. military, as it indicated that the Soviet Union (the USSR) had made greater advances in booster technology than previously thought. This made "people in high places think seriously about satellite reconnaissance"; indeed, serious thinking was urgently needed to assess the ballistic-missile threat now posed by the Soviet Union.[8] As Dwayne Day and his coeditors write in *Eye in the Sky: The Story of the CORONA Spy Satellites*, "There was a sense of extraordinary urgency in getting good pictures of the entire USSR; a satellite reconnaissance system was the obvious way to do so."

Although the Air Force was working on such a system, the President's Board of Consultants on Foreign Intelligence Activities, which was asked to review the program, was skeptical that the WS-117L could provide the needed capability. Its hundred-foot resolution was too crude to provide strategic intelligence, the program was running late, and it was encountering technical difficulties.[9] Instead of continuing with WS-117L, President Dwight D. Eisenhower directed the CIA to develop within a year a satellite reconnaissance system capable of achieving a resolution of twenty-five feet or better. Elements of the WS-117L program that promised to be ready early were transferred to the new program, under the auspices of the CIA and the newly established Advanced Projects Agency within the Department of Defense. The former would be responsible for the satellite reconnaissance effort, the latter for the boosters and upper-stage spacecraft.

A program office was soon established in Los Angeles, staffed with five Air Force officers commanded by Col. Lee Battle. It was supported by a small group of CIA officers in Washington directed by the CIA's Richard Bissell. The program group decided to use the Thor IRBM as the first stage of the rocket needed to place the reconnaissance satellite, to be known as "Corona," in Earth orbit. Thor had been flying successfully since 1957 and was further along in development than Atlas. An additional rocket stage was needed, however, to lift Corona into orbit. Although it had yet to be flown, the logical choice for the upper stage was Lockheed's Agena spacecraft, which had already been under development for two years. It was five feet in diameter and used a 16,000-pound engine that burned hypergolic propellants.

When it came time for Lockheed to select a contractor for the Agena guidance and control system, the Instrumentation Lab had to contend with a competing proposal from North American Aviation's Autonetics Division based on the latter's successful bid for Minuteman's guidance package. Draper's reputation and the availability of Dr. Joe DeLisle as project engineer—he had been largely responsible for the success of

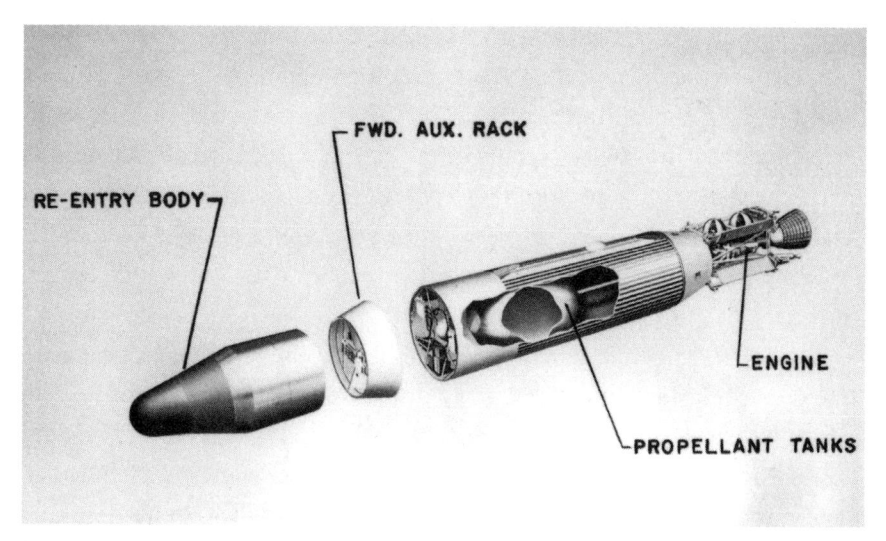

Fig. 18-2. Exploded view of Agena spacecraft. *National Reconnaissance Office*

SINS—gave the advantage to the Instrumentation Laboratory.[10] DeLisle proposed that the lab could save money by scaling down the weight and performance of the IMU from the Sperry SINS system for the Agena/Corona system. This approach would require only a small engineering staff. DeLisle's expertise and low-cost development approach was very attractive to the cash-strapped Air Force project office, which accordingly awarded a sole-source contract for the spacecraft design and on-orbit control to the Instrumentation Lab. The contract required building the first few systems. To save money, the Instrumentation Laboratory was authorized to present orally its progress reports in lieu of the more formal written reports usually required. The only documentation generated during the project were the MIT master's theses, briefing aids, engineering drawings, and manufacturing data.

Working for the Itek Corporation, which was tasked with modifying the company's Hi-AC camera for space use on Corona, was Jack Herther.[11] After graduating from MIT in 1955, Herther had been assigned to the WS-117L program office, where he oversaw the work on guidance, control, and on-orbit stabilization systems. In May 1957, he was asked to brief Secretary of the Air Force Donald Quarles and his staff on the progress being made on the guidance and control system for the SAMOS satellite—another Air Force reconnaissance satellite, intended to supply real-time data.[12] Herther asked Draper to back him up, particularly on passive gravity stabilization.* Draper accompanied him, bringing along a prop prepared by the Instrumentation Lab to explain the physical principles involved in gravity stabilization. As Coolbaugh recalled from his own experiences with attempting to obtain funds and reviewing industry proposals, good props were "very effective, important, and influential tools for understanding new physical principles and for convincing an audience of . . . the value of a project."[13]

The "visual aid" consisted of a dumbbell mounted on a rotating arm. The dumbbell was mounted slightly offset—that is, attached to the arm at a point that was not equidistant from the two weighted ends—on a jeweled pivot so that when rotated horizontally on the arm it tended to line up radically and oscillate, making clear the need for a rate-damping system. It illustrated the principle that when the center of gravity and the center of mass do not coincide, a restoring torque is created that would be similar to the vertical restoring torque on a satellite, as in the case of the Agena while in orbit. Herther pointed out the Moon an example of how the gradient principle works: the Moon's face always points to the Earth. Thus, the active flywheel damping concept on Agena, Herther explained, would settle quickly (compared to the Moon) within a few orbits, *if no other significant disturbing torques were present.*

Herther gave several other high-level briefings on the guidance system for WS-117 before deciding to leave the Air Force in order to pursue a PhD degree at MIT. He planned to start in the fall of 1958 and had obtained approval to leave the Air Force, but his orders were rescinded until October, when he was unexpectedly offered an early release from his three-year obligation.[14]

Discharged in December, Herther went to see Richard S. Leghorn, president of Itek, a newly formed company dedicated to information technology and space reconnaissance. Colonel Leghorn had cofounded Itek after leaving the Air Force the September before. In the Air Force he had served as chief

---

* Gravity gradient attitude stabilization is defined as the alignment of one axis of a satellite along the Earth's local vertical direction so that a particular end of the satellite always faces in the downward direction. Passive gravity gradient stabilization is the achievement of this orientation, including damping the resulting vibrations, without use of active control elements, such as servo systems, reaction wheels, or gas jets.

of intelligence and reconnaissance systems development at the Pentagon. He first learned of the importance of Herther's work on systems for stabilizing the Agena spacecraft during a visit to Draper to discuss one of Herther's staff briefings. Leghorn realized that Herther's approach was the key to adapting the Hi-AC (high-acuity) direct-scanning panoramic camera for high-altitude balloons that Itek was developing for Corona.[15] Back at his office, Leghorn discussed Herther's work with Dr. Duncan Macdonald, one of the company's cofounders. Both men immediately decided to offer Herther a job.[16] Jack Herther accepted their offer, becoming Itek's first employee.

Herther reported for work on December 16, 1958, and immediately began designing a three-axis stabilization system for Itek's Hi-AC camera. It was only then that he learned of the CIA's secret reconnaissance satellite project and the integral camera-vehicle concept for recovering the spacecraft's camera and film. Herther realized that to reduce drag on the Agena, which would be inserted into orbit at a much lower altitude than SAMOS, it would have to be flown horizontally. This orientation would have inherently smaller aerodynamic torques than the vertical nose-down three-axis vehicle orientation employed for SAMOS. Unlike the passive payloads on the WS-117L's programs, the larger torques produced by the Hi-AC oscillating panoramic scanning camera would require an active control system to maintain on-orbit stabilization.

Because of his MIT thesis and his previous work on the Agena spacecraft stabilization, Herther was able to come up with a suitable control system for Corona. He accomplished this by placing Itek's cross-track oscillating panoramic camera into an Agena spacecraft that would fly horizontally using the three-axis gyro stabilization system the Instrumentation Lab was developing for it. To maintain horizontal stabilization, which was essential for high-resolution photography at that altitude, he added a horizon sensor to correct for any gyro drift. The gas jets that were already a part of Agena's ascent guidance system would be used to provide active control without the weight penalty of the passive stabilization system planned for SAMOS (which would operate for months, as opposed to a Corona satellite's life of only a few days).* This approach provided three-axis Earth-viewing stabilization based on the horizontal

---

* If the mission ran longer than this, there would not be enough fuel to maintain attitude control.

sensor's updating of Agena's ascent inertial control system.[17] Any motions imparted to the spacecraft by internal equipment, such as the moving camera, could be automatically sensed and damped out with the gas jets. The vehicle would move so little that there would be minimal blurring of the images.

Itek was awarded the contract for the camera in the spring of 1959, and Herther traveled to Lockheed's "Skunk Works" (advanced-development office) with Walter Levison, head of Itek camera development, to meet with representatives of Lockheed and its subcontractors: General Electric, responsible for the recovery capsule, and Fairchild, in charge of fabricating Itek's camera design. Herther, who took on the role of system integration engineer, understood that the Agena's active stabilization system would only work if the camera design did not cause the spacecraft's attitude disturbances to blur the excellent static lens film resolution of the Hi-AC camera. He worried that the moving camera parts and film supply spools would impose inertial forces on the spacecraft.

To make sure that they did not, he spent $50,000 of CIA's "black budget" to purchase a Pace analog computer. As he later explained, "I just ordered it. I hired two engineers and a technician to run the damn thing. We set up a very elaborate analog simulation of the Agena vehicle and our reaction to it at full and empty film spool conditions in order to verify that the extremely small rates did not cause image smear during exposure."[18] Running the simulation, Herther discovered that the first Corona design created more movement of the spacecraft than the active gas stabilization system could cope with. "We had to whack down the inertia of the reciprocating scan arm." They also added several systems to reduce the effects of the various moving parts of the camera.

The three-axis control system Herther proposed was implemented by the Instrumentation Laboratory, which had been contracted by Lockheed to design Agena's guidance system. It had for ascent guidance a three-axis IMU that used a pitch/yaw gimbaled engine to maintain the ascent trajectory plane, with gas jets for vehicle roll control. On orbit, three-axis stabilization for horizontal flight kept the camera axis pointed earthward for image motion compensation, while minimizing blur from pitch, roll, and yaw using the pitch and roll gyros, updated continuously using infrared horizon sensors for determining the vertical to better than 0.1 degree accuracy.[19]

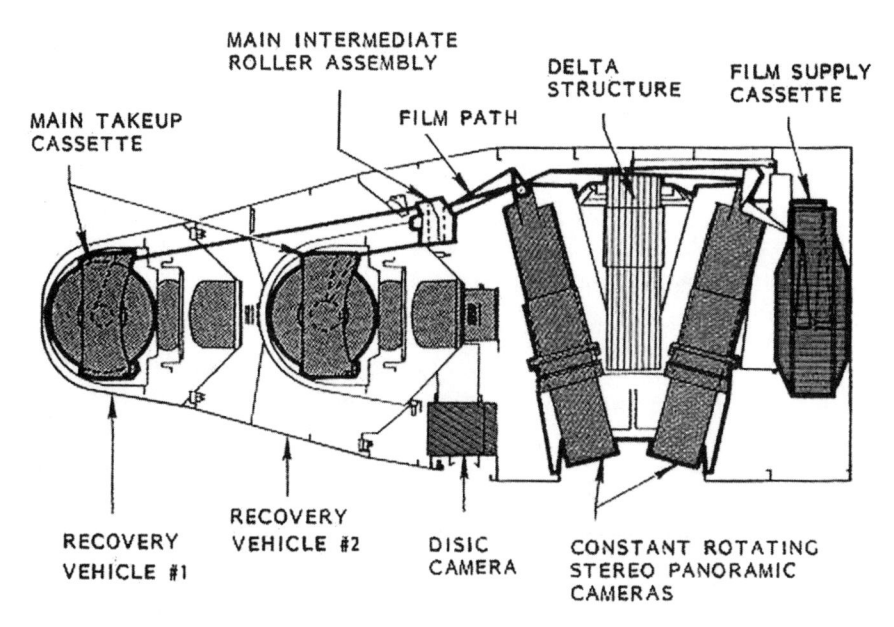

Fig. 18-3. Inboard profile of the Corona J-1 model camera system, which had a second panoramic camera and a second recovery capsule added.
National Reconnaissance Office

Herther's work in three-axis control was considered routine at the time, but in hindsight and with the declassification of national security records it becomes clear that he played a significant role in the success of the nation's efforts to develop highly specialized satellite reconnaissance systems. It was a role that he would not have undertaken, had it not been for the experience and knowledge gained during his training at the Instrumentation Laboratory. As Herther and Coolbaugh were to write in their definitive paper on the genesis of three-axis control and stabilization, "The three-axis satellite stabilization, which allowed Corona's Itek cameras to capture four hundred miles of high-resolution photos, played a crucial role in the geopolitical stability and maintaining a balance of power during the Cold War because arms control treaties could now be verified."[20] During the operational life of Corona, which lasted from 1960 to 1972, over 140 successful missions were flown and over 2.1 million feet of film were delivered to U.S. intelligence agencies.[21] This imagery gave American leaders unprecedented insights into critical national threats, particularly from the Soviet Union.

When Alan B. Shepard Jr. was blasted into space on a Mercury-Redstone rocket on May 5, 1961, in the first of Project Mercury's manned space flights, the Instrumentation Laboratory was already working on guidance systems for manned space vehicles with wings. At the time, Draper saw little future in the Project Mercury type of space flight, which shot a capsule into space like a bullet, with no wings and very little control.[22] Like all consummate pilots, Draper wanted a man in the control loop during the critical phases of flight.[23] As he told the editors of *Time* magazine, Project Mercury was "like going over Niagara Falls in a barrel. You don't expect to find many people making a career of it."[24]

More to Draper's mind was Dyna-Soar, a winged spaceplane that could be used for a variety of military missions. Its origins dated to 1952, when the Bell Aircraft Corporation proposed the development of a boost-glide bomber-missile, dubbed "Bomi," that evolved into an intercontinental reconnaissance bomber.[25] Bomi led to the System 118P project for the design of a two-stage Mach 15 reconnaissance vehicle. The 118P project was followed by a reconnaissance system named "Brass Bell." These studies, to which the Air Force was generally receptive, led to a number of Air Force–funded investigations of reconnaissance and strike boost-gliders. In November 1956, the Air Force asked NACA to review the service's boost-glide aircraft studies. In response, NACA's director, Hugh L. Dryden, formed a steering committee that evaluated the service's various projects and recommended development of a flat-bottomed, hypersonic delta glider. On October 10, 1957, six days after the Russians launched Sputnik, the Air Force consolidated research on boost-gliders into a single three-phase research program based on "dynamic soaring"—"Dyna-Soar." Two months later, the Air Research and Development Command issued a directive defining the Dyna-Soar project as a conceptual test to obtain data on the boost-glide flight regime in support of future weapon system development.[26]

By the end of January 1958, an Air Force selection board had screened a list of 111 contractors to choose potential bidders for the Phase I design. Before the year was out, ARDC had requested proposals from industry for the development of a delta-wing, single-seat boost-glider that was to serve as a technology demonstrator. Boeing won the competition and became the prime contractor for the Dyna-Soar Weapon System.[27] The

Martin Company was named associate contractor, with responsibility for the booster development.

On September 30, 1961, the Instrumentation Laboratory began work on an Air Force contract for the Pace inertial navigation and guidance system to be used on the Dyna-Soar advanced-boost glide weapon systems. These hypervelocity glide vehicles were to be boosted into orbit or near-orbit and released into a flight regime covering intercontinental ranges for bombing or reconnaissance missions.

Two years later, funding for the Dyna-Soar program was seriously undermined after Secretary of Defense Robert McNamara, believing that the Air Force had placed too much emphasis on controlled reentry, suggested that a space station serviced by a ferry vehicle would be a better approach to the reconnaissance mission.[28] As a result, the Dyna-Soar vehicle was never built, and funding for the Pace I system was discontinued at the end of the year.[29] By then the Instrumentation Laboratory had designed, built, and successfully tested an inertial navigation and guidance system that was capable of accurate long-time-of-flight performance. Funding resumed in June, when the Air Force issued a contract for Pace II, which was essentially an extension of the Pace I program.[30] Its goal was to increase accuracy by an order of magnitude while minimizing size, weight, and power requirements. A prototype was completed and successfully tested in the summer of 1967, but it never entered production. It did, however, accomplish one of the main objectives of the program, which was to design and test a prototype system capable of being produced by industry with minimum delay from start-up to production model.

19

To the Moon and Beyond

The Instrumentation Lab's involvement in deep-space missions began with a project known as the "Mars Probe," a response to the Soviet Union's launch of Sputnik, the first artificial satellite, in October 1957. Hal Laning and Dick Battin were given the responsibility of determining the probe's trajectory and for developing the navigation and guidance system needed to get it to Mars and back.

In 1956, Dick Battin, after helping to develop Q-guidance, had sought greener pastures and left the Instrumentation Laboratory to work in the operations research group of the Arthur D. Little Corporation. As Battin later admitted, it was the worst mistake he ever made: "There was an awful lot of travel, which I hated. When I looked at the people who were farther advanced than [I] in the company, they traveled more. So the only thing I had to look forward to was more travel and working on projects that I really didn't enjoy."[1] Fortunately, he had stayed in contact with Hal Laning, his mentor at the Instrumentation Lab. A few months after Sputnik was launched, Battin learned that Laning had a simulation of the solar system running on an IBM 650 computer and was "flying" round trips to Mars.[2] The two had discussed it on the telephone and Laning's work made Battin "homesick for the Lab." He wound up his affairs at Arthur D. Little and returned to the Instrumentation Laboratory. Once there, he joined the small group of scientists that was being formed to "flesh out" the proposed system for a reconnaissance mission to Mars.

Battin's return, in the spring of 1959, coincided with the publication of the lab's report on the technical feasibility of sending an unmanned probe to photograph Mars. This work, as well as the follow-on project to design the spacecraft and its guidance system, was funded through a clause in the Air Force ballistic-missile guidance contract that permitted the lab to work on anything else of interest that was relevant to guidance.[3] As Battin recalled, there was not a lot of money, but it was just enough to fund the work of three or four people.

The project, the Mars Probe, centered around a spacecraft that, launched from Earth, would coast to Mars, take a photograph at its closest approach, and return to Earth for splashdown and recovery. The flight path required the probe to make two trips around the sun during a voyage that would take more than three years.[4] The Mars probe had to be completely self-sufficient, as there was no way to uplink commands to the spacecraft; navigation and guidance would all have to be done on board. The astronomers they talked to thought they were crazy. "How [are] you going to go to Mars?," they asked. "You don't even know where Mars is."[5]

Nevertheless, Laning and his small band of engineers, which included Milt Trageser, went ahead with the project. Trageser, a physicist by training, designed an optical telescope with which the probe would orient itself relative to the Moon and the stars. He was also responsible for the design of a camera to capture the image of Mars as the probe flew past.[6] By the summer of 1959, Laning's team had completed a preliminary design, which they planned to present to the contractor for approval. By then, however, the Air Force was no longer in the business of flying spacecraft. That task had been relinquished to a new government agency called the National Aeronautics and Space Administration (NASA). So, Batten and Laning went to Washington to see Hugh L. Dryden, the deputy director of NASA.

They happened to arrive on the same day that the Soviet premier, Nikita Khrushchev, was visiting Washington. They gave their presentation and it was well received, but the high-level NASA audience they had expected was busy attending to the protocol mandated by Khrushchev's visit.[7] The NASA people they did talk to thought the team had done a nice piece of work and offered a small amount of study money to refine some of their ideas presented. The two Instrumentation Laboratory researchers were very disappointed: "We were young and naïve," recalled Battin, "and expected that they would say, 'Hey, this is great. How much do you need? Go ahead and do it.' But, of course, that didn't happen."[8] Nevertheless, the study contract issued by NASA for $50,000 allowed them to continue the work that had begun under Air Force auspices.

Because an onboard computer was needed, Laning joined forces with Ramon L. Alonzo to design a small control computer with unique characteristics for space applications, dubbed simply the "Mod 1."[9] The programming task was assigned to Hugh Blair-Smith, a young software engineer from Harvard recruited by Dan Goldenberg, the head of the lab's computing group.[10]

The stated purpose of NASA's study contract was to aid the Jet Propulsion Laboratory (JPL) in its efforts to conduct unmanned space missions to Mars, Venus, and the Moon.[11] Due to differences in the guidance philosophies of the two organizations, however, the relationship between the MIT Instrumentation Laboratory and JPL never materialized. The Instrumentation Lab's onboard self-sufficiency approach was at odds with JPL's ground-based control, which relied on large tracking antennas and telemetry systems. When the laboratory's four-volume NASA-funded report on the Mars study appeared in April 1960, it fell on deaf ears.

In the late fall of 1960, the Instrumentation Laboratory's inability to find a place for itself in NASA's unmanned deep-space missions led Draper to approach Harry J. Goett to discuss the possibility of the lab's participation in the agency's manned space missions. Goett, then director of NASA's Goddard Laboratories, was also chairman of the NASA Research Steering Committee on Manned Space Flight. By then NASA had already stated its intention to fund a six-month feasibility study for the Apollo Moon mission.

Draper's conversation with Goett led to a meeting with the Space Task Group (STG) at the Goddard Space Flight Center, in Beltsville, Maryland, on November 22. There they discussed a proposed contract with Draper's Instrumentation Laboratory to conduct a preliminary study of the navigation and guidance system for Project Apollo.[12] During the meeting Milt Trageser presented a draft statement of the proposed work. It was divided into three parts: midcourse guidance, reentry guidance, and an experimental feasibility study using the Orbiting Geophysical

Observatory satellite. The laboratory's work was clarified the next day in talks between Trageser and Robert G. Chilton of STG. The two men agreed that the purpose of the Apollo program was the development of manned space flight capability, not simply the circumnavigation of the Moon by an encapsulated man. As David Hoag writes in his history of that aspect of Apollo, Trageser and Chilton

> determined the system should consist of a general purpose digital computer, a space sextant, and inertial guidance unit (gyro stable platform with accelerometers), a control and display console for the astronauts, and supporting electronics. The inflight autonomy of the earlier Air Force and NASA studies seemed appropriate to the manned mission, particularly since some urged that the mission should not be vulnerable to interference from hostile countries. It was judged important to utilize the man in carrying out his complex mission rather than merely to bring him along for the ride.[13]

In late December, the Instrumentation Lab submitted its proposal for review by NASA; it was approved on February 7, 1961.[14] A month later, Chilton visited MIT and toured the Instrumentation Lab's facilities.[15] Although Chilton had graduated from MIT in 1949 and had worked with Draper on his thesis, he did not know anyone else now in the lab. But he was impressed with the staff selected to work on the Apollo project.[16] In particular, "he was amazed to learn about the role MIT had played in Polaris—performing 'overall system management' for the guidance system, and then integrating it into the missile itself."[17] This visit and the impression it left on Chilton would be instrumental when the time came for NASA to select a contractor for Apollo's guidance and navigation system.

The Apollo program itself took on new meaning when President John F. Kennedy urged the nation to "commit itself to achieving the goal, before the decade is out, of landing a man on the moon and returning him safely to the earth."[18] As Hoag tells us, "With the impetus of the presidential challenge, the efforts at the Instrumentation Laboratory changed character."[19] What began as a small research project to study interplanetary navigation quickly became the laboratory's largest and most important project.[20]

The Instrumentation Lab had yet to receive the NASA contract, with its anticipated $100,000, but Draper began preparing it by shifting key people from Polaris, which was winding down, to Apollo. These included David G. Hoag, heretofore technical director for Polaris, whom he appointed as technical director for the Apollo project.[21] Hoag, another of Draper's students, had joined the laboratory in 1950 after receiving his master's degree in instrumentation from MIT.[22]

In July 1961, Ralph R. Ragan, who had been director of development for the Polaris guidance program, was called back from summer vacation to work on Apollo.[23] He joined with Trageser to define the navigation and control system, in conjunction with NASA's Chilton, Robert O. Piland, and Robert Seamans.[24] The latter had worked for Draper for almost fifteen years before joining RCA in 1955. Now, he was the associate administrator at NASA, having joined that agency in 1960.[25] Together this group completed a study of what had been done on the Polaris program to design a guidance and navigation system, along with the documentation necessary for putting such a system into production on an extremely tight schedule. Using this as a basis, the group worked out a rough schedule for a similar program on Apollo. Seamans' participation ensured that the lab would have little or no competition when NASA put the work out for bids. As Mindell writes in *Digital Apollo*, "He [Seamans] wanted to see some Apollo contracts to New England and announced the guidance contract early, so when [NASA] went out for bids on the prime contracts for the spacecraft itself, it would be clear that they did not include guidance."[26]

On August 4, 1961, the Instrumentation Laboratory submitted a proposal to NASA for the Apollo guidance system. Six days later, the two entered into a contractual agreement—the first issued by NASA for the Apollo program—for the guidance and navigation system.[27] The laboratory would use the same methodology that had been applied on Polaris: it would do only the technical design and prototype development; when the manufacturing phase commenced, industrial contractors would take over.

At the end of August, James Webb, NASA's chief administrator, invited Draper and members of his Apollo team to Washington, D.C.[28] Webb, having served as treasurer of the Sperry Gyroscope Company during World War II, was already familiar with Draper and the Instrumentation

Lab's capabilities and must have had in mind the 85,000 highly profitable Mark 14 gunsights that Sperry had produced for the Navy. Nevertheless, he had several concerns regarding the guidance system. Most importantly, he wanted to know when it would be delivered, and he wanted assurances that it would work. The discussions that took place on August 31 at NASA headquarters continued at Webb's house over dinner that evening. Webb and his high-ranking guests from NASA asked Draper's team if it would be feasible to provide the guidance needed for a lunar mission before the end of the decade. "We said 'Yes.' When we were asked if the Instrumentation Laboratory would take responsibility for the navigation and guidance system, we again said 'Yes.' They asked when the equipment would be ready. We said, 'Before you need it.' Finally, they asked, 'How do we know you're telling the truth?'" This one Draper answered personally: "I said, 'I'll go along and run it.'"[29] Draper was so confident that his team could deliver that he was willing to put himself at risk to prove it.

Although others later made light of his desire to "float around in space," Draper had been dead serious.[30] In November, he wrote Robert Seamans to volunteer formally for service as a crew member on the Apollo mission to the Moon: "I fully realize my limitations as a test pilot, but I feel that my qualifications in scientific and engineering fields should be considered as a worthy background for a crew member primarily concerned with non-flight operation matters such as making observations, checking system performance, interpreting events, and helping to maintain communications."[31] "Seamans," as Mindell tells us, "wrote back to his former mentor that he would forward the request to the appropriate authorities."[32]

Draper worried that his age might weigh against his selection. After all, he was sixty years old. Draper's age would not bar his selection, according to Brig. Gen. Donald D. Flickinger, the Air Force flight surgeon on NASA's Life Science Committee, responsible for selecting astronauts.[33] One gets the sense that Flickinger was being kind, even reverential, to this esteemed scientist who was the director of MIT's Instrumentation Laboratory and the Apollo Guidance and Navigation Program. But in fact, whatever his qualifications, Draper was simply too old to be considered. Nevertheless, Draper sincerely believed that his work of the past thirty years had uniquely qualified him for the spacecraft's crew.

The guidance and navigation system proposed by the Instrumentation Laboratory consisted of a three-axis inertial measurement unit with gyros and accelerometers; an optical system to align the inertial system periodically to the stars and to make navigation measurements in a sextant configuration, by observing the direction of the Earth and the Moon against background stars; and a general-purpose digital computer to handle the data. Unlike the autonomous guidance systems developed to guide Polaris, and the one proposed for the Mars mission, NASA demanded in Apollo the active involvement of its astronauts. This placed new demands on the design for Apollo's computer that required the addition to the computer design of a display and keyboard for inputting data and commands—what would come to be known as the "DSKY" (pronounced "disky"), for display keyboard.

The digital computer, in fact, was the most challenging aspect of the system. Its design was undertaken by Eldon C. Hall's computer group. Now that the second-generation Polaris computer was undergoing flight tests, he had a sizable number of designers looking for a new challenge. The functional requirements for the Apollo guidance system's computer were unknown at the time, but Hall's group knew it would have to operate in a small spacecraft. Its size, weight, and power consumption would have to be limited. When NASA asked how big the computer was going to be, Dick Battin advised his colleagues, "Well, we've got to give them a number; just tell them it's a cubic foot"—and that is what it became.[34] Had Hall's computer-design team known then what they learned later, however, "they would probably have concluded that there was no solution with the technology of the early sixties."[35]

To meet the design characteristics of the Apollo Guidance Computer (AGC), Hall's group chose the magnetic core transistor logic circuits that they had developed and tested. This decision lasted for about one year, until it became clear to Hall that a computer using this technology would have limited speed and be difficult to fabricate, costly to manufacture, and less reliable than the application demanded.[36] By then Hall had begun investigating the potential of semiconductor integrated circuits; he now decided that these devices should be incorporated into the ACG design in place of the magnetic core transistor logic circuits. He then went about convincing everybody, including NASA program management, that the change was not only feasible but highly desirable. Fortunately for all concerned, NASA acquiesced in December 1962 and granted approval to use this new technology.[37]

The integrated circuit offered great increases in capability in a very small package. But as David Mindell notes, "the little chip was a black box, seemingly immune to human understanding and repair—and nobody had ever tried it in such a life-critical application."[38] Hall spent much of the next two years making sure that the computer would operate flawlessly during the eight-day mission to the Moon and back. He spent hours, days, and weeks in design reviews with both the Instrumentation Lab's engineers and representatives from Raytheon (the manufacturer), the spacecraft contractor's engineers, and NASA technical teams.

The guidance computer came under intensive fire from the NASA review teams, which focused on mission success and crew safety. According to Hall,

> Any statistician could prove the guidance computer fell short of the requirements for mission and crew safety. Experience with digital computers and the unknowns, added by the introduction of integrated circuits into the design, fueled the fires. The individuals who made up these review teams were like wolves nipping at our heels. Their motivation did not seem to be pure. At the time, their actions seemed to be attempts to bite off a piece of the action.
>
> Responding to the attacks and maintaining an organized design effort required constant vigilance on my part.[39]

These attacks subsided only when the first computers came off the production line and operated beyond expectation.

Three years into the project, however, NASA began to have doubts about the Instrumentation Laboratory's ability to deliver the AGC on schedule. Various committees were formed to examine the progress of the hardware and software designs. Because of reliability concerns, some in NASA wanted to replace the AGC with the launch vehicle's computer—known by the acronym LVDC—because of latter's triple redundancy. The LVDC, which was the autopilot for the Saturn launch vehicle, was composed of three identical logic systems. Each logic system was split into a seven-stage pipeline. At each stage in the pipeline, a "voting system" would compare the three results and pass the most popular on to the next stage in all pipelines. This meant that for each of the seven stages, one module in any one of the three pipelines could fail and the LVDC would still produce the correct results.

One of the committees formed to investigate the computer problem was a team of engineers from the Instrumentation Laboratory that included Hugh Blair-Smith. He was one of the key software engineers working on the Apollo computer, having helped develop the instruction set for the Mod 3C computer which became the baseline design for the AGC. After studying the LVDC concept, he determined that it would be too "clumsy and inefficient," for the nimble multitasking that would be required of the Apollo computer.[40] His input and the findings of the MIT team convinced NASA that the LVDC was inadequate for the new task, and the AGC remained as one of the critical elements in the Apollo guidance system.

On July 20, 1969, the computer helped make history when it guided the Apollo 11 spacecraft and its lunar orbiter to a soft landing on the Moon's surface. This historic event marked the first time that a human being set foot on a heavenly body. The success of the Apollo Guidance Computer can ultimately be attributed to the dedication, motivation, and confidence of the people involved, as well as to the Instrumentation Laboratory's culture, which fostered a tremendous concern for and attention to detail. "Every integrated circuit, and every line of code had a reason to be there, and by the time it was delivered, it had been examined and tested until there was no doubt that it worked."[41]

20

The Road to Divestiture

The Apollo program was conducted concurrently with Sabre, the Instrumentation Lab's second-largest project in terms of contract dollars. Both projects, along with Poseidon guidance, were undertaken while the United States was heavily engaged in the very unpopular war in Vietnam. As the war continued unabated, there was a growing resistance to what many American citizens believed was an unjust, unwinnable war. The ongoing conflict in Vietnam sparked student protests on college and university campuses throughout the United States. Students and non-student radicals were against what they considered to be the misguided policies of the U.S. government.

In January 1969, a group of MIT faculty and students shocked the academic community by calling for a research stoppage on March 4 as a symbolic gesture to provoke "a public discussion of problems and dangers related to the present role of science and technology in the life of our nation."[1] The idea to conduct this day of reflection—later dubbed a "strike" by the press—originated with the newly formed Science Action Coordinating Committee (SACC), a group of students activists who opposed the war, academic credit for classified research, and what they considered to be the militarization of research at MIT.[2]

The seeds for this dissidence had been sown during the 1960s when MIT became both the largest university contracting for defense research in the United States and a major supplier of top scientific advisors to the

Department of Defense.[3] In 1965, faculty efforts led by Noam Chomsky laid the groundwork for the creation of an organization called RESIST, to resist the taxes supporting the Vietnam War.[4] But it was student activism beginning in 1967 that led to the March 4 call for a research stoppage. The idea, according to Professor Murray Eden, originated in "a dinnertime discussion among physics graduate students on the possible impact of an anti-war science-research strike."[5] The students quickly began reaching out to other students and to professors in their own and other departments. The result was the creation of SACC and the release of a draft statement signed by the forty-eight senior faculty members calling on scientists and engineers at MIT and throughout the country's universities to take concerted action and exercise leadership in specific ways:

- Devise means for turning research applications away from the present overemphasis on military technology towards the solution of pressing environmental and social problems.
- Convey to our students the hope that they will devote themselves to bringing the benefits of science and technology to mankind, and to ask them to scrutinize the issues raised here before participating in the construction of destructive weapons systems.
- Express our determined opposition to ill-advised and hazardous projects such as the ABM system, the enlargement of our nuclear arsenal, and the development of chemical and biological weapons.[6]

The interruption of research was to be on a voluntary basis; no effort was to be made to render it compulsory.

On March 4, 1969, most MIT students and faculty spent the day in their classrooms and laboratories. Less than 20 percent of the student body (about 1,400 people, not all of whom were MIT students) crowded into Kresge Auditorium to listen to panel discussions about converting military research into peaceful pursuits, academic-government relations, and the responsibility of intellectuals. The "strike" touched off a national debate on the military's presence on campuses and classified research. It inspired sit-ins, teach-ins, and similar events at many other universities and plunged MIT into an intense discussion about the role of military research at the institute.[7] Because it ranked first among university defense contractors, critics of the administration and antiwar activists began calling MIT "Pentagon East" or the "Pentagon on the Charles." They wanted the school's administration to reexamine the university's priorities in light of the social responsibility of scientists and engineers.

At the center of this debate was the question of what to do with the Instrumentation Laboratory and the Lincoln Laboratory. The two labs were considered "special," because they were substantially larger than the others. Their combined budget of $114 million in 1969 ($779 million in current dollars) represented over half of MIT's annual budget. Unlike the Instrumentation Laboratory, whose funding was split between the Defense Department and NASA, the Lincoln Laboratory was funded completely by the military. This should have made it a prime target for the antiwar demonstrators, but Lincoln Laboratory was on the grounds of Hanscom Air Force Base, twenty miles west of Cambridge and inaccessible to demonstrators. Instead they focused on the Instrumentation Laboratory, whose main office at 68 Albany Street in Cambridge was within the main campus.

On April 22, SACC organized a morning rally of about fifty radical students in the courtyard in front of the Maclaurin Building to protest Poseidon missile research and the development of the MIRV, which many claimed would further escalate the arms race and destabilize the present balance of power by leading to a "first-strike capability."[8] At noon, several demonstrators began addressing the crowd of onlookers.[9] Using a bullhorn, speakers urged fellow students to march on the Instrumentation Lab two blocks away. The students, according to Draper's recollection of events, gathered on the steps of the Rogers Building, "put themselves in a frenzy," walked down Massachusetts Avenue, and turned left onto Albany Street. It took them just two or three minutes to arrive in front the laboratory.[10] Draper caught them by surprise, meeting them at the door and attempted to explain his position. The demonstrators threw epithets at Draper, calling him a "warmonger" and "murderer." Draper, according to the press, later claimed he had been able "to give them hell on their own bullhorn several times."[11] "The students," according to the account that appeared in *Business Week*, "demanded to see the lab. Draper astounded them when he agreed. The students wanted to go in all at once, but Draper wanted more manageable groups of two or three."

After picketing the Instrumentation Lab, the protestors walked back to the Rogers Building to present their demands to MIT president Howard W. Johnson. He was at lunch, but when he returned he invited the students to join him and chairman James R. Killian Jr. in an adjoining lecture room to discuss the issues.[12]

Three days later, on April 25, 1969, Johnson, anxious to forestall further confrontation and to address the issue of military research at MIT, appointed a twenty-two-member panel under the chairmanship of Dean William F. Pounds of the Sloan School of Management to review the status of the special laboratories. The Review Panel on Special Laboratories, usually referred to as the "Pounds Panel," was given six months to review the MIT's relationship to the special laboratories and the appropriateness of its sponsorship of them. The panel included various MIT professors, group leaders from the special labs, an undergraduate business student, graduate students from engineering and biology, President Johnson, former MIT president Julius Stratton, Yale historian Elting Morison, Peter Gray of the MIT Alumni Advisory Council, Gregory Smith (alumni term member of the MIT Corporation), and Dr. Richard Wurtman, associate professor of endocrinology at Harvard University.[13]

The panel met for its first working meeting on Sunday, April 27. After a rather lengthy introduction, Pounds got down to the business of determining what people they needed to interview or work with in order to make an informed decision on what direction to take with regard to the special laboratories. As a starting point, Pounds intended "to go around the table," inviting the members to share their views about MIT as a university and its purpose.[14] "Let's begin," he asked after a lengthy introduction, "with the most straightforward question, which people do we want to hear from, or categories of people?"[15]

Jonathan Kabat, a graduate student representative, was the first to speak up. "I have been thinking about this overnight," he said, "and I have tried to identify several general questions. The first is [to] think of . . . the various things we might recommend and . . . the questions we have to ask to get the information."[16] Kabat, one of a handful of energetic organizers of the SACC, revealed that he had only one thing in mind: "that the Instrumentation Laboratory and the Lincoln Laboratory be severed from MIT." Kabat's statement, made that first day, seemed to the panel to make the fate of the Instrumentation Laboratory a foregone conclusion.[17] That was obvious from David Hoag's reaction the following day when he alluded to the victimization of the Instrumentation Lab with the words, "the patient is sick and the doctor may kill the patient."[18]

The Pounds Panel met twenty times from the end of April through the beginning of May to hear testimony from a hundred people—faculty members, special laboratories staff, administrators, politicians, and students. One of its first tasks was to visit the special laboratories, to hear presentations, tour the facilities, and solicit statements from researchers and administrators. "At both labs, the panel's questions and lab staffers' responses emphasized the same key issues: the ratio of basic to applied research, the autonomy of the labs, the importance of the labs' work to MIT's prestige, the inevitability of weapons development, and the morality of demonstrating the feasibility of specific weapons systems."[19]

The panel, according to the prestigious journal *Science*, concluded that with their current mix and scale of research activities, the special laboratories would not over the long term fulfill the ultimate objectives of MIT. The panel offered alternative strategies for conversion of mission-orientated work, but fundamentally "it also called for a major institutional effort to carry through a reappraisal" of the laboratories' research.[20]

On May 31, after the spring semester had ended and most undergraduates had left the campus for the summer break, the Pounds Panel published its preliminary report, recommending that:

1. The Laboratories and MIT should energetically explore new projects to provide a more balanced research program. . . .
2. The educational interactions between the Special Laboratories and the campus should be expanded. . . .
3. There should be intensive efforts to reduce classification and clearance barriers in the Special Laboratories. . . .
4. A standing committee on the Special Laboratories should be established . . . as a means of providing the President with the considered advice of students, faculty and laboratory staff.[21]

The panel also concluded "that the Poseidon program at this stage of its development is inappropriate for MIT sponsorship," adding a fifth point: "The MIT administration will review the Institute's future commitments to this program. It recognizes, however, that MIT must be prepared to honor its existing contractual obligations."[22]

A *New York Times* editorial published on June 4 lauded MIT's decision not to sever ties to the labs, arguing that "the sole effect of such a spinoff in the absence of a basic redirection of Government research activities, would be to force the armed services to set up more laboratory complexes of their own or to give huge new contracts to aerospace and other corporations. The result would be precisely opposite to the one desired by the student and faculty dissenters—a strengthening of the military-industrial complex and a diminution in the capacity of university scientists to exert any useful influence in the shaping of public policy on military matters."[23]

Although the Pounds Panel's report was praised by the *New York Times* and well received by the MIT community, Draper, according to Jack Ruin, MIT vice president for special laboratories, considered it to be an inquisition, not an investigation.[24] Draper, who had testified before the panel on more than one occasion, felt this way "because there was no discussion of the background, of our motives, of what the laboratory did, what it was trying to do. The only thing . . . they wanted to do was to prove that I had worked on military projects."[25] This Draper readily agreed to (he had also worked on civilian projects). He was not ashamed of having done things that helped the country. As he told the Pounds Panel, his objective was "to carry out the developments of technology so that nobody in the world [would] think that our country was an easy mark."

Draper, as William Leavitt, senior editor for *Air Force Space Digest*, noted, was impatient with the arguments of those who opposed military research on the grounds that it must inevitably lead to an uncontrollable arms race and oblivion:

> I think it is sheer bubble-headed nonsense to feel that somehow human beings have all of a sudden gotten to the point where the proper approach is to throw away your own weapons, your own tools, in the hope that somebody else will throw theirs away, too.

I believe that the maintenance of a balance of respect between the entities of great power is the only way in which [mankind] can hope to get enough time to begin to work things out. . . . Anybody who feels that the prohibition of technological development in the weapons field is going to do anything except put this country further behind the eight ball than it already is, I think, is insane.[26]

Some members of the press did not think there was much of a chance that MIT would back away from military-funded research. It was too lucrative. John Walsh, writing in *Science,* believed that MIT was not a hospitable environment for radical activists. "By tradition," he wrote, "MIT is an engineering school, and engineering students have been a conspicuously inert group in most universities during the current upheavals. Most MIT faculty members practice in those disciplines in which research has drawn heavily on government funding, both military and nonmilitary, and faculty members serve as consultants for industry and as advisers in the government."[27]

Walsh's ideas about the MIT community might have been accurate, but they did not take into account the antiwar protests and radical demonstrations taking place in other universities across the country. Nor did they account for the MIT faculty's growing concern over the imbalance in the institute's research activities. "The Pounds Panel," Walsh wrote, "was troubled by [this] question, and it was the subject of a statement added to the panel report by chemistry Professor Edwin R. Gilliland, and graduate student Marvin A. Sirbu Jr., who called for the eventual divestiture by MIT of special laboratories."[28]

The pressure for President Howard Johnson to do something continued to mount, and it intensified as students and faculty returned to campus after the summer recess. On September 18, 1969, the Executive Committee of the Corporation, which was responsible for general administration and superintendence of all matters relating to the school, issued a statement accepting the recommendations of the Pounds Panel and urging the president to give high priority to their implementation: the committee believed "that it would be inappropriate for the Institute to incur new obligations in the design and development of systems that are intended for operational deployment as military weapons. That is not to mean that with its unique qualities the Institute should not continue to be involved in advancing the state of technology in areas which had defense applications."[29]

In September, Johnson established an oversight committee as recommended by the Pounds Panel, naming John C. Sheehan, Dreyfus Professor of Organic Chemistry, as chairman. Properly the Standing Committee on Special Laboratories, the "Sheehan Committee" was to evaluate the acceptability of contract proposals submitted by the special laboratories of the basis of nine criteria set down by the Pounds Panel. As Sarah Bridger notes in *Scientists at War,* the establishment of "this committee challenged notions of 'academic freedom' and fostered resentment among lab staffers some of whom reportedly referred to it as the 'moral committee.'" Meanwhile, SACC complained about the committee's lack of procedural transparency and small student representation. Nevertheless, student support for halting MIRV research and genuine efforts at reconversion remained strong.[30]

Johnson was well aware that Draper, as the charismatic leader of the Instrumental Laboratory and a polarizing force to many on the faculty, would be hard to control.[31] When asked on one occasion by the faculty council to describe the current activities of his laboratory, he, according to Johnson's memoirs, "came on initially as a modest little man from Missouri, and ended up as the fire-eating rocket scientist. He alarmed several of the department heads. One of my colleagues called him Dr. Strangelove."[32] Draper had also threatened to leave the institute, having told the *Boston Globe* in June that a policy change would force him to cut ties with MIT. Not surprisingly, Johnson concluded that the administration "would do better in dealing with the laboratory situation without Draper as the active head."[33]

On September 24, 1969, Johnson dropped a bombshell. Without previous discussion he announced that on January 1, 1970, Professor Charles L. Miller, director of the Urban Systems Laboratory, would assume responsibility for the direction of the Instrumentation Laboratory.[34] Professor Draper, he added, would continue as senior advisor and technical director of major projects in the laboratory. Johnson never discussed this change with Draper before it happened, nor did he give any reasons.[35] Draper was simply informed that as of January 1 he would no longer be director of the MIT Instrumentation Laboratory.

Three years earlier, Draper, then sixty-five, had in accordance with MIT rules become a professor emeritus and had stopped teaching, but he had continued as the director of the Instrumentation Laboratory. Shortly thereafter, President Johnson asked him to stay on as director of the laboratory for three more years so that he could continue to supervise the important work being on the Apollo project. As a result, Draper was expecting to continue as head of the laboratory for another six months when he was summarily replaced. He did not resign, he did not request retirement. As he explained in a later interview documented in his papers at the Library of Congress, "I didn't ask to be relieved of my job. I didn't resign, and I was informed that I didn't have it [the job] anymore and in the place where I come from in Missouri when that happens to you, why[,] you say you were fired."[36]

As a sop to Draper's feelings, Johnson also announced that the name of the facility would be changed to the Charles Stark Draper Laboratory, "in recognition of the enduring contributions that Dr. Draper has made to MIT and the country."[37] Nevertheless, Draper's achievements in the field of space-age guidance and technology had not been enough to keep him in his position as the Instrumentation Laboratory's director. This was a time, according to William Leavitt, when "a coterie of self-appointed zealots . . . at MIT and at other prestigious universities around the country [had] taken it upon themselves to 'cleanse' these institutions of what is called war research."[38]

On October 22, Johnson issued a statement signaling the future direction of MIT's special laboratories by endorsing the recommendations of the Pounds Panel, whose final report had been issued that month. First, he began, "I want very much to see our work applying high technology to domestic and social problems. . . . Second, I would like very much to see our work in basic technology related to defense continue."[39] The special laboratories would continue to do fundamental research in various fields related to national defense and to do exploratory work on new concepts with important defense applications. "However," he continued, "the laboratories will not assume responsibility for developing operational weapon systems based on these concepts, nor will either laboratory assume responsibility for the field testing or production of specific weapon systems."[40]

Although Draper was willing to work with those who wanted to change the direction of the Instrumentation Lab's research, he felt that they were devoid of ideas. "They're not about to do anything," he told William Leavitt in an interview shortly after his dethronement. "They're only making noises about how other people should do it."[41] They have no idea, he explained, how to obtain financing, what problems to concentrate on, or how to organize research. Draper was seriously worried about the survival of the laboratory and its impact on national defense. Who, he asked, would do this work if the Instrumentation Laboratory—the only facility of its kind in the free world—went out of business?

Neither President Johnson's statement to the faculty, the publication of the Pounds Panel final report, nor the administration's decision to set up the Sheehan Committee satisfied the radical elements on and off MIT's campus. They continued to push for a faster reordering of MIT's priorities and even talked of shutting down the Instrumentation Lab by force.[42] On October 29, fifty members of the radical November Action Committee (NAC) disrupted a General Electric recruiting session. Chanting "Six, five, four, three, organize to smash GE!," the group marched to the door of the recruiter's office, where they were stopped by Dean Kenneth R. Wadleigh of MIT, other members of the MIT administration, and campus policemen.[43] The incident made Draper so concerned for the security of the Instrumentation Laboratory and the safety of its employees that he issued a memorandum to all of his personnel that same day, warning them to prepare for an impending "time of troubles."[44]

The "trouble" began shortly after 7 a.m. on the morning of November 5, 1969, when 375 chanting members of the NAC, most of them not MIT students, set up an obstructive picket line outside the five entrances to the Instrumentation Lab and its parking lot.[45] The demonstrators, waving Vietcong flags and shouting "Shut it down!," scuffled with the laboratory workers arriving for the morning shift.[46] Those inside the lab tried to encourage Draper not to go there, but as his daughter, Martha, explains, he was stubborn.

"OK, I'm going in," he told her. "I'm going to go in and drop you at your place and I'll park the car there."[47]

As Draper dropped off his daughter, she told him, "I'll go upstairs and alert the lab that you're coming." He parked the car and started walking toward the Instrumentation Lab.

Martha's offices were on the eighth floor of a building located on Albany Street; once there she looked out and could see the demonstrators in front of the lab—and there was her father, walking down the street wearing the French beret and trench coat that he always wore. The demonstrators had absolutely no idea he was there. Martha, of course, thought this was hysterically funny. He walked along all by himself until some people came out of the lab and surrounded him, as an escort. Avoiding the main entrance, he went into the lab through the delivery dock. The demonstrators were totally unaware of who had just walked into the lab.

At around 9 a.m. the pickets were deemed to be unlawfully assembled, and Cambridge police wearing riot gear were sent into action. The police promptly routed the protesters down the back alleys of Cambridge. Within an hour the street had been cleared and the laboratory reopened.

Most of those working for the Instrumentation Laboratory blamed the laboratory's troubles on outside agitators who had no sense of the laboratory's real mission or its accomplishments. Nevertheless, the die had already been cast. The road to divestiture had been made unmistakable when the Executive Committee had ruled out new weapon-related design or development projects.[48]

At the time, the MIT Instrumentation Laboratory was preparing to design the guidance system for the Trident I missile (chapter 17). With the Apollo program (chapter 19) winding down, Trident was considered to be a key to the laboratory's future, but the release of the lab's proposal had to be delayed because of doubts that it would meet MIT's new standard for research projects.

The tide against the laboratory began to turn during a general faculty meeting on March 11, 1970, when Ascher H. Shapiro, head of the Department of Engineering, delivered an unequivocal call for divestiture.[49] In 1966, Shapiro and the members of his department's Fluid Mechanics Laboratory had resolved to convert the bulk of its research to socially oriented problems.[50] Previously the Fluid Mechanics Laboratory, like many of the MIT laboratories, had received the bulk of its funding from military applications, research on jet engines and on reentry physics for spacecraft and ballistic missiles. Shapiro had become increasingly interested in bioengineering, which was a newly emerging field, and turned his focus to fluid flows in the human body. The idea of converting the special laboratories, he argued, was futile. Even if they could somehow be "converted," he argued, they would compete for scarce civilian funding with other MIT laboratories, like his own.[51] The faculty tabled Shapiro's motion for divestiture, but Instrumentation Laboratory members left the meeting convinced that it was inevitable.

In fact, the administration, according to Professor Miller, had decided on divestiture before he took over the laboratory in January.[52] This is somewhat at odds with Johnson's account, according to which he came up with the plan to divest and discussed it with his closest advisors ("Jerry Weisner, Jim Killian, and some of the deans—at great length") in his isolated house in New Hampshire sometime that spring.[53] That latter seems more logical, especially Killian's involvement in the decision; he had gone through a divestiture in 1958, when the Mitre Corporation was created. At that time, Killian was president of MIT and Lincoln Laboratory was becoming overwhelmed by its responsibilities for starting up the nation's first air-defense system SAGE. The next phase, integrating interceptor weapons into the software, would be so massive that the laboratory would have to double in size. MIT, however, had been unwilling to let Lincoln Laboratory grow any larger. Furthermore, the original purpose of that laboratory—research and development—had nothing to do with system implementation, let alone on such a vast scale. When no other contractors could be persuaded to take over the project, the Secretary of the Air Force suggested that part of Lincoln Laboratory, the Digital Computer Division, be spun off to continue the systems engineering for SAGE on its own. MIT had agreed, and the Mitre Corporation had established a precedent for the divestiture of the Draper Laboratory.[54]

Whenever the decision to divest was made, the axe fell on May 20, 1970. At a special faculty meeting, President Johnson made it official: starting June 1, 1970, the Charles Stark Draper Laboratory would be an independent division of MIT under its own board. The new board would be headed by the MIT vice president for research, Dr. Albert G. Hill, and would consist of MIT senior officials plus representatives from industry, banking, and foundations.[55] The board would determine the best way to carry out full divestiture.

Draper was bitter over the decision of the administration, which he believed had given into student demands instead of "facing down those who, in his view, had finally found a scapegoat for all their frustrations and guilt over Vietnam."[56] In his opinion the Instrumentation Laboratory was that scapegoat, the elimination of which would expiate the collective guilt of the MIT community. "Only then," he was convinced, "will the students and professors feel cleansed and purified."[57]

The survival of Lincoln Laboratory gives credence to Draper's scapegoat theory. Despite its total reliance on military spending, the administration chose to uphold its standing within the MIT community, claiming that its programs were in compliance with the new policy against operational weapons development.[58] Yet the Instrumentation Laboratory—part of an academic department, with a long tradition of faculty/staff interchange, dozens of course offerings, and a balance of civilian and military funding better than Lincoln's—got the axe.

For Johnson and the MIT administration, making the Instrumentation Laboratory an independent corporation solved two ongoing problems. First, it placated the faculty members and activists advocating the elimination of classified research on the MIT campus. Second, it consolidated the administration's control over the types of research conducted by MIT's major laboratories while enabling oversight of their policies and operations. Unlike the Lincoln Laboratory, which was closely supervised by the MIT Corporation and the administration, the Instrumentation Laboratory was under Draper's complete control. Neither of the MIT past presidents (Killian and Stratton) nor Johnson had ever been in full control of Draper, and they knew it.[59]

There were other reasons behind the administration's decision for divestiture as well. As Bridger points out, "Outside observers speculated that the administration had failed to locate nonmilitary projects necessary to keep the lab open in accordance with the Pounds Panel's recommendations."[60] This was partly the result of a sharp decrease in the percentage of R&D funds in the federal budget for fiscal year 1971, which as Albert Hill pointed out in his vice president's report for research, was the lowest since 1959.[61] Draper himself expected the funding and people for the Instrumentation Laboratory to decrease by 10 percent and 12 percent, respectively, for the year ahead.[62]

But it was also the result of the lab's inability to secure the kind of nonmilitary work that its critics demanded. Conversion was a "misnomer,"

Johnson proclaimed in an interview reported by the *Boston Globe*.[63] "We cannot decree that people change the direction of their research." It appears likely that President Johnson or his supporters were counting on a large contract from the Department of Transportation for an air traffic control system. Perhaps this is why Professor Miller was selected to replace Draper. As it happened, however, any money for this purpose dried up on March 25, when the Department of Transportation announced that it would take over the NASA center in Cambridge and rename it the "Transportation Systems Center," driving the final nail in the Instrumentation Laboratory's coffin.[64]

Notwithstanding Draper's claim that it was a scapegoat, the Instrumentation Lab, like the Mitre Corporation before it, had simply outgrown its role within the university. Draper's insistence on perfecting the single-degree-of-freedom gyro above all else, along with his demand—because of his experiences with the Sperry Gyroscope Company—that his laboratory oversee the production of the lab's inventions, made divestiture inevitable.

After June 1, 1970, when the Charles Stark Draper Laboratory (CSDL), as Johnson had announced, became an independent division reporting to the MIT Corporation through its own board of directors (whose president was Draper and its chair Albert G. Hill). The laboratory continued to function as before. However, board of directors rather than Draper alone was in charge. As the historian Dorothy Nelkin observes, "The administrative decision, intended to accommodate as many interests as possible, pleased few. A major distraction was removed; but the questions raised by the critics were not resolved."[65]

In July 1973, after a three-year transition period to sort out contractual obligations, the Draper Laboratory—still housed in buildings adjacent to the MIT campus but no longer part of the university—became an independent nonprofit research corporation. Nevertheless, it continued to have a close working relationship with the institute. Although Draper was terribly hurt (some claim he was left a "broken man") by the divestiture of "his beloved creation," he continued to devote himself to the laboratory's success and, as professor emeritus, its ongoing relationship with MIT.[66]

After separating from MIT, the laboratory kept teaching as part of its own mission. "MIT graduate students still worked at the lab as 'Draper Fellows,' undergraduates attended seminars there, and the facility's resources brought together academic researchers, corporate scientists and military personnel."[67] The laboratory continued to work on Trident, Minuteman, and the Apollo programs throughout the 1970s. In the 1980s it applied its expertise on the Space Shuttle and in the 1990s moved into robotics, the Global Positioning System, and advanced guidance systems. As Stuart Leslie observes, "In every way that mattered, nothing had changed except on paper."[68]

Draper died in the summer of 1987, on July 25. As Robert Duffy, a retired Air Force brigadier general who served as the CSDL's first president, notes, "Draper's passing took from us an innovative, insightful, productive, leader of very rare qualities. His warmth and humor lightened many a heavy discussion. He could get to the nugget of an argument rapidly and he saw elements of an issue most of us would miss."[69]

After his death, the board of directors of the Draper Laboratory authorized an annual award in Draper's name to be administered by the National Academy of Engineering. The award honors the engineer who has contributed most to engineering in the opinion of the National Academy of Engineering–appointed selection committee. The award approximates, in its field, the Nobel Prize in prestige, and it is permanently endowed.

21

A Heterogeneous Engineer

Heterogeneous engineer: Term used by the sociologist John Law to describe a technical expert who can move beyond the technical aspects of engineering and mobilize humans as well as material resources to achieve a goal.[1]

Looking back on Draper's long and distinguished career, one cannot fail to notice the similarities between his approach to problem solving and the concepts of Thomas P. Hughes in his 1998 *Rescuing Prometheus*. Hughes emphasizes the synergistic effect of bringing together smart, creative people to solve problems. Under the right circumstances, they coalesce into intellectual communities that become influential far out of proportion to their numbers. This appears to be the case with Draper and the "gyro culture" he established at the Instrumentation Laboratory.

As Professor John Krige argues in his analysis of the winner of the 1984 Nobel Prize in physics, "Only a few, generally outstanding, but always well-connected individuals, situated in appropriate nodes in the social network, are able to weave together the scientific, technological, institution, and political threads that together constitute their project."[2] Charles Stark Draper was one of those rare individuals. From his laboratory's earliest days Draper knew how to mobilize the human and material resources needed to achieve the objectives of the laboratory and its varied programs. He also had a masterful understanding of the personal, professional, and institutional advantages of close relationships—with students, staff, and the laboratory's funders. The Instrumentation Laboratory, aside from its scientific and technological expertise, was a well-run social, economic, and political enterprise, due largely to Draper's personality and management skills. He also had an uncanny ability to sustain the military and agency alliances from which long-term support for projects derived.

Draper, to paraphrase Bill Trimble's description of Jerome C. Hunsaker, was the archetypal heterogeneous engineer: "A heterogeneous engineer is comfortable not only in his or her technical field but also in the wider world and able to understand and deal with technology's social, cultural, political, and economic ramifications."[3] Draper's understanding of the personal, professional, and institutional advantages of close relationships outside the confines of the research laboratory, in the worlds of business and government, shaped much of his career. His lifelong objective, that for which he "mobilize[d] human and material resources" needed to obtain his objective. For Draper, this was the lifelong pursuit of inertial navigation and guidance.

Draper "understood human beings and he understood how to challenge them. . . . [He] knew how to lead and how to get people to follow towards a common goal."[4] He believed that one of the primary goals of higher education was identifying and developing individuals who showed sparks of creativity. The young engineers trained in the Instrumentation Lab became Draper's disciples and carried the technological torch of inertial navigation and guidance.

Although many of the Draper's supporters, like Horacio Rivero, were not directly involved with any of Draper's projects, they had all taken Draper's courses and studied instrumentation at MIT or worked for the Instrumentation Laboratory. Rivero, assigned to the U.S. Navy's Bureau of Ordnance Anti-Aircraft Production Section, had already studied the gunsight problem when he unexpectedly discovered Draper's lead-computing sight. He was abundantly qualified to understand the significance of Draper's invention and ideally placed to bring it to the attention of those in the Bureau of Ordnance who had the clout to get it adopted by the Navy. Whether the Mark 14 would ever have seen the light of day without Rivero's intervention remains open to speculation.

Another early disciple who was to play a significant role in the development of inertial navigation was Leighton Davis. Unlike Draper's later acolytes, who were almost always introduced via their work on gyros or inertial navigation, Davis was initially brought to Draper's attention by their mutual interest in engine indicators. This interest brought Davis to MIT, where he studied under Draper, learned of the lead-computing gunsight, and arranged for Draper to be permitted to test it at the Watertown Arsenal. When Davis discovered the Army Air Forces lacked an effective dive-bombing sight, he contacted Draper, who sketched out a preliminary design that led to a collaborative program between the Instrumentation Laboratory and Wright Field's Armament Laboratory to develop the A1 gunsight. The cooperative effort during the next two years led to a close friendship that in turn opened the door to Draper's initial government contract to study the inertial navigation problem. That study ultimately led to the development of SPIRE and finally to Draper's initial involvement with the guidance system for Atlas.

Paul Blasingame (chapter 14) was yet another Draper protégé who became an influential supporter of the Instrumentation Laboratory's work on ballistic-missile guidance. Blasingame spent three years at MIT learning under Draper's tutelage the nuances of inertial navigation and graduating with a doctor of science degree in 1950. He was subsequently appointed chief guidance and control project officer for the Air Force's Western Development Division and became its in-house advocate for equipping ICBMs with inertial guidance. This issue put him at odds with many experts, who considered inertial guidance too experimental, as well as too heavy compared to radio guidance. But Blasingame was convinced that AC Spark Plug Division of General Motors Corporation, backed by MIT, could produce a workable inertial guidance system for the ICBM. He won approval for using inertial guidance as a backup for radio guidance on the Atlas ICBM and as the primary guidance system on the intermediate-range Thor.

Draper's quintessential disciple was undoubtedly Robert Seamans, another the graduate student who worked closely with him for fifteen years before moving on to greener pastures. It was Seamans, then an associate administrator at NASA, who made sure that Draper's Instrumentation Laboratory received the guidance contract for the Apollo spacecraft.

In addition to his roles as a scientist, teacher, and mentor, Draper was also a superb manager whose leadership of the MIT Instrumentation Laboratory was unquestioned for thirty-five years. When Draper established the laboratory in 1934 with the help of a few graduate students, he had no idea how big an enterprise it would become. When his tenure as director abruptly ceased in 1969 the MIT Instrumentation Laboratory had a budget in excess of $50 million, was involved in over thirty research projects, employed 790 people, and had 196 industrial residents and approximately 375 student associates.[5] Today, the laboratory that bears Draper's name is still one of the cutting-edge research institutions in the United States, with yearly revenue in excess of $500 million. It is a fitting memorial to one of the most talented engineer/scientists of the twentieth century.

Notes

Prologue. Hot Spot of Innovation and Invention

1. Arthur Mollela, "Welcome to a Hot Spot of Invention," *Hot Spot of Invention, People Places, and Spaces,* New Perspectives Invention & Innovation Symposium, Lemelson Center, National Museum of American History, November 6–7, 2009. Dr. Mollela is the director of the Lemelson Center.
2. Antiaircraft Operations Research Group, Headquarters, Commander in Chief, United States Fleet, Study No. 4, June 1, 1945.

Chapter 1. A Milestone in Aviation History

1. Instrumentation Laboratory, Massachusetts Institute of Technology, "R-63 Project SPIRE Flight Test Program," November 1953, CSDL Historical Collection [hereafter CSDL-HC], MIT Museum, Cambridge, Mass. [hereafter MIT-MC]; Charles Stark Draper, "The Evolution of Aerospace Guidance Technology at the Massachusetts Institute of Technology, 1935–1951: A Memoir," *Essays on the History of Rocketry and Astronautics* 2 (September 1977): 249.
2. Charles Stark Draper, "Fifty Years of Flight Technology," *IEEE Transactions on Aerospace and Electronic Systems* AES-14, no. 4 (July 1978): 715; Walter Wrigley, "The History of Inertial Navigation," *Journal of Navigation* 30, no. 1 (January 1977): 65.
3. Charles Stark Draper, "Origins of Inertial Navigation," *Journal of Guidance and Control* 4, no. 5 (September–October 1981): 457–58.
4. Draper, "On Course to Modern Guidance," *Astronautics and Aeronautics* 18 (February 1980): 51.
5. J. Scott Ferguson, "CSDL Historical Collection Projects" [hereafter CSDL-HC Finding Aid], s.v. SPIRE, Charles Stark Draper Laboratory, August 3, 1979. (Copy obtained from MIT-MC.)
6. "Laboratory Aviator Extraordinaire: Chip Collins Reflects on a 40-Year Draper Career," *D-notes,* May 20, 1988, 3.
7. Charles "Chip" Collins, interview by Thomas Wildenberg, October 16, 2012, Westford, Mass.
8. Collins interview, October 16, 2012.
9. CSDL-HC Finding Aid, s.v. SPIRE.
10. "Laboratory Aviator Extraordinaire," 3.
11. Collins interview, October 16, 2012.
12. Christopher Morgan, with Joseph O'Connor and David Hoag, *Draper at 25: Innovation for the 21st Century* (Cambridge, Mass.: Charles Stark Draper Laboratory, 1998), 11.

Chapter 2. The Formative Years

1. Draper described it as "a community where the farmers came to do their trading and sell their products." Charles S. Draper, "Transcript of a Tape-Recorded Interview with C. Stark Draper," conducted by Barton Hacker, 1, MIT Oral History Program, January 19, 1976; February 2, 1976; March 1, 1976; and April 5, 1976, Cambridge, Mass [hereafter CSD Oral History].
2. Howard I. Conrad, ed., *The Encyclopedia of the History of Missouri* (New York: Southern History, 1901), s.v. Windsor, available at http://tacnet.missouri.org/history/encycmo/towns.html#Windsor.
3. Martha Stark Draper Scrapbook 1881–1925, Charles Stark Draper Papers, Library of Congress [hereafter CSDP].
4. General Stark was responsible for establishing the New Hampshire Militia, which arrived at the battle of Bunker Hill just in time to save the day. His pronouncement "Live Free Or Die: Death Is Not the Worst of Evils" was written in a letter penned in 1809 as part of a toast he sent to his wartime comrades who were attending an anniversary commemoration that he was unable to attend.
5. Quoted by Harold K. Banks, "She Chose to Walk with Him," *Pictorial Review* [section] Boston *Advertiser,* September 15, 1962, 12. (Clippings in CSDP.)
6. CSD Oral History, 3.
7. CSD Oral History, 4.
8. Robert A. Duffy, "Charles Stark Draper, 1901–1987: A Biographical Memoir," in *National Academy of Sciences Biographical Memoirs* (Washington, D.C.: National Academy of Sciences, 1994), 123.
9. *American Miller and Processor* 38, no. 2 (February 1, 1910): 151.
10. Dr. Draper to "Dear Isabelle," postmarked October 21, 1926, Family Papers File, CSDP.
11. "He Has Struck Oil," clipping, unnamed newspaper, hand-dated December 19, 1916, Family Papers File, CSDP.
12. "May He Strike It Rich," clipping, unnamed newspaper, undated, Family Papers File, CSDP.
13. Clipping, unnamed newspaper, undated, Family Papers File, CSDP.
14. CSD Oral History, 7.
15. Stephen Birmingham, *Real Lace* (Syracuse, N.Y.: Syracuse University Press, 1997), 159.
16. James S. Draper [hereafter JSD], interview by Thomas Wildenberg, November 15, 2012, Newton, Mass. [hereafter JSD interview].
17. Fiorello La Guardia, quoted in Ellen Lawson, *Smugglers, Bootleggers, and Scofflaws: Prohibition and New York City* (Albany, N.Y.: Excelsior, 2013), 102.
18. JSD interview.
19. JSD to the author, Cambridge, Mass., October 3, 2017.
20. Per Debbie Douglas, director of MIT History Museum, based on Doc's letters held by James Draper.
21. The quotation itself is from CSD Oral History, 7; for the authorial insertion, CSD, "Fifty Years of Flight Technology," 708.
22. CSD Oral History, 61.
23. CSD Oral History, passim.
24. CSD Oral History, 21.
25. JSD, email to author, July 29, 2015.
26. CSD to "Dear Folks," April 2, 1925, Correspondence 1911–1940, CSDP.
27. Banks, "She Chose to Walk with Him," 13.

28. JSD, email to author, June 25, 2018, based on the letters in his possession.
29. U.S. Congress, Air Corps Act (ch. 721 44 Stat. 780), July 2, 1926.
30. CSD Oral History, 7. See also the captioned picture of the aircraft taken by Draper in the collection of his papers at the Library of Congress [hereafter LC].
31. CSD, "Fifty Years of Flight Technology," 709.
32. Military Service 1925–1942 File, CSDP.
33. For this and the other details of the intake process at Brooks Field, see Rebecca Hancock Cameron, *Training to Fly: Military Flight Training, 1907–1945* (Washington, D.C.: Air Force History and Museums Program, 1999), 246.
34. For example, if the instructor pushed his control stick forward, the car would begin to turn nose down just as an airplane would. The student could undo this motion by bringing the control stick slightly toward him.
35. "The Ruggles Orientator," *U.S. Air Services*, March 13, 1928, 20.
36. Duffy, "Charles Stark Draper," 125.
37. CSD, "Fifty Years of Flight Technology," 709.
38. Faculty Board of the Air Corps Primary Flying School, November 8, 1926, Military Service 1925–1942 File, CSDP. Duffy ("Charles Stark Draper") states that Doc exhibited "a tendency towards air sickness" when in the Ruggles Orientator and implies that this was the reason for his rejection from further training. This theory does not fit the chronological record, which shows a one-month gap between the time Draper was authorized to undergo flight training and the date that he was washed out. Nor does it agree with Draper's statement in his article.
39. CSD, "Fifty Years of Flight Technology," 709.
40. CSD, "On Course to Modern Guidance," 56.
41. CSD, "On Course to Modern Guidance," 56; Thomas Hughes, *Elmer Sperry: Inventor and Engineer* (Baltimore, Md.: Johns Hopkins University Press, 1971), 213.
42. Hughes, *Elmer Sperry*, 167.
43. *U.S. Naval War College Information Service for Officers* 2, no. 10 (June 1950): 1.
44. *U.S. Naval War College Information Service for Officers*, 1. On Gillmor's health, see Hughes, *Elmer Sperry*, 213.
45. CSD Oral History, 132.
46. CSD Oral History, 134.
47. CSD, "On Course to Modern Guidance," 56.
48. H. M. Goodwin to C. S. Stark [at the Electro-Physical Laboratories], April 28, 1927, General Correspondence File 1927–1957, CSDP.
49. JSD, email to author, May 4, 2015.
50. "H. M. Crane Dead: Auto Pioneer, 81," *New York Times*, June 22, 1956.
51. Jim Donnelly, "Henry M. Crane," *Hemmings Classic Car*, June 2012, http://www.hemmings.com/hcc/stories/2012/06/01/hmn_feature12.html.
52. Robert D. Edwards, foreword to *Building Technology Transfer within Research Universities*, ed. Thomas J. Allen and Rory O'Shea (Cambridge, U.K.: Cambridge University Press, 2014), xvi.

Chapter 3. Back to MIT

1. "C. F. Taylor Sr., Retired MIT Professor Who Pioneered Development of Aircraft Engines, Dies at 101," *MIT News*, June 27, 1996, n.p., http://newsoffice.mit.edu/1996/taylor; "Taylor's Career Took Wing with WWI Plane Engine Testing," *MIT News*, September 14, 1994, n.p., http://newsoffice.mit.edu/1994/taylor-0914.

2. Barnes W. McCormick et al., *Engineering Education during the First Century of Flight* (Reston, Va.: American Institute of Aeronautics and Astronautics, 2004), 34.
3. "Taylor's Career Took Wing with WWI Plane Engine Testing."
4. CSD Oral History, 13, 15.
5. For a detailed explanation of detonation, see Allen W. Cline, "Engine Basics: Detonation and Pre-ignition," *Contact!*, no. 54 (January 2000): http://www.contactmagazine.com/Issue54/EngineBasics.html.
6. CSD Oral History, 36.
7. CSD Oral History, 28.
8. *President's Report* 65, no. 3 (1928–29): 33.
9. *President's Report* 65, no. 3 (1928–29): 89.
10. CSD Oral History, 30.
11. CSD Oral History, 64; Duffy, "Charles Stark Draper," 125.
12. CSD Oral History, 64; CSD, "On Course to Modern Guidance," 58.
13. Robert R. Osborn, "The Curtiss Robin," *Aviation*, May 21, 1928, available at http://www.airminded.net/curtrob/robinarticle.html.
14. Banks, "She Chose to Walk with Him," 13.
15. CSD Oral History, 37.
16. "The History of the MIT Department of Physics: New Labs and New Frontiers—1916–1939," *MIT Department of Physics*, http://web.mit.edu/physics/about/history/1916-1939.html.
17. Christophe Lécuyer, "Patrons and a Plan," in *Becoming MIT*, ed. David Kaiser (Cambridge, Mass.: MIT Press, 2010), 71; Philip M. Morse, *In at the Beginning: A Physicist's Life* (Cambridge, Mass.: MIT Press, 1977), 100–101.
18. Peter Galison and Bruce W. Hevly, *Big Science: The Growth of Large-Scale Research* (Stanford, Calif.: Stanford University Press, 1992), 159.
19. Quoted in "History of the MIT Department of Physics."
20. Morse, *In at the Beginning*, 101.
21. Lécuyer, "Patrons and a Plan," 71.
22. "History of the MIT Department of Physics."
23. *President's Report* 67, no. 3 (1930–31): 42.
24. CSD Oral History, 37.
25. CSD Oral History, 38.
26. Roslyn R. Romanowski, "Peacetime to Wartime: Transitions in Defense Policy at MIT," BS thesis, Massachusetts Institute of Technology, 1982, 76.
27. CSD Oral History, 27, and chronological listing and transcript.
28. CSD, "The Instrumentation Laboratory of the Massachusetts Institute of Technology: Remarks by Dr. C. S. Draper, Director of the Laboratory from Its Beginnings until the Present Time (1969)," 6, Profession File 1925–1974, CSDP.
29. See E. S. Taylor and C. S. Draper, "A New High-Speed Engine Indicator," *Mechanical Engineering* 55, no. 3 (March 1933): 169–71.
30. CSD, "A Method for Detecting Detonation Waves in the Internal Combustion Engine," MS thesis, Massachusetts Institute of Technology, 1928, 144–45.
31. CSD, "Method for Detecting Detonation Waves," 37.
32. CSD, "Method for Detecting Detonation Waves," 1. See also Edgerton notebook entries for April 29, May 28, and July 5, 1931, Harold E. Edgerton Digital Collection, MIT Museum.
33. CSD, "Method for Detecting Detonation Waves," 123.

34. See CSD and D. G. C. Luck, "A Fast and Economical Type of Photographic Oscillograph," *Review of Scientific Instruments* 5 (August 1933): 440–43.
35. CSD and Luck, "Fast and Economical Type of Photographic Oscillograph."
36. Michael Aaron Dennis, "A Change of State: The Political Cultures of Technical Practice at the MIT Instrumentation Laboratory and the Johns Hopkins University Applied Physics Laboratory, 1930–1945," PhD diss., Johns Hopkins University, 1990, 44.
37. *President's Report* 69, no. 3 (1932–33): 68.
38. Clayton Draper to "Mother and Dad," November 13, 1934, Family Papers, CSDP.
39. See CSD, "The Physical Effects of Detonation in a Cylindrical Chamber," Report No. 493, National Advisory Committee for Aeronautics, Washington, D.C., January 1, 1935.
40. Dennis, "Change of State," 52.
41. Dennis, "Change of State," 56.
42. William F. Trimble, *Jerome C. Hunsaker and the Rise of American Aeronautics* (Washington, D.C.: Smithsonian Institute Press, 2002), 136–37.
43. "The IAS: Early Years (1932 to 1945)," *American Institute of Aeronautics and Astronautics*, https://www.aiaa.org/SecondaryTwoColumNo.aspx?id=1521.
44. Trimble, *Jerome C. Hunsaker*, 145.
45. *President's Report* 69, no. 3 (1932–33): 69.
46. *President's Report* 70, no. 1 (1933–34): 69.
47. "The Coming Decade in Aeronautics and Astronautics at M.I.T.," February 1960, 12, Folder: Aeronautics and Astronautics, School of Engineering, Office of the Dean 1943–1989, AC-12, MIT Archives, Cambridge, Mass.
48. CSD, "The New Instrument Laboratory at the Massachusetts Institute of Technology," *Journal of the Aeronautical Sciences* 3, no. 4 (March 1936): 151–53.
49. Dennis, "Change of State," 124–25, 127.
50. Dennis, "Change of State," 127.
51. Duffy, "Charles Stark Draper," 126.
52. John C. Slater, in *President's Report* 71, no. 1 (October 1935): 102, and 72, no. 1 (October 1936): 132.
53. Slater, in *President's Report* 72, no. 1 (October 1936): 102.
54. Duffy, "Charles Stark Draper," 124.
55. Undergraduate courses:

    Academic Year
1922–23	6 courses during three 10-week semesters
Summer 1923	5 courses
1923–24	24 courses during three 10-week semesters
1924–25	23 courses during three 10-week semesters
Summer 1925	2 courses
1925–26	20 courses in two 15-week courses

    Source: JSD, List of C. S. Draper M.I.T. Credit, Courses and Grades 1922–1935, courtesy James Draper.

Chapter 4. Aircraft Instruments and the Beginnings of the Instrumentation Lab

1. Certificate of Graduation, Commercial Pilot's Course, Curtiss-Wright Flying Schools, Boston, Massachusetts, April, 13, 1930, CSDP.
2. Martha Draper, interview by Thomas Wildenberg, December 30, 2015.
3. Log of trip to California in OX-5 Robin NC9247, CSDP.
4. *Air Corps News* 13, no. 10, July 20, 1929, 241.

5. Log of trip to California.
6. CSD Oral History, 63.
7. CSD, "On Course to Modern Guidance," 58.
8. CSD and A. F. Spilhaus, "Power Supplies for Suction-Driven Gyroscopic Aircraft Instruments," *ASME Transactions* 56, no. 5 (May 5, 1934): 289.
9. Athelstan Spilhaus, interview by Ronald E. Doel, November 10, 1989.
10. Kaiser, *The Development of the Aerospace Industry on Long Island*, 53; Wilson, "Sperry Industrial Giant of the Air Age," 70.
11. Hughes, *Elmer Sperry*, 167.
12. Spilhaus Oral History, 3.
13. Spilhaus Oral History, 4.
14. Spilhaus quoting Elmer A. Sperry Jr., in Spilhaus Oral History.
15. Spilhaus Oral History, 2.
16. Spilhaus Oral History.
17. Howard W. Johnson, "Draper Chair Dedication," MIT Faculty Club, April 25, 1978, courtesy Debbie Douglas, MIT Museum, 5–6.
18. K. O. Lange, "On the Technique of Meteorological Airplane Ascents," in *The Meteorological Airplane Ascents of the Massachusetts Institute of Technology*, Papers in Physical Oceanography and Meteorology III, no. 2 (Cambridge Mass.: Massachusetts Institute of Technology and Woods Hole Oceanographic Institution, 1934), 5.
19. John Laconture, "Early Aviation on Nantucket," *Historic Nantucket* 49, no. 1 (Spring 1992): repr. *Nantucket Historical Association*, https://www.nha.org/library/hn/HN-fall92-aviation.htm.
20. "[I] designed and largely installed the instrument loaded in this ship." Handwritten notation on newspaper clipping titled "Flying Laboratory Pries into Weather Conditions over Hub," CSDP.
21. CSD, "Aircraft Instrruments in Meteorological Flying," in *The Meteorological Airplane Ascents of the Massachusetts Institute of Technology*, Papers in Physical Oceanography and Meteorology III, no. 2 (Cambridge, Mass.: Massachusetts Institute of Technology and Woods Hole Oceanographic Institution, 1934), 56.
22. U.S. Department of Transportation, Federal Aviation Administration, *Instrument Flying Handbook*, FAA-H-8083-15B, updated October 10, 2014 (Oklahoma City, Okla.: Testing Standards Branch, 2014), 5–12.
23. See for example, CSD, W. H. Cook, and Walter McKay, "Northerly Turning Error of the Magnetic Compass for Aircraft," *Journal of the Aeronautical Sciences* 5, no. 9 (1938): 345–454.
24. CSD and Walker McKay, "Magnetic Compass," U.S. Patent No. 2,248,748, filed January 21, 1938, issued July 8, 1941, 5.
25. Draper on the General Method for Determining Characteristics of All Aircraft Instruments, n.d.; Draper to Sperry, August 19, 1934, Scientific Papers, CSDP.
26. William G. Denhard, "The Start of the Laboratory: The Beginnings of the MIT Instrumentation Laboratory," *IEEE AES Systems Magazine* (October 1992): 8.
27. See Elmer Sperry Jr.'s handwritten note described in note 25.
28. Scheduled to take place at Columbia University at the end of January 1935. See CSD, "Dynamic Characteristics of Aircraft Instruments," *Journal of Aeronautical Sciences* 2 (March 1935): passim; also "Three to Be Honored for Aiding Aviation: Aeronautical Institute Will Make Awards on Wednesday at Annual Meeting Here," *New York Times*, January 28, 1935, online archives.

29. Clayton Draper letter to "Mother and Dad," November 13, 1934.
30. Clayton Draper letter to "Mother and Dad," November 13, 1934.
31. Henry Etzkowitz, *MIT and the Rise of Entrepreneurial Science* (London: Routledge, 2002). (Accessed unpaginated Google Books edition.)
32. Kimble D. McCutcheon, "No Short Days: The Struggle to Develop the R-200 'Double Wasp' Crankshaft," *AAHS Journal* 46, no. 2 (Summer 2001): 3. Available at http://www.enginehistory.org.
33. CSD and G. Bentley, "Measurement of Aircraft Vibrations during Flight," *Journal of the Aeronautical Sciences* 3, no. 4 (March 1936): 1; CSD Oral History, 13.
34. CSD, "On Course to Modern Guidance," 58.
35. CSD and Bentley, "Measurement of Aircraft Vibrations during Flight," passim.
36. Jerome C. Hunsaker, "Aeronautical Engineering," *President's Report* 73, no. 1 (October 1937): 87.
37. Their results were published a month later; see CSD and Bentley, "Measurement of Aircraft Vibrations during Flight."
38. AnnaLee Saxenian, *Regional Advantage: Culture and Competition in Silicon Valley and Route 128* (Cambridge, Mass.: Harvard University Press, 1994), 13; Etzkowitz, *MIT and the Rise of Entrepreneurial Science*, 43.
39. Etzkowitz, *MIT and the Rise of Entrepreneurial Science*, 45.
40. Lécuyer, "Patrons and a Plan," 74.
41. Dennis, "Change of State," 78.
42. Dennis, "Change of State," 94.
43. Hunsaker, "Aeronautical Engineering," *President's Report,* 70, no. 1 (October 1934), 69.
44. Hunsaker, "Aeronautical Engineering," *President's Report* 73, no. 1 (October 1937): 87.
45. CSD, "The Sonic Altimeter for Aircraft," Technical Note No. 611, National Advisory Committee for Aeronautics, Washington, D.C., 1937.
46. CSD, "Sonic Altimeter for Aircraft," 99.
47. E. S. Taylor et al., "A New Instrument Devised for the Study of Combustion," *SAE Transactions* (1934): 59–62.
48. Dennis, "Change of State," 53.
49. David A. Mindell, *Between Human and Machine: Feedback, Control, and Computing before Cybernetics* (Baltimore, Md.: Johns Hopkins University Press, 2002), 81.
50. Mindell, *Between Human and Machine,* 99–101. Joseph Lancor graduated from MIT with a BSEE in 1936.
51. Vannevar Bush to Bureau of Aeronautics, April 21, 1938, as quoted in Dennis, "Change of State," 103.
52. Dennis, "Change of State," 104.

Chapter 5. From Turn Indicator to Gunsight

1. CSD, "Innovation in Aircraft Instruments Assists Blind Flying," *Tech Engineering News* 17, December 1936, is listed in the *President's Report* of October 1937 (p. 203).
2. CSD, "On Course to Modern Guidance," 58–59. Personal correspondence reviewed by Draper's son James indicates that this trip took place between July 14 and August 14, 1938; JSD, email to author, July 21, 2015.
3. CSD, "On Course to Modern Guidance," 59.
4. CSD, "On Course to Modern Guidance."
5. Stark's 1-2-3 method: (1) turn, bank, and rate of climb indicators to maintain level flight; (2) altimeter for safe height; and (3) compass for direction.

6. Howard C. Stark, *Blind or Instrument Flying* (Newark, N.J., 1931), 4, 9, 14.
7. Dennis, "Change of State," 360; JSD, email to author, July 21, 2015.
8. Banks, "She Chose to Walk with Him," 13. The 1929 date recalled by Ivy Willard in Banks' article appears to be in error, in light of other chronological information.
9. Ivy Willard Draper, as quoted by Banks, "She Chose to Walk with Him," 12.
10. Banks, "She Chose to Walk with Him," 12.
11. Ivy Willard Draper, as quoted by Banks, "She Chose to Walk with Him," 13.
12. JSD, email to author, July 21, 2015.
13. CSD to George Brady, Chief Engineer, Curtiss Propeller Division, Curtiss-Wright Corp., September 14, 1938, Consulting File, CSDP-LC. Draper was Dent's advisor for the latter's 1938 MS thesis on aircraft vibration.
14. Brady to CSD August 23, 1938, Curtiss Consulting Folder, CSDP-LC.
15. Frederick B. Dent Jr., typewritten note to "Dear Stark"; Brady to Dent, August 23, 1938; both Curtiss Consulting Folder, CSDP-LC.
16. CSD, "Fifty Years of Flight Technology," 710–11. See also Karl Compton, memorandum of conversation with Prof. Draper, September 30, 1939, Folder: C. S. Draper 1939–1958, Office of the President, AC 4, MIT Archives, Cambridge, Mass. [hereafter Compton, memo of conversation].
17. Clayton to CSD, February 21, 1939, Waltham Watch Folder, CSDP-LC.
18. CSD Oral History, entry for March 9, 1939; for details see CSD to Dumaine, March 15, 1939, Waltham Watch Folder, CSDP.
19. Compton, memo of conversation.
20. Handwritten manuscript, Waltham Watch Folder, CSDP.
21. Draper's promotion to professor was listed in Compton's *President's Report* for 1939–40; see *President's Report* 75, no. 1 (October 1939): 20.
22. Compton, memo of conversation.
23. CSD, "Fifty Years of Flight Technology," 711. In "On Course to Modern Guidance," Draper gives the amount as a thousand dollars.
24. CSD, "On Course to Modern Guidance," 59.
25. CSD, "On Course to Modern Guidance."
26. CSD, "Fifty Years of Flight Technology," 711.
27. CSD, "On Course to Modern Guidance," 59.
28. Horacio Rivero Jr. and Lloyd M. Mustin, "Servo Mechanism for a Rate Follow-up System," 2, cited by Mindell, *Between Human and Machine*, 178.
29. CSD, "Turn-Indicator," U.S. Patent No. 2,291,612, was applied for on April 29, 1940.
30. Lt. N. P. Bentley, memorandum to Capt. E. E. Hermann, February 9, 1944, 2, Subj: Gunsight, Public Affairs Department, FLI–Gun Box, Sperry Gyroscope Collection, Manuscripts and Archives Department, Hagley Museum and Library, Wilmington, Del. [hereafter HAG-SP].
31. Bentley memorandum. See also Dennis, "Change of State," 363.
32. Sage to Compton, November 23, 1940, Folder: C. S. Draper, 1939–1958, Office of the President, AC 4, MIT Archives, Cambridge, Mass.
33. For Bentley's position, see John Burchard, *Q.E.D.: MIT in World War II* (Cambridge, Mass.: Technology, 1948), 151.
34. F. R. House, "The Navy Mark 14 Gunsight to July 1943," unpublished history, 1, Sperry Gyroscope Collection, Hagley Museum and Library, Wilmington, Del.
35. CSD et al., "Lead Angle Computer for Gun Sights," U.S. Patent No. 2,690,014, March 29, 1941.

36. Ivan A. Getting, *All in a Lifetime: Science in the Defense of Democracy* (New York: Vantage, 1989), 166.
37. Dennis, "Change of State," 363.
38. CSD, "On Course to Modern Guidance," 59–60. Draper recalled forty years later that the tests were conducted in the spring. Archival records and other documents, including Hacker's timeline, provide evidence that his memory failed him.
39. Although the exact timing of Fowler's visit is not known, a letter from Compton jointly addressed to Draper and Sage of December 10, 1940, is a "smoking gun." The letter can be found in Folder 4: C. S. Draper, 1939–1958, Office of the President, AC 4, MIT Archives, Cambridge, Mass.
40. David Zimmerman, *Top Secret Exchange: The Tizard Mission and the Scientific War* (Stroud, Gloucestershire, U.K.: Alan Sutton, 1996), 111.
41. Zimmerman, *Top Secret Exchange,* 124.
42. CSD, "On Course to Modern Guidance," 60.
43. CSD, "The Instrumentation Laboratory of the Massachusetts Institute of Technology Remarks by Dr. C. S. Draper, Director of the Laboratory from Its Beginnings until the Present time (1969)," 15, CSDP; CSD, "On Course to Modern Guidance," 60; Dennis, "Change of State," 369.
44. CSD, "The Instrumentation Laboratory," 15, CSDP. Barton Hacker asked about the Sperry units during his oral history interview; Draper replied that they "were not satisfactory."
45. CSD to Bassett, undated handwritten draft, Series III, HAG-SP.
46. Leighton I. Davis, "Interview Conducted by Maj. Lyn R. Officer and Hugh Ahmann, Burbank, California, April 26, 1973," 11, U.S. Air Force Oral History Program, IRIS No. 00904753, Air Force Historical Research Agency, Maxwell Air Force Base, Montgomery, Ala.; Bassett to CSD, February 7, 1941, Sperry Folder, CSDP-LC [hereafter Davis interview].
47. CSD, "On Course to Modern Guidance," 60. Sperry was then heavily involved in developing a fire-control director, then called a "predictor," for use with the Army's SCR-584 radar.

Chapter 6. War Work

1. "Horatio Rivero, Admiral USN (Retired)," official biography, Navy Department Library, Washington, D.C.; Re4 to Files, memorandum, May 1, 1944, in U.S. Navy, *Administrative History of World War II: Bureau of Ordnance,* vol. 79, *Fire Control (Except Radar and Aviation Ordnance),* 310–16, Special Collections, Navy Department Library, Washington, D.C., 1947 [hereafter Admin. Hist. of Fire Control].
2. Admin. Hist. of Fire Control, 310–16.
3. Inspector of Ordnance in Charge [David I. Hedrick] to Chief BuOrd, "Test of Sperry (Draper) Sight on Oerlikon Gun," August 15, 1941, S74-1(13), Entry 1002 (BuOrd G & CC Corr. 1915–1944), RG 74, National Archives, College Park, Md. [hereafter NA-CP]. Among those attending were future admirals W. A. Lee, E. E. Hermann, J. L. Halloway Jr., M. E. Murphy, E. M. Eller, and L. M. Mustin.
4. Sperry Gyroscope Co., Inc., O. B. Whitaker, interoffice memo from to P. R. Bassett, 25 July 1941, Folder No. 4: Gunsight Navy, Advertising Dept. Vertical File, HAG-SP.
5. Re4 to Files, 313.
6. Dennis, "Change of State," 376.

7. Robert Seamans, interview by Martin Collins and Michael Dennis, February 25, 1987, MIT, NASA Oral History Project, Tape 1, Side 1.
8. W. F. Raborn and John P. Craven, "The Significance of Draper's Work in the Development of Naval Weapons," in *Air Space and Instruments,* ed. Sidney Lees (New York: McGraw-Hill, 1963), 16.
9. Killian, telegram to Compton, October 31, 1941, Folder 4: Draper, C. S. 1939–1958, AC-4 MIT archives, Cambridge, Mass.
10. CSD, "The Instrumentation Laboratory," 16, CSDP.
11. Draper to Bassett, handwritten draft, n.d., HAG-SP.
12. P. D. Gallery to Commander McDowell, February 12, [1942], Subj. File: Gun Directors and Sights, CinCUS Fleet Readiness Div., RG 38, NA-CP.
13. Just 225 Mark 14 gunsights were scheduled for delivery to the fleet during the three-month period of July–September 1942. See Cominch to Chief of BuOrd, July 5, 1942, File S71, Entry 1002 (BuOrd G & CC Corr. 1915–1944), NA-CP.
14. U.S. Navy, *Fire Control,* 79:163–64.
15. Mindell, *Between Human and Machine,* 221.
16. Re4 to Files, 314; U.S. Navy, *Fire Control,* 79:164.
17. This figure was derived by adding the 12,800 units delivered through July 1943 to the 2,700 produced every month thereafter, as cited in House, "Navy Mark 14 Gunsight to July 1943," 3.
18. Killian to Draper, December 20, 1943, Folder 4: Draper, C. S. 1939–1958, MIT, AC 4.
19. CSD Oral History, 72.
20. Re4 to Files, 314.
21. Inspector of Ordnance in Charge [Hedrick] to Chief of BuOrd, "Tests of Mark 51 Gun Director," May 29, 1942, S71(2), Entry 1002 (BuOrd G & CC Corr. 1915–1944), RG 74, NA-CP.
22. Re4 to Files, 314.
23. Re4 to Files, 315.
24. "MIT Servomechanisms Laboratory," *MIT Libraries: MIT History,* https://libraries.mit.edu/mithistory/research/labs/mit-servomechanisms-laboratory/.
25. Fernando Elichirigoity, *Planet Management: Limits to Growth, Computer Simulation and the Emergence of Global Spaces* (Evanston, Ill.: Northwestern University Press, 1999), 41.
26. Dennis, "Change of State," 383.
27. Draper to Walter, Gunsight File, CSDL-HC, as cited by Dennis, "Change of State," 384.
28. Inspector of Ordnance in Charge to Chief of BuOrd, "Tests of Mark 51 Gun Director."
29. Re4 to Files, 315.
30. Inspector of Ordnance in Charge to Chief of BuOrd, "Tests of Mark 51 Gun Director," 3–4, 7.

Chapter 7. Directors and Gun Fire-Control Systems

1. Commanding Officer [hereafter CO] to Chief of BuOrd, December 29, 1942, Subj: Anti-Aircraft Guns, Comments on, Source Book on the History of Fire Control Radar, Office of Scientific Research and Development [hereafter OSRD], RG 74, NA-CP.
2. Robert R. Wallace, *From Dam Neck to Okinawa: A Memoir of Antiaircraft Training in World War II* (Department of the Navy, Naval Historical Center, 2001), 10.
3. U.S. Navy, Headquarters of the Commander in Chief, U.S. Fleet, *Gun Sight Mark 14, Gunner's Operating Bulletin No. 2*, 22, Bureau Ordnance Pamphlets, RG 74, NA-CP.

Notes to Pages 72–82   239

4. CO to Chief of BuOrd, December 29, 1942, 1.
5. CO to Chief of BuOrd, December 29, 1942, 10.
6. CO to Chief of BuOrd, December 29, 1942, 11.
7. CO to CinCUSPacFlt, "The Battle of Santa Cruz, October 26, 1942: Report Of," November 10, 1942, WWII Action and Operation Reports, RG 38, NA-CP.
8. Vannevar Bush, memorandum to Dr. Tuve and Dr. Hazen, January 11, 1943, Tuve File, Office Files of Harold Hazen, OSRD, RG 227, NA-CP [hereafter Tuve File].
9. Admin. Hist. of Fire Control, 181.
10. Seamans interview, February 25, 1987, Tape 1, Side 1; Draper Oral History, 71.
11. Seamans interview, February 25, 1987.
12. Hazen to Bush, February 24, 1943, Tuve File.
13. Edward J. Poitras, diary, March 5, 1943, Tuve File.
14. Stewart, memorandum to Gettings, April 5, 1943, Tuve File.
15. Dennis, "Change of State," 342.
16. Dennis, "Change of State."
17. Bush, memorandum to various members of Section T and Divisions 7 & 14, 13 April 1943, Tuve File.
18. Marion R. Kelly, "VT Fuse," draft manuscript, 9 July 1962, 70, copy in Navy Department Library, Washington, D.C. Admiral Kelly was to serve as head of the Administrative Section of the Bureau of Ordnance in the early 1950s.
19. Norman Friedman, *U.S. Naval Weapons* (Annapolis, Md.: Naval Institute Press, 1982), 86, 244.
20. Seamans interview, February 25, 1987.
21. Seamans never made clear when this occurred—it might have been during the testing of the Mark 52—and there is no clear evidence in the historical record as to when this visit to Dam Neck took place.
22. Robert C. Seamans Jr., "Past, Present, and Future," in *Air Space and Instruments*, ed. Sidney Lees (New York: McGraw-Hill, 1963), 12.
23. Duffy, "Charles Stark Draper," 140–41.
24. Duffy, "Charles Stark Draper," 140–41; Seamans, "Past, Present, and Future," 12.
25. Seamans interview, February 25, 1987.
26. CSD Oral History, 44.
27. CSD Oral History, 78-79; Massachusetts Institute of Technology, *Instrumentation Laboratory*, brochure, April 1968, Early Chronology of Instrumentation Laboratory Projects [hereafter *IL* Early Chronology].
28. CSD Oral History, 78-79; U.S. Navy, Bureau of Ordnance, *Gunar Systems Marks 1, 2, and 3 Mod 1 and Gun Fire Control System Mark 69 Mod 1 Operation*, OP 2076, Vol. 6, May 20, 1958, Bureau Ordnance Pamphlets, RG 74, NA-CP, 1.
29. CSDL-HC Finding Aid, Project Summary: Gunar Fire Control System.
30. U.S. Navy, *Gunar Systems*.
31. Draper statement in CSDL-HC Finding Aid.
32. Seamans interview, February 25, 1987.

Chapter 8. The A1-C(M) Gunsight

1. CSD, "Fifty Years of Flight Technology," 711; CSD, "The Instrumentation Laboratory," 18, CSDP; CSD, notebook entry, "Theory of Dive Bomb Sight Using Damped Gyroscopic Turn Indicators," March 25, 1943, Gyroscope Reports 1937–1943, CSDL-HC.

2. "Lieutenant General Leighton I. Davis," *U.S. Air Force,* www.af.mil/.
3. Davis interview, 6.
4. For Davis' introduction to the oscilloscope, see Thomas Wildenberg, "The A-1C(M) Gunsight: A Case Study of Technological Innovation in the United States Air Force," *Air Power History* 56, no. 2 (Summer 2009): 30–31.
5. Davis interview, 6.
6. Davis interview.
7. Davis interview, 8.
8. Leighton I. Davis, "Military Significance of Draper's Work for the Air Force," in *Air Space and Instruments,* ed. Sidney Lees (New York: McGraw-Hill, 1963), 6; Davis interview, 9; CSD, J. H. Lancor, and Leighton Davis, "The Applications of an Electromagnetic Indicator to Internal Combustion Engine Problems," *Journal of Aeronautical Science* 8, no. 1 (1940): 7–16.
9. Davis, "Military Significance of Draper's Work for the Air Force," 6.
10. Davis interview, 13. The transcript lists the airplane as an A-31, which is obviously an error, as the Vultee A-31 did not see service in World War II.
11. Davis interview, 13.
12. When this meeting took place is unclear, as Draper alternately lists it as occurring in the fall of 1942 and around Thanksgiving 1943. However, the dated entry in his notebook coincides well with the timeline for Davis' change of orders.
13. CSD, "The Instrumentation Laboratory," 18, CSDP.
14. Davis, "Military Significance of Draper's Work for the Air Force," 7.
15. Davis interview, 14.
16. Davis, "Military Significance of Draper's Work for the Air Force," 7.
17. Air Force Systems Command, Aeronautical Systems Division, *Development of Airborne Armament 1910–1961* [hereafter *Airborne Armament*], vol. 3, *Fighter Fire Control,* 387 (Wright Field, Ohio: Historians Office, Air Force Materiel Command), 387; CSDL-HC Finding Aid, s.v. A-1 and A-4 Gun, Bomb, and Rocket Sight.
18. CSD, "New Instrumentation Laboratory," 18.
19. Seamans interview, February 25, 1987, Tape 1, Side 1.
20. Seamans interview, February 25, 1987.
21. Air Technical Service Command, U.S. Army Air Forces, "A-1 Combination Gun-Bombing Sight," Classified Project No. MX-402, in *Research and Development Projects of the Engineering Division* [hereafter R&D Projects], 6th ed., January 1, 1945.
22. Davis, "Military Significance of Draper's Work for the Air Force," 7.
23. "Type A-1 Sight Installation," Classified Project No. MX-402, in R&D Projects, 7th ed., July 1, 1945.
24. *Airborne Armament,* 3:388.
25. *Airborne Armament,* 3:339, 394–95.
26. *Airborne Armament,* 3:397.
27. U.S. Air Force, "United States Air Force Operations in the Korean Conflict 1 July 1952–27 July 1953," 65, USAF Historical Study No. 127, Historical Division, Air University, 1956.
28. Frank P. Robinson Jr., telephone interview conducted by the author, October 10, 2008. Robinson, a Korean War F-86 pilot, served as the armament officer for the 336th Fighter Squadron.
29. Kenneth Werrell, *Sabres over MiG Alley: The F-86 and the Battle for Air Supremacy in Korea* (Annapolis, Md.: Naval Institute Press, 2013), 28.

30. The F-86E was equipped with the J-1 fire-control system, consisting of the A-1C(M) sight and the improved AN/APG-30A ranging radar.
31. CSD, "The Instrumentation Laboratory," 18, CSDP.
32. Robert Seamans, interviewed by Martin Collins, April 9, 1987, National Academy of Science, NASA Oral History Project, Tape 1, Side 1; Entry for Robert Channing Seamans, Jr., *The Guggenheim Medal Site*.
33. Massachusetts Institute of Technology, *Instrumentation Laboratory*, Early Chronology of Instrumentation Laboratory Projects, (unpaginated).
34. Seamans interview, February 25, 1987, Tape 1, Side 1.
35. Seamans interview, February 25, 1987, Tape 2, Side 1.
36. Seamans interview, February 25, 1987, Tape 1, Side 2.
37. Seamans interview, February 25, 1987, Tape 2, Side 1.
38. Seamans interview, April 9, 1987, Tape 1, Side 2.
39. "Chip Collins Remembers Doc Draper as Dynamic Leader of Lab," *D-notes*, October 29, 2001, 6.
40. Seamans interview, April 9, 1987, Tape 1, Side 2.
41. Collins interview, October 16, 2012, 10.
42. Collins interview, October 16, 2012, 11.
43. "Chip Collins Remembers," 6.
44. The Instrumentation Lab's Douglas A-26 is frequently confused with the B-26, built by Martin.

Chapter 9. The "Immaculate Interception" and Other Air-Defense Activities

1. Lincoln Laboratory, Massachusetts Institute of Technology, "Lincoln Laboratory Origins," *Lincoln Laboratory*, https://www.ll.mit.edu/about/History/origins.html [hereafter "Lincoln Laboratory Origins"], Part 1; Kent C. Redmond and Thomas M. Smith, *From Whirlwind to Mitre: The R&D Story of the SAGE Air Defense Computer* (Cambridge, Mass.: MIT Press, 2000), 22.
2. Redmond and Smith, *Whirlwind to Mitre*, 25; "Lincoln Laboratory Origins," Part 1.
3. "Lincoln Laboratory Origins"; Bruce S. Old, "Return on Investment in Basic Research: Exploring a Methodology," 21, DTIC AD A111283, Department of the Navy, Office of Naval Research, November 1981.
4. Redmond and Smith, *Whirlwind to Mitre*, 22.
5. Redmond and Smith, *Whirlwind to Mitre*, 26.
6. ADSEC, "Air Defense Systems," Report of October 24, 1950, 9–10, as quoted by Redmond and Smith, *Whirlwind to Mitre*, 27.
7. Redmond and Smith, *Whirlwind to Mitre*, 31.
8. Forrester, "49 JWF84," dictated February 1, 1950, MIT Computation Book No. 49, 84, as quoted by Redmond and Smith, *Whirlwind to Mitre*, 33.
9. "Lincoln Laboratory Origins," Part 1.
10. Redmond and Smith, *Whirlwind to Mitre*, 90.
11. Redmond and Smith, *Whirlwind to Mitre*, 91–92.
12. See note 44, chapter 8. It is not clear whether or not this was an A-26 or a B-26. The *President's Report* for 1948 states that the laboratory was provided with one B-25, one C-47, two A-26s, and one B-29.
13. Redmond and Smith, *Whirlwind to Mitre*, 1.
14. Wieser to Forrester, "Experimental Interceptions with Bedford MEW Radar, M-2092, April 23, 1951," 1, as cited by Redmond and Smith, *Whirlwind to Mitre*, 92.

15. Vandenberg to Killian, December 15, 1950, as cited by "Lincoln Laboratory Origins," Part 1.
16. "Lincoln Laboratory Origins," Part 2.
17. Matthew Farish, *The Contours of America's Cold War* (Minneapolis: University of Minnesota Press, 2010), 159.
18. Richard M. Freeland, *Academia's Golden Age: Universities in Massachusetts, 1945–1970* (New York: Oxford University Press, 1992), 141.
19. Freeland, *Academia's Golden Age.*
20. Killian, "Report of the President," *President's Report* 89, no. 2 (November 1953): 14.
21. Farish, *Contours of America's Cold War,* 159.
22. Lincoln Laboratory Division 6 Quarterly Progress Report, June 1, 1952, 6–11, as quoted by Redmond and Smith, *Whirlwind to Mitre,* 254.
23. Donald L. Clark, "Early Advances in Radar Technology for Aircraft Detection," *Lincoln Laboratory Journal* 12, no. 2 (2000): 167.
24. Redmond and Smith, *Whirlwind to Mitre,* 254–55.
25. This was one of the Douglas A-26s loaned to MIT in 1948 and later reclassified as a B-26 when the Martin B-26 was withdrawn from service; see U.S. Air Force Fact Sheet, "B-26 Bomber, Martin or Douglas?" This change in designation has caused a great deal of confusion with regard to the manufacturer of the aircraft used in the testing by the Flight Test Facility.
26. C. Robert Wieser, "From World War II Radar to Sage," *Computer Museum* 22 (1988): 16.
27. Wieser, "World War II Radar to Sage."
28. Jefferson M. Koonce, *Human Factors in the Training of Pilots* (New York: Taylor and Francis, 2002), 286.
29. Bernd Ulmann, *AN/FSQ-7: The Computer That Shaped the Cold War* (Berlin: De Gruyter Oldenbourg, 2014), 71; Lincoln Laboratory, "Cape Cod SAGE Prototype," Part 2.
30. Redmond and Smith, *Whirlwind to Mitre,* 257.
31. Ulmann, *AN/FSQ-7,* 72.
32. CSDL-HC Finding Aid, s.v. "Dummy Gun Fire Control."
33. CSDL-HC Finding Aid, s.v. "Dummy Gun Fire Control"; Jay Miller, *Convair B-58 Hustler: The World's First Supersonic Bomber* (Leicester, U.K.: Midland/Aerofax, 1997), 104–5.
34. "Convair B-58 Hustler: USA," *Aviation History Online Museum,* http://www.aviation-history.com/convair/b58.html.
35. CSDL-HC Finding Aid, s.v. Air-to-Air Missile Project.

Chapter 10. Inertial Navigation

1. Quoted in Francis E. Wylie, *MIT in Perspective: A Pictorial History of the Massachusetts Institute of Technology* (New York: Little, Brown, 1975), 169.
2. CSD Oral History, 79.
3. CSD, "The Instrumentation Laboratory," 19, CSDP.
4. JSD interview with author, November 15, 2012.
5. Davis suggested that some of the funds allocated to purchase bomb shackles might not have to be used. See CSD, "On the Evolution of Accurate Inertial Guidance Instruments," in *The Eagle Has Returned: Proceedings of the Dedication Conference of the International Space Hall of Fame, Held at Alamogordo, New Mexico, from 5 through 9 October 1976,* ed. Ernst A. Steinhoff, part 2, 24, Alexandria, Va.: American Astronautical Society, 1976.

6. CSD, "New Instrumentation Laboratory," 19. CSDP.
7. CSD, "Origins of Inertial Navigation," 455.
8. CSD, "The Instrumentation Laboratory," 19, CSDP.
9. Donald MacKenzie, "Stellar-Inertial Guidance: A Study in the Sociology of Military Technology," in *Science, Technology and the Military,* eds. Wiebe E. Bijker, Thomas Hughes, and T. J. Pinch (Cambridge, Mass.: MIT Press, 1989).
10. CSDL-HC Finding Aid, s.v. FEBE. Because this endeavor involved navigation, it was assigned not to the Armament Laboratory, where Davis was, but to another organization at Wright Field. See CSD, "On the Evolution of Accurate Inertial Guidance Instruments," 24.
11. Massachusetts Institute of Technology, *Instrumentation Laboratory,* Biographies Section, "Walter Wrigley," 5, Cambridge, Mass.
12. See Donald MacKenzie, *Inventing Accuracy: A Historical Sociology of Nuclear Missile Guidance* (Cambridge, Mass.: MIT Press, 1990), 66–67.
13. F. K. Richtmyer and E. H. Kennard, *Introduction to Modern Physics,* 3rd ed. (New York: McGraw-Hill, 1952), 51, as cited by MacKenzie, *Inventing Accuracy,* 67.
14. Walter Wrigley, "An Investigation of Methods Available for Indicating the Direction of the Vertical from a Moving Base," PhD diss., Massachusetts Institute of Technology, 1940.
15. Wrigley, "An Investigation of Methods Available," 86.
16. CSDL-HC Finding Aid, s.v. FEBE.
17. A. G. Bogosian, "An Experimental Automatic Long Range Guidance System, Project FEBE," preface of U.S. Air Force, Scientific Advisory Board, "Seminar on Automatic Celestial and Inertial Long Range Guidance Systems at Massachusetts Institute of Technology," Cambridge, Mass., February 1, 2, and 3, 1949 [hereafter MIT Guidance Seminar], 1:141.
18. CSD, "Evolution of Aerospace Guidance Technology," 232–33.
19. Donald MacKenzie, "Missile Accuracy: A Case Study in the Social Process of Technological Change," in *Social Construction of Technological Systems: New Directions in the Sociology and History of Technology,* ed. Wiebe E. Bijker, Thomas Hughes, and T. J. Pinch (Cambridge, Mass.: MIT Press, 1989), 201.
20. MacKenzie, *Inventing Accuracy,* 76.
21. MacKenzie, *Inventing Accuracy,* 70.
22. See MacKenzie, *Inventing Accuracy,* 71n121.
23. CSD, "Origins of Inertial Navigation," 455. These values appear to have been obtained from a then-classified analysis conducted by Draper and his colleagues at the Instrumentation Laboratory in 1947; see MacKenzie, *Inventing Accuracy,* 75n130.
24. See the biographical entries for these individuals in Massachusetts Institute of Technology, *Instrumentation Laboratory,* Biographies, 7, 16.
25. These programs included Northrop's Snark and North American Aviation's Navaho air-breathing missiles. Hughes Aircraft was also working on an Air Force contract to develop a celestial navigation and guidance system.
26. Eric Roston, *The Carbon Age: How Life's Core Element Has Become Civilization's Greatest Threat* (New York: Walker, 2008), characterizes George Gamow as "a hard drinking Soviet émigré to the United States with a brilliant mind, [and] an ebullient sense of humor" (17).
27. George Gamow to R. E. Gibson, dated "Black Friday" [February 13?] 1948, Cabinet 5, Drawer 4, File: Gamow, Applied Physics Laboratory Archives, as cited by Dennis, "Change of State," 416–18.

28. MacKenzie, *Inventing Accuracy*, 75n130.
29. Ronald E. Doel and Thomas Söderqvist, *The Historiography of Contemporary Science, Technology, and Medicine: Writing Recent Science* (London: Routledge, 2006), 176.
30. Dennis, "Change of State," 419–20.
31. MIT Guidance Seminar, 1:passim.
32. MIT Guidance Seminar, 1:2. The politics behind Gamow's memorandum, which are beyond the scope of this monograph, are summarized by Dennis, "Change of State," 420–21.
33. MacKenzie, *Inventing Accuracy*, 72.
34. MIT Guidance Seminar, foreword, 1:2.
35. MIT Guidance Seminar, 1:184.
36. CSD, "Origins of Inertial Navigation," 457.
37. CSDL-HC Finding Aid, s.v. FEBE.
38. CSD, "Origins of Inertial Navigation," 456.
39. "Laboratory Aviator Extraordinaire," 3.
40. Collins interview, October 16, 2012, 1.
41. Collins, as quoted in "Laboratory Aviator Extraordinaire," 3.
42. Collins interview, October 16, 2012, 2.
43. Collins interview, October 16, 2012.
44. Collins interview, October 16, 2012.
45. Collins interview, October 16, 2012, 4.
46. "Laboratory Aviator Extraordinaire," 3; "Chip Collins Remembers," 5; Collins interview, October 16, 2012, 4.
47. Collins interview, October 16, 2012, 5.
48. "Chip Collins Remembers," 5.
49. "Laboratory Aviator Extraordinaire," 3.
50. Concord Free Public Library, Finding Aid: Massport Hangar 24 Collection, 1948–2010.
51. Denhard, "Start of the Laboratory," 11.
52. *President's Report* 84, no. 1 (October 1948): 92.

Chapter 11. Floated Gyros and SPIRE

1. Duffy, "Charles Stark Draper," 137.
2. CSD, "Evolution of Aerospace Guidance Technology," 237.
3. "Hot Claims," *Time*, April 29, 1957, 63.
4. R. E. Hopkins, Fritz K. Mueller, and Walter Haeussermann, "The Pendulous Integrating Gyroscope Accelerometer (PIGA) from the V-2 to Trident D5, the Strategic Instrument of Choice." CSDL-P-3923, 4, paper presented at the AIAA Guidance, Navigation, and Control Conference & Exhibit, August 6–9, 2001, Montreal, Canada.
5. Claude L. Emmerich, "Advances in Gyro Performance," in *Air Space and Instruments*, ed. Sidney Lees (New York: McGraw-Hill, 1963), 462.
6. William G. Denhard, "Floating Single-Degree-of-Freedom Integrating Gyros," in *Air Space and Instruments*, ed. Sidney Lees (New York: McGraw-Hill 1963), 464.
7. Sidney Lees, "Investigation of the Performance of Experimental Shear Damper," master's thesis in aeronautical engineering, Massachusetts Institute of Technology, Cambridge, Mass., 1948.
8. Sidney Lees, "Integration by Viscous Sear Process," ScD thesis in aeronautical engineering, Massachusetts Institute of Technology, Cambridge, Mass., 1950.

9. Denhard, "Floating Single-Degree-of-Freedom Integrating Gyros," 466–67.
10. Denhard, "Floating Single-Degree-of-Freedom Integrating Gyros," 469.
11. Denhard, "Floating Single-Degree-of-Freedom Integrating Gyros," 475.
12. Duffy, "Charles Stark Draper," 131.
13. Ernst A. Steinhoff, "Charles Stark Draper Biography Part II," 13, in *Eagle Has Returned*; Steinhoff, "Comments on Dr. Stevens's Introduction," in *Eagle Has Returned*, 31; CSD Oral History, 113. According to J. D. Hunley (*Preludes to U.S. Space-Launch Vehicle Technology: Goddard Rockets to Minuteman III* [Gainesville: University Press of Florida, 2008, 82–83]), this system was developed too late to be used operationally.
14. Mackenzie, *Inventing Accuracy*, 80.
15. CSD, "Origins of Inertial Navigation," 457.
16. See CSD and Claude L. Emmerich, U.S. Patent No. 2,853,287, "Motion Sensor," filed March 13, 1952.
17. CSDL-HC Finding Aid, s.v. FEBE.
18. CSDL-HC Finding Aid, s.v. FEBE; CSD, "Origins of Inertial Navigation," 457.
19. CSD, "On Course to Modern Guidance," 62.
20. CSD, "On Course to Modern Guidance," 61; CSDL-HC Finding Aid, s.v. SPIRE.
21. MIT Instrumentation Laboratory, "R-63 Project SPIRE Flight Test Program," fig. 1B-1.
22. CSD, "Origins of Inertial Navigation," 457.
23. CSD, "Origins of Inertial Navigation," 458.
24. Instrumentation Laboratory, Massachusetts Institute of Technology, "Report R-193 The SPIRE, Jr. Long-Range Inertial Guidance System," 17.
25. CSD, "Origins of Inertial Navigation," 459; Massachusetts Institute of Technology, *Instrumentation Laboratory*, Biographies, 2nd opening page (unpaginated).
26. CSD, "Origins of Inertial Navigation," 460.
27. I have been able to identify neither the new sensors mentioned here nor the models of those used for SPIRE.
28. "*Conquest* Shows SPIRE Flight," *Niagara Falls Gazette*, April 13, 1958, T-3.
29. On the Air Force missile programs, see for example, Jacob Neufeld, *The Development of Ballistic Missiles in the United States Air Force 1945–1960* (Washington, D.C.: Department of the Air Force, Office of Air Force History, 1990), 146–77.
30. CSDL-HC Finding Aid, s.v. SPIRE Jr.
31. "*Conquest* Shows SPIRE Flight."
32. Mackenzie, "Missile Accuracy," 206.
33. Mackenzie, "Missile Accuracy," 206–7.
34. Mackenzie, *Inventing Accuracy*, 79.
35. John Slater and Walter Wrigley, "Flight Control System," U.S. Patent No. 2,649,264, filed March 15, 1947, issued August 18, 1953.
36. Mackenzie, "Missile Accuracy," 209.
37. Mackenzie, *Inventing Accuracy*, 85.
38. Mackenzie, *Inventing Accuracy*, 89.
39. Marshall William McMurran, *Achieving Accuracy: A Legacy of Computers and Missiles* (Bloomington, Ind.: Xlibris, 2009), 228.
40. Anthony Lawrence, *Modern Inertial Technology: Navigation, Guidance, and Control* (New York: Springer-Verlag, 1998), 92.
41. Massachusetts Institute of Technology, *Instrumentation Laboratory*, manpower history graph (unpaginated).
42. Seamans interview, February 25, 1987, Tape 1, Side 1.

Chapter 12. SINS

1. CSDL-HC Finding Aid, Project MAST Section.
2. CSD, "On the Evolution of Accurate Inertial Guidance Instruments," 24.
3. CSD, Walter Wrigley, and John Hovorka, *Inertial Guidance* (New York: Pergamon, 1960), 22.
4. CSD, Wrigley, and Hovorka, *Inertial Guidance*; Chief of Naval Research to Chief of Bureau of Ships, June 15, 1950, Bureau of Ships January thru June 1950, ONR Gen. Corr. 1950, RG 298, NA-CP.
5. Walter Wrigley, preface to "Report R-9 [Part 1], Theoretical Background of Inertial Navigation for Submarines," by Forrest E. Houston and John Hovorka, DTIC, March 1951, iii; Robert H. Cook and Boyd E. Gustafson, "Applications of Inertial Navigation to Undersea Warfare," MS thesis in mechanical engineering, Massachusetts Institute of Technology, Cambridge, Mass., May 18, 1951, iv.
6. Cook and Gustafson, "Applications of Inertial Navigation to Undersea Warfare." See also, Mackenzie, *Inventing Accuracy*, 415n45.
7. Mackenzie, *Inventing Accuracy*, 415.
8. Forrest E. Houston and John Hovorka, "Report R-9, Part II, Characteristics of Systems Feasible for Inertial Navigation of Submarines," DTIC, August 1951, 1.
9. Raborn and Craven, "Significance of Draper's Work in the Development of Naval Weapons," 25.
10. Inspector of Naval Material [Cambridge office] to Massachusetts Institute of Technology Instrumentation, Attn. Mr. F. E. Houston, "Requirements for Submarine Inertial Guidance System," November 6, 1951, Correspondence Reports, SINS, CSDL-HC.
11. Forrest E. Houston obituary, *Boston Globe*, September 22, 2010.
12. Massachusetts Institute of Technology, *Instrumentation Laboratory*, Biographies, F. E. Houston, 4.
13. Wrigley, preface to Houston and Hovorka, "Theoretical Background of Inertial Navigation for Submarines," iii.
14. CSDL-HC Finding Aid, s.v. SINS; CSD, "Origins of Inertial Navigation," 460.
15. Raborn and Craven, "Significance of Draper's Work in the Development of Naval Weapons," 25.
16. CSD, "Origins of Inertial Navigation," 460.
17. Raborn and Craven, "Significance of Draper's Work in the Development of Naval Weapons," 25.
18. Charles D. LaFond, "FBM Accuracy Starts with SINS," *Missiles and Rockets*, July 25, 1960, 24.
19. R. A. Fuhrman, "The Fleet Ballistic Missile System: Polaris to Trident," *Journal of Spacecraft* 15, no. 5 (September–October, 1978): 272.
20. Graham Spinardi, *From Polaris to Trident: The Development of U.S. Fleet Ballistic Missile Technology* (New York: Cambridge University Press, 1994), 47.
21. McMurran, *Achieving Accuracy*, 231.
22. Marvin May, "SINS Mk2 Mod 6," *Institute of Navigation Museum,* http://www.ion.org/museum/item_view.cfm?cid=2&scid=4&iid=4.
23. U.S. Congress, Senate Appropriations Committee. "Department of Defense Appropriations Fiscal Year 1974," 93rd Cong. 1st Sess., p. 618.
24. Naval History and Heritage Command: DANFS [Dictionary of American Naval Fighting Ships], s.v. *Compass Island,* https://www.history.navy.mil/.

25. LaFond, "FBM Accuracy Starts with SINS," 24.
26. Spinardi, *From Polaris to Trident*, 48n68.
27. B. McKelvie and H. Galt Jr., "The Evolution of the Ship's Inertial Navigation System for the Fleet Ballistic Missile Program," *Navigation: Journal of the Institute of Navigation* 25, no. 3 (Fall 1978): 315, available at onlinelibrary.wiley.com/.
28. Hunley, *Preludes to U.S. Space-Launch Vehicle Technology*, 304.
29. CSD, "Submarine Inertial Navigation: A Review and Some Predictions," paper presented to the Polaris Steering Task Group, October 22, 1959, CSD-107, courtesy CSDL, 1.
30. CSD, "Submarine Inertial Navigation," 47.
31. CSDL-HC Finding Aid, s.v. SINS.
32. CSD, "Origins of Inertial Navigation," 461.
33. McKelvie and Galt, "Evolution of the Ship's Inertial Navigation System," 315.

Chapter 13. Professor, Prodigious Worker, Family Man

1. Trimble, *Jerome C. Hunsaker*, 285.
2. CSD Oral History, 94.
3. CSD Oral History, 95.
4. Massachusetts Institute of Technology, *Policies and Procedures: A Guide for Faculty and Staff Members*, http://web.mit.edu/policies/, section 4.1, "Teaching and Research."
5. James C. Hunsaker, "Aeronautical Engineering," *President's Report* 86, no. 1 (October 1950): 100.
6. Robert G. Chilton Oral History Transcript, Johnson Space Center Oral History Project, National Aeronautics and Space Administration, 11.
7. Thomas R. Weschler, *The Reminiscences of Vice Admiral Thomas R. Weschler, U.S. Navy (Retired)*. Copy located in the Navy Department Library, Washington, D.C., 135.
8. Duffy, "Charles Stark Draper," 123.
9. CSD, "The Inertial Gyro: An Example of Basic Research," 10, presented at the National Convention of the Scientific Research Society of America, held jointly with the Annual Meeting of the American Association for the Advancement of Science, Chicago, Ill., December 29, 1959.
10. James R. Killian, *The Education of a College President* (Cambridge, Mass.: MIT Press, 1985), 36.
11. As recalled by Courtland Perkins, quoted by Killian, *Education of a College President*.
12. CSD Oral History, 93.
13. Robert Seamans, interviewed by Martin Collins, December 4, 1986, at Dr. Seamans' home in Cambridge, Massachusetts, NASA Oral History Project, Tape 1, Side 1.
14. Thomas K. Sherwood, *President's Report* 88, no. 1 (October 1952): 55.
15. CSD Oral History, 94.
16. James C. Hunsaker, "Aeronautical Engineering," *President's Report* 83, no. 1 (October 1947): 84.
17. Massachusetts Institute of Technology, *Instrumentation Laboratory*, manpower history graph, annual expenditures graph (unpaginated).
18. Ernst A. Steinhoff, ed., *The Eagle Has Returned: Proceedings of the Dedication Conference of the International Space Hall of Fame, Held at Alamogordo, New Mexico, from 5 through 9 October 1976*, Alexandria, Va.: American Astronautical Society, 1976, 14.
19. Denhard, "Start of the Laboratory," 7.
20. Duffy, "Charles Stark Draper," 123.

21. Seamans interview, February 25, 1987, Tape 1, Side 1.
22. Duffy, "Charles Stark Draper," 129.
23. Seamans interview, February 25, 1987, Tape 1, Side 1.
24. Robert C. Seamans Jr., *Aiming at Targets* (Washington, D.C.: NASA, 1996), 28.
25. Duffy, "Charles Stark Draper," 136.
26. Duffy, "Charles Stark Draper," 129.
27. Seamans interview, February 25, 1987, Tape 1, Side 2.
28. Seamans interview, February 25, 1987, Tape 1, Side 2.
29. Seamans interview, February 25, 1987, Tape 1, Side 2. See also William Leavitt, "A Triumph of Reverse McCarthyism: The Dethronement of Dr. Draper," *Air Force Space Digest* (December 1969): 49.
30. Nugent, bachelor of science in mechanical engineering, MIT '37, appears to have been an early student of Doc's who was employed by the IL.
31. Carl Machover, *Gyro Primer* (Commack, N.Y.: United Aircraft, 1958), 11. A similar anecdote appears in "Men of the Year: U.S. Scientists," the *Time* magazine cover story for its January 2, 1961, issue.
32. JSD interview.
33. Martha Stark Ditmeyer (née Draper), interview by Thomas Wildenberg, December 30, 2015, Alexandria, Va.
34. JSD interview.
35. JSD interview.
36. Collins interview, October 16, 2012, 2.
37. Martha Stark Ditmeyer interview.
38. JSD interview.
39. Martha Stark Ditmeyer interview.
40. JSD interview.
41. Martha Stark Ditmeyer interview.
42. JSD interview.
43. Collins interview, October 16, 2012, 2.
44. JSD interview.
45. JSD interview.
46. Martha Stark Ditmeyer interview.

Chapter 14. Inertial Guidance for Atlas and Thor

1. Neufeld, *Ballistic Missiles in the United States Air Force*, 74.
2. Neufeld, *Ballistic Missiles in the United States Air Force*, 64.
3. Neufeld, *Ballistic Missiles in the United States Air Force*, 74–75.
4. Eric Schlosser, *Command and Control: Nuclear Weapons, the Damascus Accident, and the Illusion of Safety* (New York: Penguin, 2013), 224.
5. Neufeld, *Ballistic Missiles in the United States Air Force*, 78.
6. Futrell, *Ideas, Concepts, Doctrine*, 1:489.
7. Herbert York, *Race to Oblivion: A Participants' View of the Arms Race* (New York: Simon & Schuster, 1970), 86.
8. York, *Race to Oblivion*.
9. York, *Race to Oblivion*, 89.
10. Philip J. Klass, "MIT Sparks Inertial Guidance Efforts," *Aviation Week,* August 10, 1959, 73.
11. Duffy, "Charles Stark Draper," 143.

12. CSDL-HC Finding Aid, Thor Section.
13. Neufeld, *Ballistic Missiles in the United States Air Force,* 108.
14. Instrumentation Laboratory, Massachusetts Institute of Technology, "Report R-299 Final Report for Contract AF 04(645)-9 WS-107A Inertial Guidance," January 1, 1955, through December 31, 1958 [hereafter IL MIT Report R-299], 2, CSDL Historical Collection, MIT Museum, Cambridge, Mass.
15. Richard H. Battin, "Space Guidance Evolutions: A Personal Narrative," *Journal of Guidance and Control* 5 (1982): 98.
16. Massachusetts Institute of Technology, *Instrumentation Laboratory,* Biographies, R. H. Battin, 6.
17. Massachusetts Institute of Technology, *Instrumentation Laboratory,* Biographies, J. H. Laning, 6. See also J. Halcombe Laning Jr. "Mathematical Theory of Lubrication-Type Flow," PhD diss., Massachusetts Institute of Technology, Cambridge, Mass., 1947.
18. Battin, "Space Guidance Evolutions," 98.
19. See Battin, "Space Guidance Evolutions," figure 1.
20. Battin, "Space Guidance Evolutions," 99.
21. Battin, "Space Guidance Evolutions," 100.
22. Lester R. Grothe et al., "The MIT 25 IRIG Inertial Reference Integrating Gyro Unit and the MIT 25 PIG Pendulous Integrating Gyro Unit," Report R-141, Instrumentation Laboratory, Massachusetts Institute of Technology, January 1958 (courtesy of the Charles Stark Draper Laboratory, Cambridge, Mass.).
23. Grothe, "MIT 25 IRIG . . . MIT 25 PIG," 4. Engineers assisting Grothe included Edward J. Hall, Michele S. Sapuppa, and Andrew E. Scoville.
24. Neufeld, *Ballistic Missiles in the United States Air Force,* 160.
25. CSDL-HC Finding Aid, Thor Section.
26. CSDL-HC Finding Aid, Thor Section; James R. Burnett, Oral History, January 19, 1989, Glennan-Webb-Seamans Project for Research in Space History, National Air and Space Museum, Smithsonian Institution, Washington, D.C.
27. Rueben Mettler, interview by Martin Collins, June 26, 1990, Los Angeles, Calif. NASM Archives, Tape 1, Side 1.
28. George Campbell, "Recollections," available at http//acdelco.org/Documents/Delco%20History%20Recollections.%20George%20Campbell, accessed February 4, 2009.
29. Hunley, *Preludes to U.S. Space-Launch Vehicle Technology,* 234–35.
30. William J. Normyle, "Milwaukeeans Guide a Missile to Perfection," *Milwaukee Journal,* November 4, 1958, 21; Hunley, *Preludes to U.S. Space-Launch Vehicle Technology,* 235.
31. Everett G. Martin, "Thor 'Gun Barrels' Cleaned," *Boston Monitor,* January 18, 1958 (news clipping, CSDL Collection, MIT Museum, Cambridge, Mass.).
32. Martin, "Thor 'Gun Barrels' Cleaned."
33. Martin, "Thor 'Gun Barrels' Cleaned."
34. Hunley, *Preludes to U.S. Space-Launch Vehicle Technology,* 235.
35. Hunley, *Preludes to U.S. Space-Launch Vehicle Technology,* 236–37.

Chapter 15. Titan, FLIMBAL, AIRS, and the MX/Peacekeeper

1. Warren E. Greene, "The Development of the SM-68 Titan," Historical Office, Deputy Commander for Aerospace Systems, Air Force Systems Command, August 1962, 22.
2. Hunley, *Preludes to U.S. Space-Launch Vehicle Technology,* 263.
3. Mackenzie, *Inventing Accuracy,* 119n79.

4. Greene, "Development of the SM-68 Titan," 24.
5. See Massachusetts Institute of Technology, *Instrumentation Laboratory*, Report R-299.
6. National Academy of Sciences (U.S.), Committee on Utilization of Scientific and Engineering Manpower, *Toward Better Utilization of Scientific Engineering Talent: A Program for Action Report* (Washington, D.C.: National Research Council, 1964), 133.
7. Greene, "Development of the SM-68 Titan," 24; MIT press release, May 13, 1959, Titan folder, CSDL-HC, MIT-MC.
8. Massachusetts Institute of Technology, *Instrumentation Laboratory*, "Report R-299," 4.
9. David K. Stumpf, *Titan II: A History of a Cold War Missile Program* (Fayetteville: University of Arkansas Press, 2000), 63.
10. Massachusetts Institute of Technology, *Instrumentation Laboratory*, "Report R-299," 1–2.
11. CSDL-HC Finding Aid, FLIMBAL Section; Morgan, *Draper at 25*, 12. See also Mackenzie, *Inventing Accuracy*, 218.
12. Massachusetts Institute of Technology, *Instrumentation Laboratory*, Biographies, P. N. Bowditch, 7.
13. Massachusetts Institute of Technology, *Instrumentation Laboratory*, Biographies, K. Fertig, 8.
14. From 1957 to 1963, see CSDL-HC Finding Aid.
15. Mackenzie, *Inventing Accuracy*, 222.
16. CSDL-HC Finding Aid, Sabre Section.
17. For details of the project funding, see Mackenzie, *Inventing Accuracy*, 222.
18. J. M. Wuerth, "Minuteman Guidance and Control," *Navigation: Journal of the Institute of Navigation* 23, no. 1 (Spring 1976): 69; Burnett Oral History.
19. Mackenzie, *Inventing Accuracy*, 224; McMurran, *Achieving Accuracy*, 156.
20. Mackenzie, *Inventing Accuracy*, 226.
21. "Advanced Inertial Navigation—The Airs FLIMBAL," www.deanspacedrive.org, accessed September 20, 2010. See also John H. Cushman Jr., "Northrop's Struggle with the MX," *New York Times*, November 22, 1987, online archives.
22. Cushman, "Northrop's Struggle with the MX."
23. *Air Force Missileers* (Paducah, Ky.: Turner, 1998), 31.

Chapter 16. Polaris

1. Fuhrman, "Fleet Ballistic Missile System," 266.
2. Paul C. Dow Jr., "History of the Trident Guidance Program at Draper Laboratory," CSDL-C-6455, September 1997, 3, Charles Stark Draper Laboratory, Cambridge, Mass.
3. Mackenzie, *Inventing Accuracy*, 146–47; Spinardi, *From Polaris to Trident*, 42.
4. "Sixtieth Graduation Anniversary of the Class of 1940," *United States Naval Academy*, https://www.usna.com/NC/History/ClassOf1940/Contents.htm/; Samuel A. Forter, "A Light-Weight Multi-Channel Telemetering System," master of science in electrical engineering thesis, MIT, 1947.
5. Willis M. Hawkins, "Levering Smith 1910–1999," *Memorial Tributes: National Academy of Engineering* 7 (1994): 257.
6. Ignatius J. Galantin, *Submarine Admiral: From Battlewagons to Missiles* (Urbana: University of Illinois Press, 1995), 231.
7. Mackenzie, *Inventing Accuracy*, 146.
8. Spinardi, *From Polaris to Trident*, 43n44.
9. Spinardi, *From Polaris to Trident*, 42.
10. Spinardi, *From Polaris to Trident*, 43n46.

11. The Polaris program officially began on December 8, 1956, with a directive from Secretary of Defense Charles E. Wilson.
12. Raborn and Craven, "Significance of Draper's Work in the Development of Naval Weapons," 27.
13. Rear Adm. Robert Wertheim, USN (Ret.), audio interview by staff October 3, 2012, Charles Stark Draper Laboratory, Cambridge, Mass. [Courtesy Draper Laboratory]
14. Duffy, "Charles Stark Draper," 129.
15. Wertheim interview.
16. Wertheim interview.
17. Hunley, *Preludes to U.S. Space-Launch Vehicle Technology*, 235.
18. Eldon C. Hall, *Journey to the Moon: The History of Apollo Guidance Computer* (Reston, Va.: American Institute of Aeronautics and Astronautics, 1996), 38.
19. Hall, *Journey to the Moon*.
20. Hunley, *Preludes to U.S. Space-Launch Vehicle Technology*, 237.
21. Eldon C. Hall, "The Apollo Guidance Computer: A Designer's View," in *Computer Museum Report* (Boston: Computer Museum, 1982), 2.
22. Eldon C. Hall, "From the Farm to Pioneering with Digital Control Computers: An Autobiography," *IEEE Annals of the History of Computing* 22 (April–June 2000): 27; Hall, *Journey to the Moon*, 43.
23. Jack Ward, "Texas Instruments R212," Transistor Museum Historic Photo Gallery, www.semiconductormuseum.com/PhotoGallery_TI_R212.htm.
24. Hall, "From the Farm to Pioneering with Digital Control Computers," 27.
25. Charles Wright, "Polaris-Poseidon-Trident: An Awesome Concept—Sea-Launched Ballistic Missile Systems; General Electric Ordnance System–United States Navy," revised September 10, 2009, *Polaris: The Free World's Most Effective Deterrent*, 2, www.navsource.org/archives/08/PolarisHistoryGEX1.htm.
26. Wright, "Polaris-Poseidon-Trident," 1.
27. Wright, "Polaris-Poseidon-Trident," 2.
28. Robert V. Gates, "Strategic Systems Fire Control," Naval Surface Warfare Center, Dahlgren Division *Technical Digest* (1995): 168.
29. Gates, "Strategic Systems Fire Control"; Raymond H. Hughey Jr., "History of Mathematics and Computing Technology at the Dahlgren Laboratory," Naval Surface Warfare Center, Dahlgren Division *Technical Digest* (1995): 17.
30. James Rife and Rodney P. Carlisle, *The Sound of Freedom: Naval Weapons Technology at Dahlgren, Virginia, 1918–2006* (Dahlgren, Va.: Naval Surface Warfare Center, Dahlgren Division, 2007), 105.
31. Lockheed Missiles and Space Company, *The Fleet Ballistic Missile System: From Polaris to Poseidon and Trident*, 20.
32. Spinardi, *From Polaris to Trident*, 66.
33. Hunley, *Preludes to U.S. Space-Launch Vehicle Technology*, 308–9.
34. Spinardi, *From Polaris to Trident*, 70.
35. Spinardi, *From Polaris to Trident*.

Chapter 17. Poseidon and Trident

1. "Men of the Year: U.S. Scientists," *Time* 77, no. 1 (January 2, 1961), 40–46.
2. Dow, "History of the Trident Guidance Program at Draper Laboratory," 8.
3. Dow, "History of the Trident Guidance Program at Draper Laboratory."
4. Lockheed, *Fleet Ballistic Missile System*, 25.

5. Spinardi, *From Polaris to Trident*, 88.
6. Dow, "History of the Trident Guidance Program at Draper Laboratory," 8.
7. Mackenzie, *Inventing Accuracy*, 263; U.S. Navy, Strategic Systems Programs, *Facts/Chronology Polaris Poseidon Trident* (Washington, D.C., 2000), 39.
8. Mackenzie, *Inventing Accuracy*, states (p. 262n23) that it was "to investigate the propriety of the Instrumentation laboratory, rather than a private firm." What he meant by "proprietary" is not clear.
9. Sam Forter, interview by Graham Spinardi, May 15, 1987, as cited in Spinardi, *From Polaris to Trident*, 94.
10. Dow, "History of the Trident Guidance Program at Draper Laboratory," 9.
11. Gates, "Strategic Systems Fire Control," 170.
12. U.S. Navy, Naval Surface Weapons Center, "The SLBM Program: Brief History of NSWC's Involvement in the FBM/SLBM Fire Control Systems, 1956–1984," Dahlgren, Va., November 1984, 12.
13. Gates, "Strategic Systems Fire Control," 171.
14. Mackenzie, *Inventing Accuracy*, 263; Dow, "History of the Trident Guidance Program at Draper Laboratory," 7–8.
15. Spinardi, *From Polaris to Trident*, 91.
16. Rear Adm. Robert Wertheim, interview by Donald Mackenzie, as cited by Spinardi, *From Polaris to Trident*, 95; see also 222n58.
17. Levering Smith, as quoted by Spinardi, *From Polaris to Trident*, 97; see also 222n62.
18. Spinardi, *From Polaris to Trident*, 97.
19. Mackenzie, *Inventing Accuracy*, 266.
20. Mackenzie, *Inventing Accuracy*.
21. Forter, as quoted by Spinardi, *From Polaris to Trident*, 98; see also 222n67. Although Olson is not mentioned by name, he was head of the technical design team responsible for Polaris and Poseidon guidance; see Massachusetts Institute of Technology, *Instrumentation Laboratory*, Biographies, O. B. Olson, 20.
22. Spinardi, *From Polaris to Trident*.
23. For the quotation, Spinardi, *From Polaris to Trident*, 99.
24. Alton Frye, *A Responsible Congress: The Politics of National Security* (New York: McGraw-Hill, 1975), 75. See also Mackenzie, *Inventing Accuracy*, 269, and Spinardi, *From Polaris to Trident*, 100.
25. Spinardi, *From Polaris to Trident*, 100.
26. See Institute for Defense Analyses, *The Strat-X Report*, as cited by Spinardi, *From Polaris to Trident*, 113.
27. Levering Smith, Robert H. Wertheim, and Robert A. Duffy, "Innovative Engineering in the Trident Missile Development," *Bridge* 10, no. 2 (Summer 1980): 11.
28. Spinardi, *From Polaris to Trident*, 113.
29. Smith, Wertheim, and Duffy, "Innovative Engineering in the Trident Missile Development," 11–12. See Spinardi, *From Polaris to Trident*, 113–14, for details on the political process that determined these characteristics.
30. Gates, "Strategic Systems Fire Control," 172; Mackenzie, *Inventing Accuracy*, 273.
31. Dow, "History of the Trident Guidance Program at Draper Laboratory," 12.
32. John Brett, as quoted by Mackenzie, *Inventing Accuracy*, 274.
33. Spinardi, *From Polaris to Trident*, 132. For a schematic of a dry tuned-rotor gyroscope, see Mackenzie, *Inventing Accuracy*, 181.
34. Spinardi, *From Polaris to Trident*, 133.

35. Robert Duffy, as cited by Spinardi, *From Polaris to Trident*, 132, 230n99.
36. Mackenzie, *Inventing Accuracy*, 276.
37. Dow, "History of the Trident Guidance Program at Draper Laboratory," 13; Spinardi, *From Polaris to Trident*, 133.
38. See Gates, "Strategic Systems Fire Control," 172–73.
39. Spinardi, *From Polaris to Trident*, 133.
40. Dow, "History of the Trident Guidance Program at Draper Laboratory," 13.
41. Smith, Wertheim, and Duffy, "Innovative Engineering in the Trident Missile Development," 13.
42. Dow, "History of the Trident Guidance Program at Draper Laboratory," 14.
43. U.S. Navy, Strategic Systems Programs, *Facts/Chronology Polaris, Poseidon, Trident*, 73.
44. Spinardi, *From Polaris to Trident*, 142.
45. Dow, "History of the Trident Guidance Program at Draper Laboratory," 17.
46. For a discussion of U.S. nuclear doctrine see Spinardi, *From Polaris to Trident*, 143–47; and Dow, "History of the Trident Guidance Program at Draper Laboratory," 17–19.
47. Dow, "History of the Trident Guidance Program at Draper Laboratory," 21.
48. U.S. Congress, Government Accountability Office [hereafter GAO], "Kearfott Guidance & Navigation Corporation," B-292895.2, May 25, 2004, Washington, D.C., 2.
49. Dow, "History of the Trident Guidance Program at Draper Laboratory," 22.
50. Dow, "History of the Trident Guidance Program at Draper Laboratory," 23–24.
51. Various sources list the D5's accuracy as ninety meters (three hundred feet); George Koutosumpos, "Submarine Launched ICBM Trident D5 Conventional Trident Modification," *Journal of Computations & Modeling* 4, no. 1 (2014): 245–65, states that flight tests have demonstrated significantly better results.
52. GAO, "Kearfott Guidance & Navigation Corporation," 2.
53. "Draper Draws in $101M Navy Contract Addition," *Boston Business Journal*, December 12, 2005; "$101.M to Develop the MK6 LE: So, What's That?," *Defense Industry Daily*, December 15, 2005, http://www.defenseindustrydaily.com/1011m-to-develop-the-mk6-le-so-whats-that-01627/.
54. "$101.M to Develop the MK6 LE."
55. "$243M to Draper Laboratory for Trident II D5 Guidance System Support," *Defense Industry Daily*, December 6, 2009, http://www.defenseindustrydaily.com/243M-to-Draper-Laboratory-for-Trident-II-D5-Guidance-System-Support-06004/.
56. Todd Jackson, "Modifying the MARK 6 Guidance System Part 1," *Evaluation Engineering* (September 2009), 4, http://www.evaluationengineering.com/articles/200909/modifying-the-mark-6-guidance-system-part-1.
57. Charles Stark Draper Laboratory, *Annual Report* (Cambridge, Mass., 2009).
58. Charles Stark Draper Laboratory, *Annual Report* (Cambridge, Mass., 2007).
59. U.S. Navy, Strategic Systems Programs, Public Affairs, "Back to the Future with Trident Life Extension," *Undersea Warfare* (Spring 2012): 10.

Chapter 18. Spy Satellites and Space Planes

1. John C. Herther and James S. Coolbaugh, "Genesis of Three-Axis Spacecraft Guidance, Control, and On-Orbit Stabilization," AIAA *Journal of Guidance, Control, and Dynamics* 29, no. 6 (November–December 2006): 1251.
2. Aeronautical Engineering Department, Weapons System Section, Report Prepared for Dean Soderberg, September 1955, 1, MIT, School of Engineering, Office of the Dean, Records, 1943–1989, AC 12, MIT Archives, Cambridge, Mass.

3. "Academic Program in Aeronautics and Astronautics (Preliminary)," MIT, School of Engineering, Office of the Dean, Records, 1943–1989, Box 1, AC 12, MIT Archives, Cambridge, Mass.
4. Alex Abella, *Soldiers of Reason: The Rand Corporation and the Rise of the American Empire* (Orlando, Fla.: Harcourt, 2008), 153; J. E. Lipp and R. M Slater, eds., "Project Feed Back Summary Report," vol. 1, March 1, 1954, RAND Corporation, Santa Monica, Calif.
5. Herther and Coolbaugh, "Genesis of Three-Axis Spacecraft Guidance," 1251.
6. Herther, John C., and James S. Coolbaugh. "Genesis of Three-Axis Spacecraft Guidance, Control, and On-Orbit Stabilization." AIAA *Journal of Guidance, Control, and Dynamics* Vol. 29, No. 6, November–December 2006.
7. William O. Covington, "Orientation Control Study," MA thesis, Massachusetts Institute of Technology, Cambridge, Mass., 1955.
8. Dwayne A. Day, John M. Logsdon, and Brian Latell, eds., *Eye in the Sky: The Story of the CORONA Spy Satellites* (Washington, D.C., Smithsonian, 1998), 32.
9. Day, Logsdon, and Latell, *Eye in the Sky,* 146.
10. Herther and Coolbaugh, "Genesis of Three-Axis Spacecraft Guidance," 1253.
11. Dwayne Day, "Square Peg in a Cone-Shaped Hole: The Samos E-5 Recoverable Satellite (Part 2)," *Spaceflight* (February 2003), 1, www.thespaceview.com/article/1419/1.
12. Herther and Coolbaugh, "Genesis of Three-Axis Spacecraft Guidance," 1258.
13. Herther and Coolbaugh, "Genesis of Three-Axis Spacecraft Guidance," 1258–59.
14. In the reference to disturbing torques in the paragraph above, emphasis has been supplied. For a detailed description of the events shaping Herther's request and discharge from the Air Force, see Herther and Coolbaugh, "Genesis of Three-Axis Spacecraft Guidance," 1259.
15. Day, "Square Peg in a Cone-Shaped Hole," 1.
16. Herther and Coolbaugh, "Genesis of Three-Axis Spacecraft Guidance," 1261.
17. Herther and Coolbaugh, "Genesis of Three-Axis Spacecraft Guidance."
18. Day, "Square Peg in a Cone-Shaped Hole," 2.
19. "AutoMATE LLC Founder's Inertial Control Background," www.automarine.com/wp-content/p-loads/Genesis-of-Inertial-Rudder-Controllers.pdf.
20. Herther and Coolbaugh, "Genesis of Three-Axis Spacecraft Guidance," 1266.
21. Neil Sheehan, *A Fiery Peace in a Cold War: Bernard Schriever and the Ultimate Weapon* (New York: Vintage, 2009).
22. "Men of the Year: U.S. Scientists," 43.
23. For the pilots role in project Mercury, see David A. Mindell, *Digital Apollo: Human and Machine in Spaceflight* (Cambridge, Mass.: MIT Press, 2008), 73–74.
24. Mindell, *Digital Apollo.*
25. Clarence J. Geiger, "The Strangled Infant: The Boeing X-20A Dyna-Soar," in *The Hypersonic Revolution,* ed. Richard Hallion (Wright-Patterson Air Force Base, Ohio: ASD Special Staff Office, 1987), II-x, http://www.wpafb.af.mil/shared/media/document/AFD-080408-031.pdf.
26. Clarence J. Geiger, "History of the X-20A Dyna-Soar," vol. I, AFSC Historical Publications Series 63-50-I, September 1963, 28.
27. Geiger, "History of the X-20A Dyna-Soar," 49.
28. Geiger, "Strangled Infant," 292.
29. CSDL-HC Finding Aid, s.v. Pace I.
30. CSDL-HC Finding Aid, s.v. Pace II.

Chapter 19. To the Moon and Beyond

1. Richard H. Battin, interview by Rebecca Wright, Lexington, Mass., April 18, 2000, NASA Johnson Space Center Oral History Project, http://www.jsc.nasa.gov/history/oral_histories/BattinRH/BattinRH_4-18-00.htm.
2. Battin, "Space Guidance Evolutions," 103.
3. Battin Oral History. See also Battin, "Space Guidance Evolutions," 103.
4. Richard H. Battin, "A Funny Thing Happened on the Way to the Moon," transcript, Fall 2008 lecture, Astrodynamics Course 16.346, MIT Open Course Ware, 2, http://ocw.mit.edu/courses/aeronautics-and-astronautics/16-346-astrodynamics-fall-2008/video-lecture/MIT16-346F08.pdf.
5. Battin Oral History.
6. Mindell, *Digital Apollo*, 99–100.
7. Battin, "Space Guidance Evolutions," 104; Battin Oral History.
8. Battin Oral History.
9. Battin, "Space Guidance Evolutions," 105.
10. Hugh Blair-Smith, "Hugh Blair-Smith's Introduction," Apollo Guidance Computer History Project, Third Conference, November 30, 2001, http://authors.library.caltech.edu/5456/1/hrst.mit.edu/hrs/apollo/public/conference3/blairsmith.htm.
11. David G. Hoag, "The History of Apollo On-Board Guidance, Navigation, and Control," Cambridge, Mass.: Charles Stark Draper Laboratory, September 1976, 2 (courtesy Charles Stark Draper Laboratory). Also published in *The Eagle Has Returned: Proceedings of the Dedication Conference of the International Space Hall of Fame, Held at Alamogordo, New Mexico, from 5 through 9 October 1976*, ed. Ernst A. Steinhoff, part 1 (Alexandria, Va.: American Astronautical Society, 1976).
12. "CSM Guidance Development Diary," *Encyclopedia Astronautica*, s.v. November 22, 1960, http://www.astronautix.com/craft/csmdance.htm.
13. Hoag, "History of Apollo On-Board Guidance, Navigation, and Control," 3–4.
14. "CSM Guidance Development Diary," s.v. February 7, 1961.
15. See Mindell, *Digital Apollo*, 104n27.
16. Mindell, *Digital Apollo*, 104. See also Robert G. Chilton and Alfred R. Seville, "Distance Measurement by *Inertial* Means," MS thesis, Dept. of Aeronautical Engineering, Massachusetts Institute of Technology, Cambridge, Mass., 1949.
17. Mindell, *Digital Apollo*, 104.
18. President John F. Kennedy, Address to Congress on Urgent National Needs, May 25, 1961, John F. Kennedy Presidential Museum and Library, http://www.jfklibrary.org/JFK/Media-Gallery.aspx.
19. Hoag, "History of Apollo On-Board Guidance, Navigation, and Control," 4.
20. By 1965, funding for Apollo, in excess of $19 million, was twice as large as for the next-largest project, SABRE. See Massachusetts Institute of Technology, *Instrumentation Laboratory*, Major Program Funding 1955 thru 1970, unpaginated.
21. "Astronavigation: The First Apollo Contract," *NASA History Program Office*, http://www.hq.nasa.gov/pao/History/SP-4205/ch2-4.html.
22. Massachusetts Institute of Technology, *Instrumentation Laboratory*, Biographies, D. G. Hoag, 11.
23. Mindell, *Digital Apollo*, 105.
24. Hoag, "History of Apollo On-Board Guidance, Navigation, and Control," 4; "CSM Guidance Development Diary," s.v. July 1961.
25. Robert C. Seamans Jr., NASA, http://history.nasa.gov/Biographies/seamans.html.

26. Mindell, *Digital Apollo,* 108.
27. Mindell, *Digital Apollo*; "Astronavigation."
28. Hoag, "History of Apollo On-Board Guidance, Navigation, and Control," 5.
29. John Tylko, "MIT and Navigating the Path to the Moon," *MIT AeroAstro* (2008–2009), available at http://web.mit.edu/aeroastro/news/magazine/aeroastro6/mit-apollo.html. Of the many versions of this anecdote and of what Doc said, Tylko's version is the most plausible. Unfortunately, he failed to cite his source.
30. Hugh Blair-Smith, as quoted by Elwin C. Ong, "Profile Apollo: Hugh Blair-Smith," *NASA Office of Logic Design,* http://klabs.org/history/bios/hugh_blair_smith/elwin_mit_report.htm.
31. CSD to Seamans, November 21, 1961, Robert C. Seamans Jr. Papers, MIT Archives.
32. Mindell, *Digital Apollo,* 107.
33. CSD to Seamans, November 21, 1961.
34. As quoted by Donald C. Fraser, "Richard Battin 1925–2014," *Memorial Tributes: National Academy of Engineering* 20 (2016): 34.
35. Hall, "Apollo Guidance Computer," 3.
36. Hall, "From the Farm to Pioneering with Digital Control Computers," 28.
37. Hall, "From the Farm to Pioneering with Digital Control Computers."
38. Mindell, *Digital Apollo,* 123.
39. Hall, "From the Farm to Pioneering with Digital Control Computers," 29.
40. Hugh Blair-Smith, as quoted by Ong, "Profile Apollo: Hugh Blair-Smith."
41. Ong, "Profile Apollo: Hugh Blair-Smith."

Chapter 20. The Road to Divestiture

1. Bryce Nelson, "Scientists Plan Research Strike at M.I.T. on 4 March," *Scientist* 163 (January 24, 1969): 373.
2. Stuart W. Leslie, "'Time of Troubles' for the Special Laboratories," in *Becoming MIT,* ed. David Kaiser (Cambridge, Mass.: MIT Press, 2010), 126.
3. Bernard Dixon, "Science and Politics at MIT," review of *The University and Military Research,* by Dorothy Nelkin, *New Scientist,* July 20, 1972, 161.
4. Sarah Bridger, *Scientists at War: The Ethics of Cold War Weapons Research* (Cambridge, Mass.: Harvard University Press, 2015), 163.
5. Bridger, *Scientists at War.*
6. Nelson, "Scientists Plan Research Strike at M.I.T.," 373.
7. Stuart W. Leslie, *The Cold War and American Science: The Military-Industrial-Academic Complex at MIT and Stanford* (New York: Columbia University Press, 1993), 235.
8. Dorothy Nelkin, *The University and Military Research: Moral Politics at M.I.T.* (Ithaca, N.Y.: Cornell University Press, 1972), 49.
9. "Can Defense Work Keep a Home on Campus?" *Business Week,* June 7, 1969, 69.
10. CSD Oral History, 120.
11. "Can Defense Work Keep a Home on Campus?," 69.
12. John Walsh, "M.I.T.: Panel on Special Labs Asks More Nondefense Research," *Science* 164, June 13, 1969, 1264; Leslie, "Time of Troubles," 128.
13. Harvey Baker, "Review Panel Members Named," *Tech* 89, no. 20 (April 29, 1969): 3.
14. William F. Pounds transcript, https://infinitehistory.mit.edu/video/william-f-pounds.
15. William Pounds, Transcript of Meeting Sunday, April 27, 1969, Review Panel on Special Laboratories [hereafter Pounds Panel], Records, 1969–1971, 12, AC 054, MIT Archives.

16. Jonathan Kabat, Pounds Panel, April 27, 1969, 12.
17. Kabat would later characterize MIT as the "mother and father" of the MIRV. See Robert Elkin, "SACC-NAC Rally Begins—Afternoon Research Project," *Tech* 89, no. 43 (November 7, 1969), 1.
18. David Hoag, Pounds Panel, April 28, 3. Hoag was an associate director of the laboratory at the time.
19. Bridger, *Scientists at War*, 174.
20. Walsh, "Panel on Special Labs," 1264.
21. Nelkin, *University and Military Research*, 80. A condensed version of these statements can also be found in Jack P. Ruin, Report of the Vice President, Special Laboratories, "Report of the President 1969," *Massachusetts Institute of Technology Bulletin* 105, no. 3 (December 1969): 676.
22. Nelkin, *University and Military Research*.
23. Bridger, *Scientists at War*, 187; "Sense on Defense Science," *New York Times*, June 4, 1969, online archives.
24. Ruin, Report of the Vice President, Special Laboratories, 676; William Leavitt, "A Triumph of Reverse McCarthyism: The Dethronement of Dr. Draper," *Air Force Space Digest* (December 1969): 49.
25. Leavitt, "Triumph of Reverse McCarthyism," 49.
26. Leavitt, "Triumph of Reverse McCarthyism."
27. Walsh, "Panel on Special Labs," 1265.
28. Walsh, "Panel on Special Labs."
29. Albert G. Hill, Report of the Vice President for Research, "Report of the President for the Academic Year 1969–1970," *Massachusetts Institute of Technology Bulletin* 106, no. 2 (December 1969): 428.
30. Sarah Bridger, "Scientists and the Ethics of Cold War Weapons Research," PhD diss., Columbia University, New York, N.Y., 2011, 308.
31. Howard W. Johnson, *Holding the Center: Memoirs of a Life in Higher Education* (Cambridge, Mass.: MIT Press, 1999), 169.
32. Johnson, *Holding the Center*, 169.
33. Johnson, *Holding the Center*.
34. Howard W. Johnson, Memorandum to Members of the Faculty and Staff, September 24, 1969, MIT, School of Engineering, Office of the Dean, Records, 1943–1989, Box 1, AC 12, MIT Archives.
35. CSD, Airport Transcript, Box 17, CSDP-LC.
36. CSD, Airport Transcript.
37. Johnson, Memorandum.
38. Leavitt, "Triumph of Reverse McCarthyism," 46.
39. Howard W. Johnson, "Statement by Howard W. Johnson on the Special Laboratories, October 22, 1969," as reproduced in Nelkin, *University and Military Research*, app. 2, 168–77.
40. Johnson, "Statement on the Special Laboratories."
41. Johnson, "Statement on the Special Laboratories."
42. Leslie, *Cold War and American Science*, 237.
43. "GE Recruiting Disrupted," *Harvard Crimson*, October 30, 1969, http://www.thecrimson.com/article/1969/10/30/ge-recruiting-disrupted-pabout-50-members/.
44. CSD to All Laboratory Personnel, October 29, 1969, HC Activism Cor., MIT-MC, and as cited in Leslie, *Cold War and American Science*, 237n32, 310.

45. Carol R. Sternhell, "Police Rout NAC Pickets in Protest at M.I.T. Lab," *Harvard Crimson*, November 6, 1969, http://www.thecrimson.com/article/1969/11/6/police-rout-nac-pickets-in-protest/.
46. Leslie, *Cold War and American Science*, 237.
47. Martha Stark Ditmeyer interview.
48. Morgan, *Draper at 25*, 36.
49. Leslie, *Cold War and American Science*, 239.
50. Matthew Wisnioski, *Engineers for Change: Competing Visions of Technology in 1960s America* (Cambridge, Mass.: MIT Press, 2012), 103.
51. Ascher H. Shapiro, "A Position Paper on Retention or Divestiture of the Special Laboratories," February 1970, as cited by Leslie, *Cold War and American Science*, 239.
52. Charles Miller, interview by Dorothy Nelkin, *University and Military Research*, 99.
53. Johnson, *Holding the Center*, 172.
54. "Sage Transition," *Lincoln Laboratory Massachusetts Institute of Technology*, https://ll.mit.edu/about/History/transition.html.
55. Morgan, *Draper at 25*, 38.
56. Athelstan Spilhaus, interview by Ronald E. Doel, November 10, 1989.
57. CSD, as quoted by Leavitt, "Triumph of Reverse McCarthyism," 50.
58. Morgan, *Draper at 25*, 38.
59. Johnson, *Holding the Center*, 173.
60. Bridger, *Scientists at War*, 190.
61. Hill, Report of the Vice President for Research, 427.
62. CSD, "Charles Stark Draper Laboratory," in *Massachusetts Institute of Technology Bulletin Report of the President for the Academic Year 1969–1970*, 434.
63. Bridger, *Scientists at War*, 193.
64. Victor K. McElheny, "MIT Administration Makes Public Its Intentions on Disposition of Draper and Lincoln Laboratories," *Science* 168 (May 29, 1970): 1975.
65. Nelkin, *University and Military Research*, 145.
66. Spilhaus interview; Duffy, "Charles Stark Draper," 153.
67. Bridger, *Scientists at War*, 191.
68. Leslie, *Cold War and American Science*, 249–50; Bridger, *Scientists at War*, 191.
69. Spilhaus interview; Duffy, "Charles Stark Draper," 155.

Chapter 21. A Heterogeneous Engineer

1. John Law, "Technology and Heterogeneous Engineering: The Case of Portuguese Expansion," in *Social Construction of Technological Systems: New Directions in the Sociology and History of Technology*, ed. Wiebe E. Bijker, Thomas Hughes, and T. J. Pinch (Cambridge, Mass.: MIT Press, 1989), passim.
2. John Krige, "The 1984 Physics Prize for Heterogenous Engineering," *Minerva* 39, no. 4 (December 2001), 426.
3. Trimble, *Jerome C. Hunsaker*, 3.
4. Duffy, "Charles Stark Draper," 123.
5. Data from Massachusetts Institute of Technology, *Instrumentation Laboratory*, passim.

Source Material

Archival Records and Manuscript Collections
Concord Free Public Library, Concord, Mass.
Massport Hangar 24 Collection.
Hagley Museum and Library, Wilmington, Del.: Manuscripts and Archives Department
Sperry Gyroscope Company Collection.
Library of Congress, Washington, D.C.: Manuscript Division
Papers of Charles Stark Draper.
Library of Congress, Washington, D.C.: Recorded Sound Division
Audio Tape No. RXB8588.
Massachusetts Institute of Technology, Cambridge, Mass.: Archives
Department of Aeronautics and Astronautics, Records, 1910–1978, AC 43.
Office of the Dean, School of Engineering, Records, 1943–1989, AC 12.
Office of the President, AC 4.
Review Panel on Special Laboratories (Pounds Panel), Records, 1969–1971, AC 054.
Massachusetts Institute of Technology, Lincoln Laboratory, Lexington, Mass.
Photo Archives.
Massachusetts Institute of Technology Museum, Cambridge, Mass.
Charles Stark Draper Laboratory Historical Collection.
National Air and Space Museum, Smithsonian Institution, Washington, D.C.
Vertical Files, Library.
National Archives, College Park, Md.
Bureau of Ordnance Confidential Correspondence 1942–47, RG 74.
Bureau of Ordnance Correspondence, Classified and Unclassified, 1905–1944, RG 74.
Bureau of Ordnance, Ordnance Data, RG 74.
Bureau of Ordnance Pamphlets, RG 74.
Bureau of Ordnance Secret Correspondence, 1942–1959, RG 74.
CNO/Secretary of the Navy Security Classified Correspondence, 1940–1947, RG 38.
Office of Naval Research, General Correspondence, 1950, RG 298.
Records of Division 7, Office of Scientific Research and Development, RG 227.
Records of Section T, Office of Scientific Research and Development, RG 227.
Records of the Research and Development Board, Department of Defense, RG 330.
Secretary of the Navy Confidential Correspondence, 1940–1941, RG 80.
Technical Publications 1901–1960, Records of the Office of the Chief of Naval Operations, RG 38.
World War II Action and Operation Reports, RG 38.

Navy Department Library, Washington, D.C.
Special Collections.

Books

Abella, Alex. *Soldiers of Reason: The Rand Corporation and the Rise of the American Empire.* Orlando, Fla.: Harcourt, 2008.

Air Force Missileers. Paducah, Ky.: Turner, 1998.

Alexander, Philip N. *A Widening Sphere: Evolving Cultures at MIT.* Cambridge, Mass.: MIT Press, 2011.

Aseltine, John A., ed. *Peaceful Uses of Automation in Outer Space: Proceedings of the Symposium.* New York: Plenum, 1966.

Bennett, Stuart. *History of Control Engineering, 1930–1955.* Stevenage, Herts., U.K.: Peregrinus, 1993.

Berhow, Mark A., and Chris Taylor. *U.S. Strategic and Defensive Missile Systems, 1950–2004.* Oxford, U.K.: Osprey, 2005.

Bilstein, Roger E., and Frank W. Anderson. *Orders of Magnitude: A History of the NACA and NASA, 1915–1990.* Washington, D.C.: NASA, Office of Management, Scientific and Technical Information Division, 1989.

Birmingham, Stephen. *Real Lace.* Syracuse, N.Y.: Syracuse University Press, 1997.

Bridger, Sarah. *Scientists at War: The Ethics of Cold War Weapons Research.* Cambridge, Mass.: Harvard University Press, 2015.

Brooks, Courtney G., James M. Grimwood, and Loyd S. Swenson. *Chariots for Apollo: A History of Manned Lunar Spacecraft.* NASA History Series SP-4205. Washington, D.C.: NASA Scientific and Technical Information Office, 1979.

Broxmeyr, Charles. *Inertial Navigation Systems.* New York: McGraw-Hill, 1964.

Burchard, John. *Q.E.D.: MIT in World War II.* Cambridge, Mass.: Technology, 1948.

Cameron, Rebecca Hancock. *Training to Fly: Military Flight Training, 1907–1945.* Washington, D.C.: Air Force History and Museums Program, 1999.

Cleveland, Reginald. *America Fledges Wings: The History of the Daniel Guggenheim Fund for the Promotion of Aeronautics.* New York: Pitman, 1942.

Color Technology in the Textile Industry. 2nd ed. Research Triangle Park, N.C.: American Association of Textile Chemists and Colorists Committee RA36, Color Measurement, 1997.

Conrad, Howard I., ed. *The Encyclopedia of the History of Missouri.* New York: Southern History, 1901. Available at http://tacnet.missouri.org/history/encycmo/towns.html#Windsor.

Converse, Elliot V., III. *Rearming for the Cold War.* Vol. 1. Washington, D.C.: Historical Office, Office of the Secretary of Defense, 2012.

Conway, Erik M. *Blind Landings: Low-Visibility Operations in American Aviation.* Baltimore, Md.: Johns Hopkins University Press, 2006.

"CSM Guidance Development Diary," *Encyclopedia Astronautica*, s.v. November 22, 1960, http://www.astronautix.com/craft/csmdance.htm.

Day, Dwayne A., John M. Logsdon, and Brian Latell, eds. *Eye in the Sky: The Story of the CORONA Spy Satellites.* Washington, D.C., Smithsonian, 1998.

Diner, Steven J. *A Very Different Age: Americans of the Progressive Era.* New York: Hill and Wang, 1998.

Doel, Ronald E., and Thomas Söderqvist. *The Historiography of Contemporary Science, Technology, and Medicine: Writing Recent Science.* London: Routledge, 2006.

Downs, Winfield Scott, ed. *Who's Who in Engineering 1941.* New York: Lewis Herford, 1941.

Draper, Charles Stark, Walter Wrigley, and Lester R. Grothe. *The Floating Integrating Gyro and Its Application to Geometrical Stabilization Problems on Moving Bases.* Sherman M. Fairchild Paper No. FF-13. New York: Institute of Aeronautical Sciences, 1955.

Draper, Charles Stark, Walter Wrigley, and John Hovorka. *Inertial Guidance.* New York: Pergamon, 1960.

Edwards, Robert D. Foreword to *Building Technology Transfer within Research Universities,* edited by Thomas J. Allen and Rory O'Shea. Cambridge, U.K.: Cambridge University Press, 2014.

Elichirigoity, Fernando. *Planet Management: Limits to Growth, Computer Simulation and the Emergence of Global Spaces.* Evanston, Ill.: Northwestern University Press, 1999.

Engineering Directory 1911. Chicago: Crawford, 1911.

Etzkowitz, Henry. *MIT and the Rise of Entrepreneurial Science.* London: Routledge, 2002.

Farish, Matthew. *The Contours of America's Cold War.* Minneapolis: University of Minnesota Press, 2010.

Freeland, Richard M. *Academia's Golden Age: Universities in Massachusetts, 1945–1970.* New York: Oxford University Press, 1992.

Freudenthal, Elisabeth E. "The Aviation Business in the 1930's." In *The History of the American Aircraft Industry an Anthology,* edited by G. R. Simons. Cambridge, Mass.: MIT Press, 1968.

Friedman, Norman. *U.S. Naval Weapons.* Annapolis, Md.: Naval Institute Press, 1982.

Frye, Alton. *A Responsible Congress: The Politics of National Security.* New York: McGraw-Hill, 1975.

Funderberg, Anne. *Bootleggers and Beer Barons of the Prohibition Era.* Jefferson, N.C.: McFarland, 2014.

Futrell, Robert F. *Ideas, Concepts, Doctrine: Basic Thinking in the United States Air Force 1907–1960.* Vol. 1, *Early Days.* Maxwell Air Force Base, Ala.: Air University Press, 1989.

Galantin, Ignatius J. *Submarine Admiral: From Battlewagons to Missiles.* Urbana: University of Illinois Press, 1995.

Galison, Peter, and Bruce W. Hevly. *Big Science: The Growth of Large-Scale Research.* Stanford, Calif.: Stanford University Press, 1992.

Getting, Ivan A. *All in a Lifetime: Science in the Defense of Democracy.* New York: Vantage, 1989.

Grewal, Mohinder S., Angus P. Andrews, and Chris Barton. *Global Navigation Satellite Systems, Inertial Navigation, and Integration.* Hoboken, N.J.: John Wiley & Sons, 2013.

Hall, Eldon C. *Journey to the Moon: The History of Apollo Guidance Computer.* Reston, Va.: American Institute of Aeronautics and Astronautics, 1996.

Hughes, Thomas. *Elmer Sperry: Inventor and Engineer.* Baltimore, Md.: Johns Hopkins University Press, 1971.

———. *Networks of Power: Electrification in Western Society, 1980–1930.* Baltimore, Md.: Johns Hopkins University Press, 1983.

———. *Rescuing Prometheus.* New York: Pantheon Books, 1998.

Hunley, J. D. *Preludes to U.S. Space-Launch Vehicle Technology: Goddard Rockets to Minuteman III.* Gainesville: University Press of Florida, 2008.

———. *U.S. Space-Launch Vehicle Technology: Viking to Space Shuttle.* Gainesville: University Press of Florida, 2008.
Jenkins, Dennis R. *X-15: Extending the Frontiers of Flight.* Washington, D.C.: NASA, 2007.
Johnson, Howard W. *Holding the Center: Memoirs of a Life in Higher Education.* Cambridge, Mass.: MIT Press, 1999.
Kaiser, David, ed. *Becoming MIT.* Cambridge, Mass.: MIT Press, 2010.
Kaiser, William K. *The Development of the Aerospace Industry on Long Island 1904–1964.* Hempstead, N.Y.: Hofstra University, 1968.
Killian, James R. *The Education of a College President.* Cambridge, Mass.: MIT Press, 1985.
Koonce, Jefferson M. *Human Factors in the Training of Pilots.* New York: Taylor and Francis, 2002.
Lannius, Roger D., and Dennis R. Jenkins, eds. *To Reach the High Frontier: A History of U.S. Launch Vehicles.* Lexington: University Press of Kentucky, 2002.
Launius, Richard D. *Innovation and the Development of Flight.* College Station: Texas A&M University Press, 1999.
Lawrence, Anthony. *Modern Inertial Technology: Navigation, Guidance, and Control.* New York: Springer-Verlag, 1998.
Lawson, Ellen. *Smugglers, Bootleggers, and Scofflaws: Prohibition and New York City.* Albany, N.Y.: Excelsior, 2013.
Lees, Sidney, ed. *Air Space Instruments.* New York: McGraw-Hill, 1963.
Leslie, Stuart W. *The Cold War and American Science: The Military-Industrial-Academic Complex at MIT and Stanford.* New York: Columbia University Press, 1993.
Machover, Carl. *Gyro Primer.* Commack, N.Y.: United Aircraft, 1958.
Mackenzie, Donald. *Inventing Accuracy: A Historical Sociology of Nuclear Missile Guidance.* Cambridge, Mass.: MIT Press, 1990.
Massachusetts Institute of Technology. *Instrumentation Laboratory.* Cambridge, Mass., April 1968.
———. *Policies and Procedures: A Guide for Faculty and Staff Members,* http://web.mit.edu/policies/.
McCormick, Barnes W., Conrad F. Jewberry, and Eric Jumper. *Engineering Education during the First Century of Flight.* Reston, Va.: American Institute of Aeronautics and Astronautics, 2004.
McMurran, Marshall William. *Achieving Accuracy: A Legacy of Computers and Missiles.* Bloomington, Ind.: Xlibris, 2009.
Mendelsohn, Everett, Merritt Roe Smith, and Peter Weingart, eds. *Science, Technology and the Military.* Dordrecht, Neth.: Kluwer Academic, 1989.
Merhav, Shmuel. *Aerospace Sensor Systems and Applications.* New York: Springer-Verlag, 1996.
Miller, Jay. *Convair B-58 Hustler: The World's First Supersonic Bomber.* Leicester, U.K.: Midland/Aerofax, 1997.
Mindell, David A. *Between Human and Machine: Feedback, Control, and Computing before Cybernetics.* Baltimore, Md.: Johns Hopkins University Press, 2002.
———. *Digital Apollo: Human and Machine in Spaceflight.* Cambridge, Mass.: MIT Press, 2008.
Morgan, Christopher, with Joseph O'Connor and David Hoag. *Draper at 25: Innovation for the 21st Century.* Cambridge, Mass.: Charles Stark Draper Laboratory, 1998.
Morse, Philip M. *In at the Beginning: A Physicist's Life.* Cambridge, Mass.: MIT Press, 1977.

National Academy of Sciences (U.S.), Committee on Utilization of Scientific and Engineering Manpower. *Toward Better Utilization of Scientific Engineering Talent: A Program for Action Report.* Washington, D.C.: National Research Council, 1964.

Nelkin, Dorothy. *The University and Military Research: Moral Politics at M.I.T.* Ithaca, N.Y.: Cornell University Press, 1972.

Neufeld, Jacob. *The Development of Ballistic Missiles in the United States Air Force 1945–1960.* Washington, D.C.: Department of the Air Force, Office of Air Force History, 1990.

Office of the Home Secretary, National Academy of Sciences. *Biographical Memoirs.* Vol. 65. Washington, D.C.: National Academies Press, 1994.

Redmond, Kent C., and Thomas M. Smith. *From Whirlwind to Mitre: The R&D Story of the SAGE Air Defense Computer.* Cambridge, Mass.: MIT Press, 2000.

Reuter, Claus. *The A4 (V2) the Russian and American Rocket Program.* Scarborough, Ont.: German-Canadian Museum of Applied History, 1998.

Rife, James, and Rodney P. Carlisle. *The Sound of Freedom: Naval Weapons Technology at Dahlgren, Virginia, 1918–2006.* Dahlgren, Va.: Naval Surface Warfare Center, Dahlgren Division, 2007.

Rockefellor, H. E. *Technique 1922.* Cambridge, Mass.: Junior Class of MIT, 1921.

Roland, Alex, and Philip Shiman. *Strategic Computing: DARPA and the Quest for Machine Intelligence, 1983–1993.* Cambridge, Mass.: MIT Press, 2002.

Roland, Buford, and William B. Boyd. *U.S. Navy Bureau of Ordnance in World War II.* Washington, D.C.: Department of the Navy, Bureau of Ordnance, 1953.

Rolfe, J. M., and Rodney Carlise. *The Sound of Freedom: Naval Weapon Technology at Dahlgren, Virginia, 1918–2006.* Dahlgren, Va.: Naval Surface Warfare Center, Dahlgren Division, 2007.

Rolfe, J. M., and K. J. Staples. *Flight Simulation.* Cambridge, U.K.: Cambridge University Press, 1986.

Roston, Eric. *The Carbon Age: How Life's Core Element Has Become Civilization's Greatest Threat.* New York: Walker, 2008.

Saxenian, AnnaLee. *Regional Advantage: Culture and Competition in Silicon Valley and Route 128.* Cambridge, Mass.: Harvard University Press, 1994.

Schlosser, Eric. *Command and Control: Nuclear Weapons, the Damascus Accident, and the Illusion of Safety.* New York: Penguin, 2013.

Seamans, Robert C., Jr. *Aiming at Targets.* Washington, D.C.: NASA, 1996.

Spinardi, Graham. *From Polaris to Trident: The Development of U.S. Fleet Ballistic Missile Technology.* New York: Cambridge University Press, 1994.

Stark, Howard C. *Blind or Instrument Flying.* Newark, N.J., 1931.

Steinhoff, Ernst A., ed. *The Eagle Has Returned: Proceedings of the Dedication Conference of the International Space Hall of Fame, Held at Alamogordo, New Mexico, from 5 through 9 October 1976.* Alexandria, Va.: American Astronautical Society, 1976.

Stumpf, David K. *Regulus: America's First Nuclear Submarine Missile.* Paducah, Ky.: Turner, 1996.

———. *Titan II: A History of a Cold War Missile Program.* Fayetteville: University of Arkansas Press, 2000.

Sturn, Thomas A. *The USAF Scientific Advisory Board: Its First Twenty Years, 1944–1964.* Washington, D.C.: Department of the Air Force, Office of Air Force History, 1986.

Trimble, William F. *Jerome C. Hunsaker and the Rise of American Aeronautics.* Washington, D.C.: Smithsonian Institute Press, 2002.

U.S. Department of Transportation, Federal Aviation Administration. *Instrument Flying Handbook,* FAA-H-8083-15B. Updated October 10, 2014. Oklahoma City, Okla.: Testing Standards Branch, 2014.
Utilizing Our Limited Human Resources. Publication No. L51-148. Washington, D.C.: Industrial College of the Armed Forces, 1951.
Wallace, Robert E. *From Dam Neck to Okinawa: A Memoir of Antiaircraft Training in World War II.* Department of the Navy, Naval Historical Center, 2001.
Werrell, Kenneth. *Sabres over MiG Alley: The F-86 and the Battle for Air Supremacy in Korea.* Annapolis, Md.: Naval Institute Press, 2013.
Wigmore, Barrie A. *The Crash and Its Aftermath: A History of Securities Markets in the United States.* Westport, Conn.: Greenwood, 1985.
Wisnioski, Matthew. *Engineers for Change: Competing Visions of Technology in 1960s America.* Cambridge, Mass.: MIT Press, 2012.
Wrigley, Walter, Walter M. Holister, and William G. Denhard. *Gyroscopic Theory, Design, and Instrumentation.* Cambridge, Mass.: MIT Press, 1969.
Wylie, Francis E. *MIT in Perspective: A Pictorial History of the Massachusetts Institute of Technology.* New York: Little, Brown, 1975.
York, Herbert. *Race to Oblivion: A Participants' View of the Arms Race.* New York: Simon & Schuster, 1970.
Zimmerman, David. *Top Secret Exchange: The Tizard Mission and the Scientific War.* Stroud, Gloucestershire, U.K.: Alan Sutton, 1996.

Articles, Chapters in Edited Collections, and Web Publications

"$101.1M to Develop the MK6 LE: So, What's That?" *Defense Industry Daily,* December 15, 2005, http://www.defenseindustrydaily.com/1011m-to-develop-the-mk6-le-so-whats-that-01627/.
"$243M to Draper Laboratory for Trident II D5 Guidance System Support." *Defense Industry Daily,* December 6, 2009, http://www.defenseindustrydaily.com/243M-to-Draper-Laboratory-for-Trident-II-D5-Guidance-System-Support-06004/.
"Advanced Inertial Navigation: The Airs FLIMBAL," www.deanspacedrive.org, accessed September 20, 2010. [Site discontinued as of April 8, 2019.]
"Advanced Inertial Reference Sphere." *Nuclear Weapons Archive,* http://nuclearweaponarchive.org/Usa/Weapons/Airs.html.
"A.E.S. Announces Free Plane Ride for Lucky Man." *Tech* 52, no. 43 (October 5, 1932): 1.
"AIAA Fellow Ragan Died in July." *AIAA Bulletin* (September 2014): B13.
"Along the Road of Progress." *Popular Science Monthly* (February 1929): 59.
"Athelstan Frederick Spilhaus Lieutenant Colonel, United States Army." *Arlington National Cemetery,* http://www.arlingtoncemetery.net/spilhaus.htm.
Baker, Harvey. "Review Panel Members Named." *Tech* 89, no. 20 (April 29, 1969): 3.
Banks, Harold K. "She Chose to Walk with Him." *Pictorial Review* [section] Boston *Advertiser,* September 15, 1962, 12–13. (Clippings in CSD Papers, Library of Congress [hereafter LC].)
Barbour, Neil M., John M. Elwell, and Roy H. Setterlund. "Inertial Instruments: Where to Now?" AIAA Guidance, Navigation and Control Conference, Hilton Head Island, S.C., August 10–12, 1992, Technical Papers, Pt. 2 (A92-55151 23-63), American Institute of Aeronautics and Astronautics, Washington, D.C., 1992, 566–74.
Battin, Richard H. "Space Guidance Evolutions: A Personal Narrative." *Journal of Guidance and Control* 5 (1982): 97–110.

Bedi, Joyce. "MIT in World War II: A Hot Spot of Invention." *Lemelson Center for the Study of Invention and Innovations,* http://invention.si.edu/mit-world-war-ii-hot-spot-invention.

Bernhardt, Greg. "Johnson Appoints Panel to Review Labs Group to Report by May 31: No Action on Research Halt." *Tech* 89, no. 20 (April 29, 1969): 1.

Brandon, Bruce. "Gone West: Col. Charles L. 'Chip' Collins." *Aero News Network,* February 25, 2015, http://www.aero-news.net/index.cfm?do=maiNo.textpost&id=b3bf1d53-d5ec-4afe-9f40-22b20fbd62cf.

Brodie, Bernard. "Our Ships Strike Back." *VQR* (Spring 1945), http://www.vqronline.org/essay/our-ships-strike-back.

Bushnell, David. "Historic Hanscom Hangar Is Potential Site for Massachusetts Technology Museum." *Boston Globe,* December 7, 2006, http://www.saveourheritage.com/news_museum.htm.

"C. F. Taylor Sr., Retired MIT Professor Who Pioneered Development of Aircraft Engines, Dies at 101." *MIT News,* June 27, 1996, http://newsoffice.mit.edu/1996/taylor.

"Can Defense Work Keep a Home on Campus?" *Business Week,* June 7, 1969, 68–70.

"Chip Collins Remembers Doc Draper as Dynamic Leader of Lab." *D-notes,* October 29, 2001, 5–6.

Chu, Jennifer. "Going Up: Doc Draper's Quest to Visit Space." *MIT Technology Review,* August 23, 2011, http://www.technologyreview.com/article/425104/going-up/.

Clark, Donald L. "Early Advances in Radar Technology for Aircraft Detection." *Lincoln Laboratory Journal* 12, no. 2 (2000): 167–80.

Cline, Allen W. "Engine Basics: Detonation and Pre-ignition." *Contact!,* no. 54 (January 2000), http://www.contactmagazine.com/Issue54/EngineBasics.html.

Connelly, Robert L., Sr., and Richard W. Harold. "Benefits of Hand-Held Color Measurement Equipment." In *Color Technology in the Textile Industry,* 2nd edition, 18–20. Research Triangle Park, N.C.: American Association of Textile Chemists and Colorists Committee RA36, Color Measurement, 1997.

"Conquest Shows SPIRE Flight." *Niagara Falls Gazette,* April 13, 1958, T-3.

"Cops Break Up Viet Protest Rally at M.I.T." *Chicago Tribune,* November 6, 1969, sec. 2, 16.

Cushman, John H., Jr., "Northrop's Struggle with the MX." *New York Times,* November 22, 1987, online archives.

Davis, Leighton I. "Military Significance of Draper's Work for the Air Force." In *Air Space and Instruments,* edited by Sidney Lees. New York: McGraw-Hill, 1963.

Day, Dwayne. "The Flight of Big Bird (Part 1): The Origins of the KH-9 Hexagon Reconnaissance Satellite," Monday, January 17, 2011, *Space Review,* http://www.thespacereview.com/article/1761/1.

———. "A Sheep in Wolf's Clothing: The Samos E-5 Recoverable Satellite (Part 1)." *Space Review,* http://www.thespacereview.com/article/1410/1. (Originally appeared in *Spaceflight,* October 2002.)

———. "Square Peg in a Cone-Shaped Hole: The Samos E-5 Recoverable Satellite (Part 2)." *Spaceflight,* February 2003, www.thespaceview.com/article/1419/1.

Denhard, William G. "Floated Single-Degree-of-Freedom Integrating Gyros." In *Air Space and Instruments,* edited by Sidney Lees. New York: McGraw-Hill, 1963.

———. "The Start of the Laboratory: The Beginnings of the MIT Instrumentation Laboratory." *IEEE AES Systems Magazine,* October 1992, 6–13.

Dennis, Michael A. "'Our First Line of Defense:' Two University Laboratories in Postwar American State." *ISIS,* 85, September 1994, 427–55.

"Distinguished Life and Career of George Gamow." *Department of Physics, University of Colorado Boulder,* http://phys.colorado.edu/public-outreach/distinguished-life-and-career-george-gamow.

Dixon, Bernard. "Science and Politics at MIT." Review of *The University and Military Research,* by Dorothy Nelkin. *New Scientist,* July 20, 1972, 161–62.

Donnelly, Jim. "Henry M. Crane." *Hemmings Classic Car,* June 2012, http://www.hemmings.com/hcc/stories/2012/06/01/hmn_feature12.html.

"Dr. Robert Channing Seaman Jr." *U.S. Air Force: Biographies,* http://www.af.mil/AboutUs/Biographies/Display/tabid/225/Article/105667/dr-robert-channing-seamans-jr.aspx.

Draper, Charles Stark. "Aircraft Instruments in Meteorological Flying." In *The Meteorological Airplane Ascents of the Massachusetts Institute of Technology.* Papers in Physical Oceanography and Meteorology III, no. 2. Cambridge, Mass.: Massachusetts Institute of Technology and Woods Hole Oceanographic Institution, 1934.

———. "Control, Navigation, and Guidance." Reprint, *Control Systems* (December 1981).

———. "Dynamic Characteristics of Aircraft Instruments." *Journal of Aeronautical Sciences* 2 (March 1935): 59.

———. "The Evolution of Aerospace Guidance Technology at the Massachusetts Institute of Technology, 1935–1951: A Memoir." *Essays on the History of Rocketry and Astronautics* 2 (September 1977): 219–52.

———. "Fifty Years of Flight Technology." *IEEE Transactions on Aerospace and Electronic Systems* AES-14, no. 4 (July 1978): 708–12.

———. "Innovation in Aircraft Instruments Assists Blind Flying." *Tech Engineering News* 17, December 1936, 170.

———. "The Instruments of Blind Flying." *Science Digest* 1, no. 1 (March 1937): 555–59. (Condensed from *Tech Engineering News.*)

———. "The New Instrument Laboratory at the Massachusetts Institute of Technology." *Journal of the Aeronautical Sciences* 3, no. 4 (March 1936): 151–53.

———. "On Course to Modern Guidance." *Astronautics and Aeronautics* 18 (February 1980): 56–62.

———. "Origins of Inertial Navigation." *Journal of Guidance and Control* 4, no. 5 (September–October, 1981): 449–63.

———. "Pressure Waves Accompanying Detonation in Engines." *Journal of Aeronautical Science* 5 (1938): 219–26.

Draper, Charles Stark, and G. Bentley. "Measurement of Aircraft Vibrations during Flight." *Journal of the Aeronautical Sciences* 3, no. 4 (March 1936): 116–21.

Draper, Charles Stark, G. Bentley, and H. H. Willis. "The M.I.T.-Sperry Apparatus for Measuring Vibration." *Journal of the Aeronautical Sciences* 4, no. 7 (May 1937): 281–85.

Draper, Charles Stark, W. H. Cook, and Walter McKay. "Northerly Turning Error of the Magnetic Compass for Aircraft." *Journal of the Aeronautical Sciences* 5, no. 9 (1938): 345–454.

Draper, Charles Stark, J. H. Lancor, and Leighton Davis. "The Applications of an Electromagnetic Indicator to Internal Combustion Engine Problems." *Journal of Aeronautical Science* 8, no. 1 (1940): 7–16.

Draper, Charles Stark, and D. G. C. Luck. "A Fast and Economical Type of Photographic Oscillograph." *Review of Scientific Instruments* 5 (August 1933): 440–43.

Draper, Charles Stark, and A. F. Spilhaus. "Power Supplies for Suction-Driven Gyroscopic Aircraft Instruments." *ASME Transactions* 56, no. 5 (May 5, 1934): 289–94.

"Draper Draws in $101M Navy Contract Addition." *Boston Business Journal,* December 12, 2005.

Droney, James F. "Mark 14 Gunsight Stopped Dive Bombers." *Boston Sunday Herald,* July 8, 1962, sec. 1, 2.

Duffy, Robert A. "Charles Stark Draper, 1901–1987: A Biographical Memoir." In *National Academy of Sciences Biographical Memoirs.* Washington, D.C.: National Academy of Sciences, 1994.

Elkin, Robert. "SACC-NAC Rally Begins—Afternoon Research Protest," *Tech* 89, no. 43 (November 7, 1969).

Emmerich, Claude L. "Advances in Gyro Performance." In *Air Space and Instruments,* edited by Sidney Lees. New York: McGraw-Hill, 1963.

Fellows, Laurence. "New ICBM Nears Trials in Flight." *New York Times,* March 16, 1958, L33.

Fischell, R. E. "Gravity Gradient Stabilization of Earth Satellites." *APL Technical Digest* (May–June 1964): 12–17.

Forant, Robert. "United American Bosch (Magneto Corporation); Springfield (MA)," originally published in the *Historical Journal of Massachusetts,* Winter 2003, reprinted in part in the Radio Museum website, https://www.radiomuseum.org/dsp_hersteller_detail.cfm?company_id=3684 [accessed April 8, 2019].

Forrester, Jay. *A Science Odyssey: People and Discoveries.* WGBH Boston, http://www.pbs.org/wgbh/aso/databank/entries/btforr.html.

Fraser, Cline W. "Digital Autopilot for Apollo: A Radical Change," http://klabs.org/mapld04/papers/g/g237_frasier_p.pdf. [The article is no longer available].

Fraser, Donald C. "Richard Battin 1925–2014." *Memorial Tributes: National Academy of Engineering* 20 (2016): 30–39.

Freeman, Ira H. "Inertial System Guided *Nautilus.*" *New York Times,* August 10, 1958. *New York Times,* online archives.

Fuhrman, R. A. "The Fleet Ballistic Missile System: Polaris to Trident." *Journal of Spacecraft* 15, no. 5 (September–October, 1978): 265–86.

Gates, Robert V. "Strategic Systems Fire Control." Naval Surface Warfare Center, Dahlgren Division *Technical Digest* (1995): 166–79.

"GE Recruiting Disrupted." *Harvard Crimson,* October 30, 1969, http://www.thecrimson.com/article/1969/10/30/ge-recruiting-disrupted-pabout-50-members/.

Geiger, Clarence J. "The Strangled Infant: The Boeing X-20A Dyna-Soar." In *The Hypersonic Revolution,* edited by Richard Hallion. Wright-Patterson Air Force Base, Ohio: ASD Special Staff Office, 1987, http://www.wpafb.af.mil/shared/media/document/AFD-080408-031.pdf.

Gescheliln, Joseph. "Thor's Brain." *Aircraft and Missiles Manufacturing* (April 1958): 24–27.

Goebel, Greg. "The North American F-86 Sabre." *AirVectors,* http://www.airvectors.net/avf86.html.

"H. M. Crane Dead: Auto Pioneer, 81." *New York Times,* January 22, 1956, online archives.

Haeussermann, Walter. "On the Evolution of Rocket Navigation, Guidance and Control (NGC) Systems." In *The Eagle Has Returned: Proceedings of the Dedication Conference of the International Space Hall of Fame, Held at Alamogordo, New Mexico, from 5 through 9 October 1976,* edited by Ernst A. Steinhoff, part 1, 258–69. Alexandria, Va.: American Astronautical Society, 1976.

Hall, Eldon C. "The Apollo Guidance Computer: A Designer's View." In *Computer Museum Report*. Boston: Computer Museum, 1982.

———. "From the Farm to Pioneering with Digital Control Computers: An Autobiography." *IEEE Annals of the History of Computing* 22 (April–June 2000): 22–31.

Hawkins, Willis M. "Levering Smith 1910–1999." *Memorial Tributes: National Academy of Engineering* 7 (1994): 214–20.

Herther, John C., and James S. Coolbaugh. "Genesis of Three-Axis Spacecraft Guidance, Control, and On-Orbit Stabilization." AIAA *Journal of Guidance, Control, and Dynamics* 29, no. 6 (November–December 2006): 1247–70.

"The History of the MIT Department of Physics: New Labs and New Frontiers—1916–1939." *MIT Department of Physics,* http://web.mit.edu/physics/about/history/1916-1939.html.

"Hot Claims." *Time,* April 29, 1957, 60, 63. Reproduced in Naval Surface Warfare Center, Dahlgren Division *Technical Digest* (1995).

Hughey, Raymond H., Jr. "History of Mathematics and Computing Technology at the Dahlgren Laboratory." Naval Surface Warfare Center, Dahlgren Division *Technical Digest* (1995): 8–21.

"The IAS: Early Years (1932 to 1945)." *American Institute of Aeronautics and Astronautics,* https://www.aiaa.org/SecondaryTwoColumNo.aspx?id=1521.

"Ivy H. W. Draper Dies at 85." *MIT News,* http://newsoffice.mit.edu/1994/draper-0518. (A version of this article appeared in *MIT Tech Talk* 38, no. 33 [May 18, 1994].)

Jackson, Todd. "Modifying the MARK 6 Guidance System Part 1." *Evaluation Engineering* (September 2009), http://www.evaluationengineering.com/articles/200909/modifying-the-mark-6-guidance-system-part-1.

———. "Modifying the MARK 6 Guidance System Part 2." *Evaluation Engineering* (October 2009), http://www.evaluationengineering.com/articles/200910/modifying-the-mark-6-guidance-system-part-2.php.

Johnson, Howard W. "Statement by Howard W. Johnson on the Special Laboratories, October 22, 1969," as reproduced in Dorothy Nelkin, *The University and Military Research: Moral Politics at M.I.T.* (Ithaca, N.Y.: Cornell University Press, 1972), app. 2, 168–77.

Klass, Philip J. "Inertial Guidance." *Aviation Week Special Report*. New York: McGraw-Hill, 1956.

———. "MIT Sparks Inertial Guidance Efforts." *Aviation Week,* August 10, 1959, 71–77.

———. "Thor Guidance Goes on Production Line." *Aviation Week,* December 30, 1957, 38–44.

Koutosumpos, George. "Submarine Launched ICBM Trident D5 Conventional Trident Modification." *Journal of Computations & Modeling* 4, no. 1 (2014): 245–65.

Krige, John. "The 1984 Physics Prize for Heterogenous Engineering," *Minerva* 39, no. 4 (December 2001), 426.

Kunin, Jay, Alex Makowski, and Greg Bernhardt. "Johnson Appoints Panel to Review Labs Group: Special Faculty Meeting Asks 'Involvement' Study," *Tech* 89, no. 20 (April 29, 1969): 1.

"Laboratory Aviator Extraordinaire: Chip Collins Reflects on a 40-Year Draper Career." *D-notes,* May 20, 1988, 3.

Laconture, John. "Early Aviation on Nantucket." *Historic Nantucket* 49, no. 1 (Spring 1992): 12–15. Reprinted *Nantucket Historical Association,* https://www.nha.org/library/hn/HN-fall92-aviation.htm.

LaFond, Charles D. "FBM Accuracy Starts with SINS." *Missiles and Rockets,* July 25, 1960, 24–25.

Lange, K. O. "On the Technique of Meteorological Airplane Ascents." In *The Meteorological Airplane Ascents of the Massachusetts Institute of Technology.* Papers in Physical Oceanography and Meteorology III, no. 2. Cambridge Mass.: Massachusetts Institute of Technology and Woods Hole Oceanographic Institution, 1934.

Laurence, William L. "'Inertial Guidance System' Can Free Planes and Missiles from Present Limitations." *New York Times,* April 21, 1957, B-9.

Law, John. "Technology and Heterogeneous Engineering: The Case of Portuguese Expansion." In *Social Construction of Technological Systems: New Directions in the Sociology and History of Technology,* edited by Wiebe E. Bijker, Thomas Hughes, and T. J. Pinch. Cambridge, Mass.: MIT Press, 1989.

Leavitt, William. "A Triumph of Reverse McCarthyism: The Dethronement of Dr. Draper." *Air Force Space Digest* (December 1969): 46–50.

Lécuyer, Christophe. "Patrons and a Plan." In *Becoming MIT,* edited by David Kaiser. Cambridge, Mass.: MIT Press, 2010.

Leslie, Stuart W. "'Time of Troubles' for the Special Laboratories." In *Becoming MIT,* edited by David Kaiser. Cambridge, Mass.: MIT Press, 2010.

"LGM-118A Peacekeeper." *Federation of American Scientists,* https://fas.org/nuke/guide/usa/icbm/lgm-118.htm [accessed April 8, 2019].

Lincoln Laboratory, Massachusetts Institute of Technology. "Cape Cod SAGE Prototype." Part 1 & 2. *Lincoln Laboratory,* https://www.ll.mit.edu/about/History/capecodprototype.html. [The article is no longer available.]

———. "Lincoln Laboratory Origins." Part 1 & 2. *Lincoln Laboratory,* https://www.ll.mit.edu/about/History/origins.html [This article is no longer available.]

MacKenzie, Donald. "Missile Accuracy: A Case Study in the Social Process of Technological Change." In *Social Construction of Technological Systems: New Directions in the Sociology and History of Technology,* edited by Wiebe E. Bijker, Thomas Hughes, and T. J. Pinch. Cambridge, Mass.: MIT Press, 1989.

———. "Stellar-Inertial Guidance: A Study in the Sociology of Military Technology." In *Science, Technology and the Military,* edited by Everett Mendelsohn, Merritt Roe Smith, and Peter Weingart, eds. Dordrecht, Neth.: Kluwer Academic, 1989.

Makowski, Alex, and Duff McRoberts. "Labs Threaten Faculty Split," *Tech* 89, no. 51 (December 12, 1969).

Malley, Ken, Vice Adm. USN (Ret.). "Trident II: Flipping a Flop," *U.S. Naval Institute Proceedings* (January 2013): 62–68.

Martin, Everett G. "Thor 'Gun Barrels' Cleaned." *Boston Monitor,* January 18, 1958. (News clipping, CSDL Collection, MIT Museum, Cambridge, Mass.)

Massachusetts Institute of Technology, Department of Aeronautics and Astronautics. "A Brief History of MIT Aeronautics and Astronautics." *MIT AeroAstro,* http://aeroastro.mit.edu/about-aeroastro/brief-history.

Mattorano, Gino. "AF Announces Space, Missile, Pioneer Awards." *Air Force Magazine,* November 2003, 18.

May, Marvin. "SINS Mk2 Mod 6." *Institute of Navigation Museum,* http://www.ion.org/museum/item_view.cfm?cid=2&scid=4&iid=4.

McCutcheon, Kimble D. "No Short Days: The Struggle to Develop the R-200 'Double Wasp' Crankshaft." *AAHS Journal* 46, no. 2 (Summer 2001): 125–46. Available at http://www.enginehistory.org.

McElheny, Victor K. "MIT Administration Makes Public Its Intentions on Disposition of Draper and Lincoln Laboratories." *Science* 168 (May 29, 1970): 1074–75.

McKay, Walter. "Dynamic Coefficients of the Aircraft Magnetic Compass." *Journal of Aeronautical Sciences* 2 (March 1932): 60.

McKelvie, B., and H. Galt Jr., "Evolution of the Ship's Inertial Navigation System for the Fleet Ballistic Missile Program," *Navigation: Journal of the Institute of Navigation* 25, No. 3 (Fall 1978), 315.

"Men of the Year: U.S. Scientists," *Time* 77, no. 1 (January 2, 1961): 40–46.

Mollela, Arthur. "Welcome to a Hot Spot of Invention." *Hot Spot of Invention: People Places, and Spaces.* New Perspectives Invention & Innovation Symposium, Lemelson Center, Museum of American History, November 6–7, 2009.

Murray, Don. " 'Doc' Draper's Wonderful Tops." Reprint *Readers Digest* (September 1957).

Nelson, Bryce. "Scientists Plan Research Strike at M.I.T. on 4 March." *Scientist* 163 (January 24, 1969): 373.

Normyle, William J. "Milwaukeeans Guide a Missile to Perfection." *Milwaukee Journal,* November 4, 1958, 21.

Norris, Robert S. "Counterforce at Sea: The Trident II Missile." *Arms Control Today* (September 1985): 5–10.

Ong, Elwin C. "Profile Apollo: Hugh Blair-Smith." *NASA Office of Logic Design,* http://klabs.org/history/bios/hugh_blair_smith/elwin_mit_report.htm.

Osborn, Kris. "Missile Improves Navigation, Targeting, Firing." *Defense Maven,* https://defensemaven.io/warriormaven/sea/new-nuclear-armed-trident-ii-d5le-missile-improves-navigation-targeting-firing-7RF-NUywTE6uziDLoEjgoA/ [accessed April 9, 2019].

Osborn, Robert R. "The Curtiss Robin." *Aviation,* May 21, 1928. Available at http://www.airminded.net/curtrob/robinarticle.html.

Palmer, Loren. "Flying without Wings Has Its Thrills." *Popular Science,* July 1919, 63.

Petosky, Henry. "The Draper Prize." *American Scientist* 82 no. 2 (March/April 1994): 114–18.

Pickens, Tom. "Doc Gyro and His Wonderful 'Where Am I?' Machine." *American Way* 5, no. 4 (April 1972): 24–27.

"Pioneer Alumnus Inducted into Space and Missile Hall of Fame." *Penn State News,* http://news.psu.edu/story/175059/2009/08/13/pioneer-alumnus-inducted-space-and-missile-hall-fame.

Raborn, W. F., and John P. Craven. "The Significance of Draper's Work in the Development of Naval Weapons." In *Air Space and Instruments,* edited by Sidney Lees. New York: McGraw-Hill, 1963.

"Regulus I RGM-6 SSM-N-6." GlobalSecurity.org, http://www.globalsecurity.org/wmd/systems/regulus1.htm.

"Ruggles Orientator." *U.S. Air Service,* March 13, 1928, 20.

Seamans, Robert C., Jr. "Past, Present, and Future." In *Air Space and Instruments,* edited by Sidney Lees. New York: McGraw-Hill, 1963.

Sherman, Elizabeth J. "Charles Stark Draper: The Man Who Set the World Straight." *Natural Science,* October 1990, available at http//www.worldandi.com.

"Sixtieth Graduation Anniversary of the Class of 1940." *United States Naval Academy,* https://www.usna.com/NC/History/ClassOf1940/Contents.htm.

Smith, Levering, Robert H. Wertheim, and Robert A. Duffy. "Innovative Engineering in the Trident Missile Development." *Bridge* 10, no. 2 (Summer 1980): 10–19.

Steinhoff, Ernst A. "Charles Stark Draper Biography Part II." In *The Eagle Has Returned: Proceedings of the Dedication Conference of the International Space Hall of Fame, Held at Alamogordo, New Mexico, from 5 through 9 October 1976,* edited by Ernst A. Steinhoff, part 1, 12–15. Alexandria, Va.: American Astronautical Society, 1976.

———. "Comments on Dr. Stevens's Introduction." In *The Eagle Has Returned: Proceedings of the Dedication Conference of the International Space Hall of Fame, Held at Alamogordo, New Mexico, from 5 through 9 October 1976,* edited by Ernst A. Steinhoff, part 2, 29–32. Alexandria, Va.: American Astronautical Society, 1976.

Sternhell, Carol R. "Police Rout NAC Pickets in Protest at M.I.T. Lab." *Harvard Crimson,* November 6, 1969, http://www.thecrimson.com/article/1969/11/6/police-rout-nac-pickets-in-protest/.

Stone, Irving. "Thor Designed for Fast Mass Production." *Aviation Week,* December 2, 1957, 26–29.

Stratton, Julius A. "Charles Stark Draper: An Appreciation." In *Air Space and Instruments,* edited by Sidney Lees. New York: McGraw-Hill, 1963.

———. "Karl Taylor Compton: A Biographical Memoir." In National *Academy of Sciences Biographical Memoirs.* Washington, D.C.: National Academy of Sciences, 1994.

Taylor, E. S., and C. S. Draper. "A New High-Speed Engine Indicator." *Mechanical Engineering* 55, no. 3 (March 1933): 169–71.

Taylor, E. S., C. S. Draper, and G. L. Williams. "A New Instrument Devised for the Study of Combustion." *SAE Transactions* (1934): 59–62.

"Taylor's Career Took Wing with WWI Plane Engine Testing." *MIT News,* September 14, 1994, http://newsoffice.mit.edu/1994/taylor-0914.

"Third Annual Meeting of the Institute of the Aeronautical Sciences." *Journal of the Aeronautical Sciences* 2 (1938): 53–60.

"Three to Be Honored for Aiding Aviation: Aeronautical Institute Will Make Awards on Wednesday at Annual Meeting Here." *New York Times,* January 28, 1935, online archives.

Trageser, Milton B., and David G. Hoag. "Apollo Space Spacecraft Guidance System." In *Peaceful Uses of Automation in Outer Space: Proceedings of the Symposium,* edited by John A. Aseltine, 435–66. New York: Plenum, 1966.

"Trident II D-5 Fleet Ballistic Missile: Recent Developments." GlobalSecurity.org, http://www.globalsecurity.org/wmd/systems/d-5-recent.htm.

Tylko, John. "MIT and Navigating the Path to the Moon." *MIT AeroAstro* (2008–2009), available at http://web.mit.edu/aeroastro/news/magazine/aeroastro6/mit-apollo.html.

U.S. Navy, Strategic Systems Public Affairs. "Back to the Future with Trident Life Extension." *Undersea Warfare* (Spring 2012): 8–11.

Walsh, John. "M.I.T.: Panel on Special Labs Asks More Nondefense Research." *Science* 164, June 13, 1969, 1264–65.

Ward, Jack. "Texas Instruments R212." *Transistor Museum Historic Photo Gallery,* www.semiconductormuseum.com/PhotoGallery/PhotoGallery_TI_R212.htm.

Wieser, C. Robert. "From World War II Radar to Sage." *Computer Museum* 22 (1988): 15–16.

Wildenberg, Thomas. "The A-1C(M) Gunsight: A Case Study of Technological Innovation in the United States Air Force." *Air Power History* 56, no. 2 (Summer 2009): 28–37.

———. "The Shoebox That Transformed Antiaircraft Firecontrol." *Naval History* 27, no. 6 (December 2013).
Wilson, Gill R. "Sperry Industrial Giant of the Air Age." *Flying* 68, no. 4 (April 1961): 46–47, 70, 72.
Wright, Charles. "Polaris-Poseidon-Trident: An Awesome Concept—Sea-Launched Ballistic Missile Systems; General Electric Ordnance System–United States Navy." Revised September 10, 2009. *Polaris: The Free World's Most Effective Deterrent,* www.navsource.org/archives/08/PolarisHistoryGEX1.htm.
Wrigley, Walter. "The History of Inertial Navigation." *Journal of Navigation* 30, no. 1 (January 1977): 61–67.
———. "Schuler Tuning Characteristics in Navigational Instruments." *Navigation* 2, no. 8 (December 1950): 282–90.
Wuerth, J. M. "Minuteman Guidance and Control." *Navigation: Journal of the Institute of Navigation* 23, no. 1 (Spring 1976): 64–75.

Dissertations, Papers, and Other Unpublished Materials

Aeronautical Engineering Department, Weapons System Section. Report Prepared for Dean Soderberg, September 1955. MIT, School of Engineering, Office of the Dean, Records, 1943–1989, AC 12, MIT Archives, Cambridge, Mass.
Air Force Systems Command, Aeronautical Systems Division. *Development of Airborne Armament 1910–1961.* Vol. 3, *Fighter Fire Control.* Wright Field, Ohio: Historical Division, Office of Information, Aeronautical Systems Division, Air Force Systems Command, 1961.
Air Technical Service Command, U.S. Army Air Forces. "A-1 Combination Gun-Bombing Sight." Classified Project No. MX-402. In *Research and Development Projects of the Engineering Division.* 6th ed., January 1, 1945. Historians Office, Air Force Materiel Command, Wright Field, Ohio.
"Background on Instrumentation Laboratory Massachusetts Institute of Technology." News Releases, Titan, Charles Stark Draper Historical Collection, MIT Museum, Cambridge, Mass.
Barbour, Neil M., John M. Elwell, and Roy H. Setterlund. "Inertial Instruments: Where To Now?" AIAA Guidance, Navigation and Control Conference, Hilton Head Island, S.C., August 10–12, 1992, Technical Papers. Part 2 (A92-55151 23-63). Washington, D.C., American Institute of Aeronautics and Astronautics, 1992, 566–74.
Biographical Register of the Officers and Graduates of USMA. Special Collections, USMA Library, United States Military Academy at West Point, http://www.library.usma.edu/archives/special.asp.
Bridger, Sarah. "Scientists and the Ethics of Cold War Weapons Research." PhD diss., Columbia University, New York, N.Y., 2011.
Bogosian, A. G. "An Experimental Automatic Long Range Guidance System, Project FEBE." Preface of Scientific Advisory Board, *Seminar on Automatic Celestial and Inertial Long Range Guidance Systems at Massachusetts Institute of Technology,* Cambridge, Mass., February 1, 2, and 3, 1949.
Charles Stark Draper Laboratory. "CSDL-P-1740 Evolution of the FBM Guidance System." Presentation to the AFFTD, April 26, 1983.
Chilton, Robert G. and Alfred R. Seville. "Distance Measurement by *Inertial* Means," MS thesis, Dept. of Aeronautical Engineering, Massachusetts Institute of Technology, Cambridge, Mass., 1949.

Concord Free Public Library. Finding Aid: Massport Hangar 24 Collection, 1948–2010.

Cook, Robert H., and Boyd E. Gustafson. "Applications of Inertial Navigation to Undersea Warfare." MS thesis in mechanical engineering, Massachusetts institute of Technology, Cambridge, Mass., May 18, 1951.

Covington, William O. "Orientation Control Study." MA thesis, Massachusetts Institute of Technology, Cambridge, Mass., 1955.

Dennis, Michael Aaron. "A Change of State: The Political Cultures of Technical Practice at the MIT Instrumentation Laboratory and the Johns Hopkins University Applied Physics Laboratory, 1930–1945." PhD diss., Johns Hopkins University, 1990.

Dow, Paul C., Jr. "History of the Trident Guidance Program at Draper Laboratory." CSDL-C-6455, September 1997. Charles Stark Draper Laboratory, Cambridge, Mass.

Draper, Charles Stark. "The Inertial Gyro: An Example of Basic Research." Presented at the National Convention of the Scientific Research Society of America, held jointly with the Annual Meeting of the American Association for the Advancement of Science, Chicago, Ill., December 29, 1959.

———. "The Instrumentation Laboratory of the Massachusetts Institute of Technology Remarks by Dr. C. S. Draper, Director of the Laboratory from Its Beginnings until the Present Time (1969)," Charles Stark Draper Papers, Manuscript Collection, Library of Congress, Washington, D.C.

———. "A Method for Detecting Detonation Waves in the Internal Combustion Engine." MS thesis, Massachusetts Institute of Technology, 1928.

———. "On the Evolution of Accurate Inertial Guidance Instruments." In *The Eagle Has Returned: Proceedings of the Dedication Conference of the International Space Hall of Fame, Held at Alamogordo, New Mexico, from 5 through 9 October 1976*, edited by Ernst A. Steinhoff, part 2, 23–28. Alexandria, Va.: American Astronautical Society, 1976.

———. "The Physical Processes Accompanying Detonation in the Internal Combustion Engine." PhD diss., Massachusetts Institute of Technology, 1938.

———. "Submarine Inertial Navigation: A Review and Some Predictions." Paper presented to the Polaris Steering Task Group, October 22, 1959. CSD-107, courtesy CSDL.

Emanuel, Kerry. "A Brief History of Meteorology and Oceanography at MIT." PowerPoint presentation, available at ftp://texmex.mit.edu/pub/emanuel/Powerpoint/PAOC_history.pptx.

Ferguson, J. Scott. "C-5249 Historical Collection Projects." Charles Stark Draper Laboratory, August 3, 1979. (Copy obtained from MIT Museum, Cambridge, Mass.)

Forter, Samuel A. "A Light-Weight Multi-Channel Telemetering System." MSEE thesis, MIT, 1947.

Greene, Warren E. "The Development of the SM-68 Titan." Historical Office, Deputy Commander for Aerospace Systems, Air Force Systems Command, August 1962.

Hall, Eldon C. "P-3357 MIT's Role: Project Apollo Vol III, Computer Subsystem." Charles Stark Draper Laboratory, MIT, August 1972. (Courtesy CSDL.)

Herther, John C., and James S. Coolbaugh. "Genesis of Three-Axis Spacecraft Guidance, Control, and On-Orbit Stabilization." AIAA *Journal of Guidance, Control, and Dynamics* Vol. 29, No. 6, November–December 2006.

Hoag, David G. "The History of Apollo On-Board Guidance, Navigation, and Control." Cambridge, Mass.: Charles Stark Draper Laboratory, September 1976. (Courtesy Charles Stark Draper Laboratory.) Also published in *The Eagle Has Returned:*

Proceedings of the Dedication Conference of the International Space Hall of Fame, Held at Alamogordo, New Mexico, from 5 through 9 October 1976, edited by Ernst A. Steinhoff, part 1. Alexandria, Va.: American Astronautical Society, 1976.

Hopkins, R. E., Fritz K. Mueller, and Walter Haeussermann. "The Pendulous Integrating Gyroscope Accelerometer (PIGA) from the V-2 to Trident D5, the Strategic Instrument of Choice." CSDL-P-3923. Paper presented at the AIAA Guidance, Navigation, and Control Conference & Exhibit, August 6–9, 2001, Montreal, Canada.

House, F. R. "The Navy Mark 14 Gunsight to July 1943." Unpublished history, Sperry Gyroscope Collection, Hagley Museum and Library, Wilmington, Del.

Jackson, Todd, with David Monti, Scot Berry, James Connelly, and Michael Murphy. "Simulation-Based Integration Approach for the Trident MK6 Life Extension Guidance System." AIAA Missile Sciences Conference, Monterey, Calif., November 18–20, 2008.

Johnson, Howard W. "Draper Chair Dedication," MIT Faculty Club, April 25, 1978. (Courtesy Debbie Douglas, MIT Museum.)

Laning, J. Halcombe, Jr. "Mathematical Theory of Lubrication-Type Flow," PhD diss., Massachusetts Institute of Technology, Cambridge, Mass., 1947.

Lees, Sidney. "Integration by Viscous Sear Process." ScD thesis in aeronautical engineering, Massachusetts Institute of Technology, Cambridge, Mass., 1950

———. "Investigation of the Performance of Experimental Shear Damper." Master's thesis in aeronautical engineering, Massachusetts Institute of Technology, Cambridge, Mass., 1948.

Lipp, J. E., and R. M. Slater, eds. "Project Feed Back Summary Report." Vol. 1, March 1, 1954. RAND Corporation, Santa Monica, Calif.

Lockheed Missiles and Space Company. *The Fleet Ballistic Missile System: From Polaris to Poseidon and Trident.*

Machover, Carl. *Gyro Primer . . . or "Selling Gyros for Fun and Profit."* Commack, N.Y.: Norden, Division of United Aircraft Corporation, 1960.

Massachusetts Institute of Technology, *Instrumentation Laboratory, Cambridge, Massachusetts.* MIT, April 1968. (Copy courtesy of Draper Laboratory.)

National Air and Space Museum. "Camera, Balloon Recon, HYAC," https://airandspace.si.edu/collection-objects/camera-balloon-recon-hyac [accessed April 10,2019].

Old, Bruce S. "Return on Investment in Basic Research: Exploring a Methodology." DTIC AD A111283. Department of the Navy, Office of Naval Research, November 1981.

"Report of the Daniel Guggenheim Fund for the Promotion of Aeronautics 1926 and 1927." Daniel Guggenheim Fund for the Promotion of Aeronautics, New York, N.Y.

"Rivero, Horacio, Admiral USN (Retired)," official biography, Navy Department Library, Washington, D.C.

Romanowski, Roslyn R. "Peacetime to Wartime: Transitions in Defense Policy at MIT." BS thesis, Massachusetts Institute of Technology, 1982.

Schmidt, George T. "Strapdown Inertial Systems: Theory and Applications—Introduction and Overview." AGARD Lecture Series No. 95, Strap-Down Inertial Systems. North Atlantic Treaty Organization Advisory Group for Aerospace Research and Development, May 1978.

Seamans, Robert C., Jr. "Comparison of Automatic Tracking System for Interceptor Aircraft." PhD diss., Massachusetts Institute of Technology, 1951.

Sheehan, Neil. *A Fiery Peace in a Cold War: Bernard Schriever and the Ultimate Weapon.* New York: Vantage, 2009.

Spinardi, Graham. "From Polaris to Trident: The Development of US Fleet Ballistic Missile Technology." PhD diss., University of Edinburgh, 1988.
Trageser, Milton B., and David G. Hoag. *R-495 Apollo Spacecraft Guidance System*. MIT Instrument Laboratory, Cambridge, Mass., 1965.
Ulmann, Bernd. *AN/FSQ-7: The Computer That Shaped the Cold War*. Berlin: De Gruyter Oldenbourg, 2014.
U.S. Air Force. "United States Air Force Operations in the Korean Conflict 1 July 1952–27 July 1953." USAF Historical Study No. 127, Historical Division, Air University, Maxwell Air Force Base, 1956.
Weschler, Thomas R. *The Reminiscences of Vice Admiral Thomas R. Weschler, U.S. Navy (Retired)*, Vol. 1. Interviewed between October 1982 and September 1984; copy located in the Navy Department Library, Washington, D.C.
Wrigley, Walter. "An Investigation of Methods Available for Indicating the Direction of the Vertical from a Moving Base." PhD diss., Massachusetts Institute of Technology, 1941.
Young, Lawrence R. "My Twenty-Five Years with the MIT Man-Vehicle Laboratory (1962–1967)." MLV@50, Man-Vehicle Laboratory, Massachusetts Institute of Technology, 2012.

Technical Reports, Public Documents, and Patents

Draper, Charles S. "The Physical Effects of Detonation in a Cylindrical Chamber." Report No. 493, National Advisory Committee for Aeronautics, Washington, D.C., January 1, 1935.
———. "The Sonic Altimeter for Aircraft." Technical Report No. 611, National Advisory Committee for Aeronautics, Washington, D.C., 1937.
———. "Turn Indicator." U.S. Patent No. 2,291,612, issued August 4, 1942.
Draper, Charles S., et al. "Detonation Detector System." U.S. Patent No. 2,275,675, issued March 10, 1942.
———. "Gunsight Having Lead Computing Device." U.S. Patent No. 2,609,606, issued September 9, 1952.
Draper, Charles S., Newton Bentley, and Edward Bentley. "Lead Angle Computer for Gun Sights." U.S. Patent No. 2,690,014, filed March 29, 1941, issued September 28, 1954.
Draper, Charles S., and Claude L. Emmerich, "Motion Sensor," U.S. Patent No. 2,853,287, filed March 13, 1952.
Draper, Charles S., and Walker McKay. "Magnetic Compass." U.S. Patent No. 2,248,748, filed January 21, 1938, issued July 8, 1941.
Geiger, Clarence J. "History of the X-20A Dyna-Soar." Vol. I. AFSC Historical Publications Series 63-50-I, September 1963.
Grothe, Lester R., Edward J. Hall, Michele S. Sapuppa, and Andrew E. Scoville. "The MIT 25 IRIG Inertial Reference Integrating Gyro Unit and the MIT 25 PIG Pendulous Integrating Gyro Unit." Report R-141, Instrumentation Laboratory, Massachusetts Institute of Technology, January 1958. (Courtesy of the Charles Stark Draper Laboratory, Cambridge, Mass.)
Grothe, Lester R., and Hugh H. McArdle. "The 10-Series of Floated Instruments." Report R-66, Report R-141, Instrumentation Laboratory Massachusetts Institute of Technology, July 1955 (revised printing October 1957). (Courtesy of the Charles Stark Draper Laboratory, Cambridge, Mass.)
Houston, Forrest E., and John Hovorka. "Report R-9 [Part I], Theoretical Background of Inertial Navigation for Submarines." DTIC. March 1951.

———. "Report R-9, Part II, Characteristics of Systems Feasible for Inertial Navigation of Submarines." DTIC. August 1951.

Instrumentation Laboratory, Massachusetts Institute of Technology. "Gunsight Mark 15 for the Control of Short-and Medium-Range Antiaircraft Fire from Naval Vessels." NavOrd Report 3-47. Cambridge, Mass., April 1949.

———. "R-63 Project Spire Flight Test Program." November 1953. CSDL Historical Collection, MIT Museum, Cambridge, Mass.

———. "Report R-193 The SPIRE, Jr. Long-Range Inertial Guidance System." CSDL Historical Collection, MIT Museum, Cambridge, Mass.

———. "Report R-299 Final Report for Contract AF 04(645)-9 WS-107A Inertial Guidance," January 1, 1955, through December 31, 1958. CSDL Historical Collection, MIT Museum, Cambridge, Mass.

Kelly, Marion R. "VT Fuse." Draft Manuscript Dated 9 July 1962. Copy located in Navy Department Library, Washington, D.C.

Leacok, Richard. *November Action* [documentary video], 2017 restored version, MIT Museum, Cambridge, Mass.

"Lieutenant General Leighton I. Davis." U.S. Air Force, www.af.mil/.

Magrath, Howard A. R&D Contributions to Aviation Progress (RADCAP) Study, Vol. II, Appendix 6 Air Vehicle Technology, August, 1972, NTIS.

Mueller, F. K. *A History of Inertial Guidance*. N.d. Army Ballistic Missile Agency, Redstone Arsenal, Ala. DTIC No. AD 419538.

Slater, John. "Self-Compensating Gyro Apparatus." U.S. Patent No. 2,999,391A, filed December 11, 1950, issued September 12, 1961.

Slater, John, and Walter Wrigley. "Flight Control System." U.S. Patent No. 2,649,264, filed March 15, 1947, issued August 18, 1953.

Sontag, Harcourt, and Daniel J. Harcourt. "Performance Characteristics of Venturi Tubes Used in Aircraft for Operating Air-Driven Gyroscopic Instruments." Report No. 624, National Advisory Committee for Aeronautics, Washington, D.C., November 1937.

Sperry, Elmer A., Jr., "Magnetic Compass." U.S. Patent No. 2,280,726, filed June 10, 1939, issued April 21, 1942.

U.S. Air Force. "A Brief History of Hanscom Air Force Base." *Hanscom Air Force Base*, https://www.hanscom.af.mil/About-Us/Fact-Sheets/Display/Article/379480/a-brief-history-of-hanscom-air-force-base/ [accessed April 6, 2019].

———. "B-26 Bomber, Martin or Douglas?" *Air Force Historical Support Division*, 2011, https://www.afhistory.af.mil/FAQs/Fact-Sheets/.../b-26-bomber-martin-or-douglas/ [accessed April 6, 2019].

———. Scientific Advisory Board. "Seminar on Automatic Celestial and Inertial Long Range Guidance Systems at Massachusetts Institute of Technology." Cambridge, Mass., February 1, 2, and 3, 1949, Vols. 1 and 3.

U.S. Congress, Government Accountability Office. "Kearfott Guidance & Navigation Corporation," B-292895.2, May 25, 2004, Washington, D.C.

U.S. Congress, Senate Appropriations Committee. "Department of Defense Appropriations Fiscal Year 1974," 93rd Cong. 1st Sess.

U.S. Congress, Senate Armed Services Committee, Strategic Forces Subcommittee. "Statement of Rear Admiral Terry Benedict, USN, Director, Strategic Systems Programs Before the Subcommittee of Strategic Forces of the Senate Armed Services Committee" FY2013 Strategic Systems, March 28, 2012, Washington, D.C.

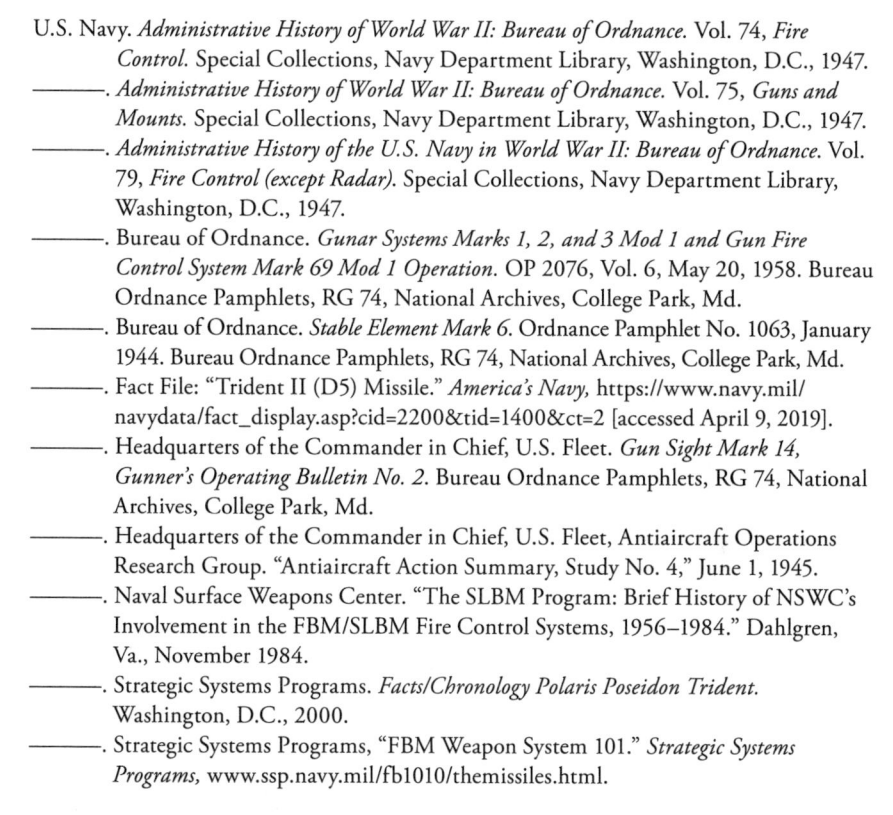

U.S. Navy. *Administrative History of World War II: Bureau of Ordnance.* Vol. 74, *Fire Control.* Special Collections, Navy Department Library, Washington, D.C., 1947.
———. *Administrative History of World War II: Bureau of Ordnance.* Vol. 75, *Guns and Mounts.* Special Collections, Navy Department Library, Washington, D.C., 1947.
———. *Administrative History of the U.S. Navy in World War II: Bureau of Ordnance.* Vol. 79, *Fire Control (except Radar).* Special Collections, Navy Department Library, Washington, D.C., 1947.
———. Bureau of Ordnance. *Gunar Systems Marks 1, 2, and 3 Mod 1 and Gun Fire Control System Mark 69 Mod 1 Operation.* OP 2076, Vol. 6, May 20, 1958. Bureau Ordnance Pamphlets, RG 74, National Archives, College Park, Md.
———. Bureau of Ordnance. *Stable Element Mark 6.* Ordnance Pamphlet No. 1063, January 1944. Bureau Ordnance Pamphlets, RG 74, National Archives, College Park, Md.
———. Fact File: "Trident II (D5) Missile." *America's Navy,* https://www.navy.mil/navydata/fact_display.asp?cid=2200&tid=1400&ct=2 [accessed April 9, 2019].
———. Headquarters of the Commander in Chief, U.S. Fleet. *Gun Sight Mark 14, Gunner's Operating Bulletin No. 2.* Bureau Ordnance Pamphlets, RG 74, National Archives, College Park, Md.
———. Headquarters of the Commander in Chief, U.S. Fleet, Antiaircraft Operations Research Group. "Antiaircraft Action Summary, Study No. 4," June 1, 1945.
———. Naval Surface Weapons Center. "The SLBM Program: Brief History of NSWC's Involvement in the FBM/SLBM Fire Control Systems, 1956–1984." Dahlgren, Va., November 1984.
———. Strategic Systems Programs. *Facts/Chronology Polaris Poseidon Trident.* Washington, D.C., 2000.
———. Strategic Systems Programs, "FBM Weapon System 101." *Strategic Systems Programs,* www.ssp.navy.mil/fb1010/themissiles.html.

Oral Histories and Interviews

Battin, Richard H. "A Funny Thing Happened on the Way to the Moon." Transcript. Fall 2008 lecture Astrodynamics Course 16.346, MIT Open Course Ware, http://ocw.mit.edu/courses/aeronautics-and-astronautics/16-346-astrodynamics-fall-2008/video-lecture/MIT16-346F08.pdf.
———. Interview by Rebecca Wright, Lexington, Mass., April 18, 2000. NASA Johnson Space Center Oral History Project, http://www.jsc.nasa.gov/history/oral_histories/BattinRH/BattinRH_4-18-00.htm.
Blair-Smith, Hugh. "Hugh Blair-Smith's Introduction." Apollo Guidance Computer History Project, Third Conference, November 30, 2001, http://authors.library.caltech.edu/5456/1/hrst.mit.edu/hrs/apollo/public/conference3/blairsmith.htm.
Burnett, James R. Oral History, January 19, 1989. Glennan-Webb-Seamans Project for Research in Space History, National Air and Space Museum, Smithsonian Institution, Washington, D.C.
Campbell, George. "Recollections." Available at http//acdelco.org/Documents/Delco%20History%20Recollections.%20George%20Campbell. [Accessed February 4, 2009. The site is no longer available.]
Chilton, Robert G. Oral History Transcript, Johnson Space Center Oral History Project, National Aeronautics and Space Administration.
Collins, Charles "Chip." Interview by Thomas Wildenberg, October 16, 2012, Westford, Mass.

Davis, Leighton I. "Interview Conducted by Maj. Lyn R. Officer and Hugh Ahmann, Burbank, California, April 26, 1973." U.S. Air Force Oral History Program, IRIS No. 00904753, Air Force Historical Research Agency, Maxwell Air Force Base, Montgomery, Ala.

Ditmeyer, Martha Stark (née Draper). Interview by Thomas Wildenberg, December 30, 2015, Alexandria, Va.

Draper, Charles S. "Transcript of a Tape-Recorded Interview with C. Stark Draper." Conducted by Barton Hacker, MIT Oral History Program, January 19, 1976; February 2, 1976; March 1, 1976; and April 5, 1976, Cambridge, Mass.

Draper, James S. Interview by Thomas Wildenberg, November 15, 2012, Newton, Mass.

Mettler, Rueben. Interview by Martin Collins, June 26, 1990, Los Angeles, Calif. NASM Archives.

Rivero, Horacio. *Reminiscences of Admiral Horacio Rivero, USN (Retired)*. U.S. Naval Institute, Annapolis, Md., 1978. Copy in Navy Department Library, Washington, D.C.

Robinson, Frank P., Jr. Telephone interview by Thomas Wildenberg, October 10, 2008. Author's collection.

Rosenblith, Walter A. Interview by Eden Miller, July 26, 2000, Marston Mills, Mass. MIT Archives and Special Collections.

Seamans, Robert. Interviewed by Martin Collins, December 4, 1986, at Dr. Seamans' home in Cambridge, Massachusetts. NASA Oral History Project.

———. Interviewed by Martin Collins and Michael Dennis, February 25, 1987, MIT. NASA Oral History Project.

———. Interviewed by Martin Collins, April 9, 1987, National Academy of Science. NASA Oral History Project.

Spilhaus, Athelstan F. Interview by Ronald E. Doel, November 10, 1989, in Middleburg, Va. Neils Bohr Library and Archives, American Institute of Physics, College Park, Md.

Wertheim, Rear Adm. Robert. Interview by staff, October 3, 2012, at Charles Stark Draper Laboratory, Cambridge, Mass. (Courtesy Draper Laboratory.)

Periodicals
Aero Digest
Air Corps News
Air Force Magazine
Boston Business Journal
Boston Globe
Charles Stark Draper Laboratory Annual Reports
Defense Industry Daily
Massachusetts Institute of Technology Bulletin, President's Report / Report of the President
Niagara Falls Gazette
Popular Aviation
The Tech

Index

Page numbers in italics indicate figures.

A1-C gunsight, 155
   advantages over existing sights, 87
   B model for F-84 and F-86, 88
   and Dummy Gun System, 101
   experimental sight, 87
   flight testing of experimental units, 87–88
   and HIG-5 gyro, 118
   idea for, 85
   introduced on F-84E, 92
   modification of, 90
   operation and maintenance, 88, 90, 91
   post–World War II testing, 90
   prototype contract awarded MIT, 85
   reticle jitter, 90
   schematic, *89*
   shown mounted in F-86, *91*
   Sperry selected to manufacture, 90
A-4 gunsight, 92
AC Spark Plug Div. General Motors Corp.
   and A1-C, 90, 103
   ICBM inertial guidance, 161, 226
   Thor guidance, 155–56
   Titan II guidance, 158, 159
accelerometers
   10 PIGA, 184
   16 PIGA, 161
   16 PIPA, 173, 178
   25 PIGA, 123, 153, 156, 173, 178
   AIRS, cost of, 162
   floated pendulum, 119
   integrating, 123
   pendulous integrating gyro accelerometer (PIGA), 18–19, *20*
   for Trident I, 178
   for V-2, 118–19
   VM4A (Autonetics), 161
Advanced Development Program for SLBMs, 184–85
Advanced Inertial Reference Sphere (AIRS). *See* AIRS IMU
Advanced Projects Agency, 192
Aeronautical Engine Laboratory (MIT), 21, 41
   Navy vibration study, 45
Aerospace Engineering during the First Century of Flight, 19
Agena spacecraft, 191, *193*
   inertial control system for ascent, 196
   three-axis control system, 195
Air Corps Act, 14
Air Corps Training School, Brooks Field, 14
air defense, U.S. Fleet, 70
Air Defense Systems Engineering Committee (ADSEC), 95, 96, 97, 98
Air Force Cambridge Research Center, 100
Air Force Cambridge Research Laboratories (AFCRL), 96, 97
Air Force reconnaissance projects, 192
Air Force/Space Digest, 214
Air Material Command (AMC), 157
   authorizes modification of A1-C, 192
Air Research and Development Command (ARDC), 147, 148
aircraft
   Beechcraft: C-45, 97, 103
   Boeing: 203 biplane, 47; B-29, 112–15; B-52, 102; C-97, 123; P-12, 82, 85; P-26, 82
   Convair: B-58, 102
   Curtiss: JN-4, 14; Robin, 22, 30–31, 35–36
   Douglas: A-24, 85; A-26, 93–94, 115; B-26, 97, 100; C-47, 114, 115; SBD, 85
   Lockheed: F-94, 90, 102, 160; Hudson, 51; P-38, 87; P-80A, 88
   MIG-15, 91

North American: A-36, 85; B-25, 93–94, 97, 115; F-86, 88, 90, 92; P-51, 85; T-6, 97
  Republic F-84, 88, 90
AIRS IMU, 162
Alcore (AK 256), 131
Alexander, Philip N., 22
Ames Aircraft Co., 51
Angell, Frank, 12
antiaircraft directors, 67
antiaircraft guns
  1.1-inch gun, 67
  3-inch/50, 73, 80
  3-inch/70, 80
  5-inch/38, 70, 72, 74, 76
  5-inch/54, 80
  20-mm Oerlikon, 60, 64, 70–72
  40-mm Bofors, 61–62, 68–69, 70
Apollo, 219, 223
  guidance computer, 203, 207
  guidance system, 177, 203, 204, 206
  Launch Vehicle Digital Computer (LVDC), 207–8
Applied Physics Laboratory, John Hopkins Univ., 76
Arma Div. American Bosch Arma Corp., 106, 158
"Armament Engineering and Fire Control Sequence," Aero. Dept. MIT, 189
armament laboratory, Wright Field, 104
  source of Instrumentation Lab's funds, 110
Armed Forces Policy Council, 149
Army-Navy Ballistic Missile Committee, 164
Arnold, Henry H. "Hap," 85
Aronson, Joseph, 3
artificial gun line, 101
Ashworth, Harry, 51, 53, 54, 65
Atlas ICBM, 155, 157, 169, 226
  firing sequence, 148–49
  priority development, 164
  radio guidance system, 147
  specification, 150
  study contract for, 147
Autonetics Div. North American Aviation
  G6B4 gyro, 161
  G7B floated gyro, 135–36
  Minuteman II guidance, 161
  and Sabre inertial guidance, 161
  SINS: contract, 132; Mark 2, 134; primary source, 67

and two-degree-of freedom gyros, 126
VM4A accelerometer, 156
autopilot, 100
  George, 100–101
  G-system, 101
Azusa guidance system, 148

Ballistic Missile Office, Air Material Command, 161
ballistic trajectory theory, 148
barber chair director, 67
Barta Building, 97, 100
Bassett, Preston R., 33, 54, 57, 63
Battin, Richard H. "Dick"
  and Apollo guidance computer, 206
  and Atlas guidance, 151–53
  and Mars Probe, 201–2
  Q-guidance, 153, 166
  sojourn from Instrumentation Lab, 200–201
Bentley, Edward P., 54, 65, 69, 84
Bentley, George C., 42, 50, 54
beryllium, 154, 159, 168
beryllium baby, 161–62
  See also FLIMBAL
Birmingham, Stephen, 11
Black Warrior, 102
Blandy, William H. P., 59, 67
  and Section T, NDRC, 76
Blasingame, Benjamin P. "Paul," 226, 150–51
blind firing, 73, 76–77
blind flying, 46–47
Blue Angels warhead dispersal project, 175
Boeing Company
  Dyna-Soar contract, 198
  FLIMBAL computer, 161
Boeing School of Aeronautics, 46–47
Bofors. *See* antiaircraft guns: 40-mm Bofors
Bomi, 198
Bon Homme Richard (CV 31), 79
Bossart, Charles, 152
Boston Airport Corporation, 35
Boutilier, Napeen, 14
Bowditch, Philip N., 160
Bowen, E. G., 57
Box, F. M., 158
Boyd, Albert, 114
Brady, George G., 48
Brass Bell reconnaissance system, 198
Brett, John, 181
Bridger, Sarah, 216, 221

Index

British Technical and Scientific Mission to the United States, 56
Brode, Robert B., 76
Brown, Gordon S., 68
Brown, Harold, 176
Brown, William B., 24
Bureau of Aeronautics, USN
   contract for stable element, 128
   contracts vibration monitoring system, 41
   letter of intent for Mark 14 production, 62
   need for 1.1-inch gun director, 66–67
   stand-alone director for 1.1-inch gun, 67
   wants Mark 14 gunsight, 61
Bush, Vannevar, 26
   and fire-control equipment, 72
   negotiates detonation detector agreement, 45
   and Section T, NDRC, 76
Business Week, 212
Buxton, Dave, 3

C4 missile, 180
   See also Trident I (C4)
Canisteo (AO 99), 131
Cape Cod System, 100–101
Carousel warhead dispersal project, 173, 175
Carter, Leslie F., 33
Case, Theodore, 16
Challenger engine, 35
charged-coupled device (CCD), 185
Charles Stark Draper Award, 223
Charles Stark Draper Laboratory
   10 PIGA development, 184
   activities after establishment, 222–23
   creates Draper likeness (mannequin), 144
   design agent Trident guidance, 182
   established, 222
   Improved Accuracy Program, 183–84
   independent corporation, 222
   Mark 5 guidance: computer design, 182–83; design agent, 181; gyro selection, 185; stellar tracking, 182; system redesign, 187
   Mark 6 Life Extension Program, 187
Chilton, Robert G., 138, 203
Clemens, John E., 85, 194
Cockroft, John, 57
Coleman, Al, 114
Collins, Charles "Chip"
   comments on Draper's Air Force demonstration flight, 93–94
   and Draper's boxing career, 146
   evaluates MIT's B-29, 113
   first assignment, 115
   first impression of Draper, 113
   and MIT interview, 114
   pilots FEBE test flight, 112–13
   pilots SPIRE flight, 3
   recalls drinking incident, 143–44
   Whirlwind interceptions, 97–98, 100
Compass Island (EAG 153), 133, 135
Compton, Karl T.
   academic vision for MIT, 23
   affirms ties with industry, 42
   backs development of Mark 14 gunsight, 54
   and MIT research partnerships, xviii
   offered MIT presidency, 23
   reassures Draper on his status at MIT, 50
   recruits Hunsaker, 26
   recruits John Slater, 22
   Waltham Watch recommendation, 49
Confidential Instrument Development Laboratory MIT, 54
   ballistic corrections, 67
   established to design Mark 14, 65
   fabricates Mark 14s for 1.1-inch gun, 67
   Mark 51 design coordination, 68
   name change, 79
Conquest TV show, 123
Consolidated Aircraft Corporation, 147
controllable-pitch propellers, 41
Cook, Robert H., 129
Coolbaugh, James S., 189, 190, 194, 197
Corona spy satellite, 189, 192–93, 195–97
   camera system, 197
   importance of three-axis control, 197
   record of accomplishments, 197
   three-axis stabilization, 195
correlation trajectory, 156
counterforce weapon, 178, 179
Covington, William O., 17
Crane, Henry M., 17
Crane Automotive Engineering Fellowship, 17
Craven, John P., 132
Curtiss Wright Flying School, 21
Curtiss-Wright Corporation, 48–49

Dahlgren Div. Naval Surface Warfare Center, 183
damping fluid, 21, 117–21

Daniel Guggenheim Fund for the Promotion of Aeronautics, 21
data analyzer. *See* Whirlwind computer
Davis, Leighton I., 84, 226
  allows Draper to fly Air Force plane, 93
  arranges Watertown Arsenal test, 84
  assigned A1-C project officer, 85
  assigned dive-bomber squadron, 84–85
  discuss lead-computing sight with Draper, 85
  dive-bomb sight developed, 85
  and Draper's bombing system, 103–4
  and Draper's idea for inertial navigation, 103
  early career, 82
  engine indicator project, 83–84
  Gamow's threat to inertial navigation, 109
  introduces Chip Collins to Draper, 113
  makes attack passes in A-26, 93–94
  and Mark 14 prototype testing, 57
  recommends Collins for test center pilot, 114
  recommends Hanscom Field, 115
  teaching duties at U.S. Military Academy, 82
  tests A1-C prototype, 87
  Tracking Control Project, 92, 84
Day, Dwayne, 192
Defense Systems Operator (DSO), 102
Delaware (BB 28), 16
DeLisle, Joseph, 192
Delta Guidance, 152
Demonstration and Shakedown Operation, 188
Denhard, William G., 39, 109, 115
Dennis, Michael A.
  Bush's strategy for research, 45
  comments on Draper's oscillograph, 25
  on Draper's instrumentation notation, 28
  Draper's reliance on Fay Taylor, 44
  on Gamow's memo, 110
  on Taylor's membership in NACA, 26
Dent, Frederick R., Jr., 48
Department of Transportation, 36, 222
detonation (engine knock), 17
detonation detector. *See* Knockmeter
Digital Apollo, 204
direction of the vertical, 195
Discover space-exploration program, 189
disturbed reticle, 77

disturbed-line-of-sight principle, 80
Ditmeyer, Martha Stark Draper, 145, 219
dive-bombing, 226
Doel, Ronald D., 110
Dolittle, James H., 46
Dow, Paul, Jr., 181
Draper, Charles A., 8, 10
Draper, Charles Stark, 8
  A1-C gunsight: criticizes initial design, 87; notebook page, *86*; sketches design, 85, *86*
  Aero Department duties, 138, 140
  Aero Eng. Lab. research assistant, 20–21
  age concerns for Apollo crew selection, 205
  agrees to let Air Force graduate students work on reconnaissance satellite, 190
  airborne fire-control work, comments on, 91
  aircraft instruments: acoustic altimeter study, 43–44; engine indicator, 44; interest in, 30, 31; Knockmeter, 20, 44; magnetic compass study, 36; for meteorological study, 35–36; patents improved magnetic compass, 38; turn indicator, 47, 51; turn indicator patent, 52
  ancestors and family background, 8
  anti-war protests: avoids pickets, 219; confronts demonstrators, 211; polarizing force against, 216; policy changes from, 216; security concerns, 218
  Apollo crew volunteer, 205
  Apollo guidance, confidence in providing, 203; preparations for, 203
  appointed to Air Defense System Engineering Committee, 95
  approach to problem solving, 225
  arranges to hire Chip Collins, 114–15
  Atlas guidance, 150
  attends Curtiss Wright Flying School, 21
  automotive engineering fellowship, xvii, 17
  awarded PhD, 29
  birth date, 8
  blind flying, 46, 47
  bootlegging anecdote, 11
  and boxing, 14
  boxing career, 146
  Brooks Field flight training, 14–15
  and CBS production on SPIRE, 124

changes to doctorial requirements, 22
charismatic leader, 216
christens Flight Test Center, 115
collaboration and relationship with others, xviii
commissioned 2nd Lt., U.S. Army Reserve, 14
confidence in providing Apollo guidance, 205
Confidential Instrument Development Lab., comments on, 65
consulting activities of, 40
consults for Curtiss Wright, 49
credit for developing Floated Integrating Gyro, 117
"Cremation of Sam McGee, The," 145
criticism of production models, 74
criticizes Pounds Panel recommendations, 218
cross country flight, 30–31
culinary artist, 138
Davis, Leighton: allows Draper to fly Air Force plane, 93; arranges Watertown Arsenal test, 84; celebrates end of World War II with Draper, 103; coauthors paper, 84; discusses idea for inertial navigation, 103; discusses lead-computing sight, 85; house guest of, 82; introduces Chip Collins to Draper, 113; shows engine indicator, 83–84; suggests bombing system, 103–4
death of, 223
on defeating Mother Nature, 141
deputy head of Aero Dept., 137
designs data recorder, 35
develops engine indicator, 24–25
disciples and protégés, 131, 150, 225–26
discovers characteristic gyro time, 55
discovers MIT, 12
Doc's "dollar bills," 124
Doc's parties, 142
dogmatism and stubbornness, 141, 146
Dummy Gun System, 101
and education of military officers, 189–90
establishes Flight Test Center, 93
faith in single-degree-of-freedom gyro, 179
"Father of Inertial Navigation," xvii
first flight, 14
fixes Mueller's accelerometer, 119
floated gyro: design, 116; inspiration for, 80–81
and flotation in gyroscope, 116–17
flying skills, 34
fondness for drink, 127, 142; John B. Nugent Medicinal Aid Foundation, 79; late afternoon ritual, 142; liquor as a social lubricant, 143; soothing syrup, 77
Gamow's threat to Draper's ideas, 109
generalized mathematic model of dynamic error, 38–39
Gillmor (Reginald), research for, 15–17
goals of education, 138, 225
greasy-thumb mechanic, 10
growing up in Windsor, Mo., 9
holiday dinners, 145
honors course, 138
importance to U.S. ICBM program, 149
inertial navigation: arranges seminar for, 109; idea for, 103–4; smoothing to correct errors, 111; theory of, 108–9
inquisitive mind, 13
instrument lab: established, 24; funding problem, 49; lauded by Hunsaker, 27
interest in gadgetry, 12
interest in psychology, 12
laboratory titles, choice of, 131
leadership of Instrumentation Lab, 141
legend on courses taken, 29
leisure activities, 145–46
lifetime accomplishments, xvii–xviii
Link Trainer practice, 47
love of Chinese food, 145, 146
management skills, 225
Mark 14 gunsight: basic design, 55, 58; and dive-bombing problem, 85; gyros for, 62; idea for, 52–53; manufacturing problems, 68; observes acceptance tests, 62; "shoebox," 57; Sperry production problems, 63; test at rifle range, 55–56
Mark 15 gunsight, 73, 74, 75
Mark 58 director design, 77
marries Ivy Willard, 48
mathematical approach to instrumentation, 28
member of Scientific Advisory Board, 147

Index

Draper, Charles Stark *(continued)*
  member of Valley Committee, 96
  meteorological flights, 35
  MIT appointments: assistant professor, 27; full professor, 50
  and MIT Instrumentation Lab, xvii
  MIT student: admitted, 13; enrolls as doctoral candidate, 21; passes doctorial exam, 25; undergraduate difficulties, 13
  NACA report, 25–26
  nicknames: Doc, xv; Droopy Drawers, 33; Snowdrift, 34
  occupation and business interests, 10
  opinion of Project Mercury, 198
  opinion on venturi, 31
  opposition to stellar tracker, 179
  oscillograph for engine indicator, 25
  papers and thesis, 20, 24, 44, 46, 52, 134
  passion for single-degree-of-freedom gyro, 124
  patents turn indicator, 52
  perfection of single-degree-of-freedom gyroscope, *124*, 220–21
  personal relationships, 40
  personality, xvii, 224–25
  Polaris missile: A2 accuracy requirements, 134; negotiates contract, 167
  post-war programs in fire control, 89
  pressure-sensor development, 20
  public acclaim for guidance work, 174
  purchases Curtiss Robin, 22
  quest for knowledge, 21
  reaction to appointment as head of Aero Department, 137
  reaction to Pounds Panel report, 214
  recruits Spilhaus, 32
  refutes Gamow's concerns, 111
  and relation with MIT, xvii–xviii
  relationship with Elmer Sperry Jr., 40
  removed as director of Instrumentation Lab, 216
  retires as professor emeritus, 217
  SINS: improves characteristics of, 135; principle investigator, 129
  spectroscopy fuel flame study, 19–20
  and Sperry's handling of SINS, 134, 166–67
  SPIRE Jr., expected performance of, 23
  and SPIRE test flight, 1–7
  Stanford University student, 11–12
  and star trackers, 104
  statement to Pounds Panel, 214
  on Steering Task Group, SPO, 165
  Stellar Bombing System, 104–6
  Stratton demonstration flight, 34
  studies internal combustion process, 17–18, 20–21
  and Stutz Bearcat, 11–12
  taste of Army life, 10
  teaches aircraft instrument course, 24
  teaching philosophy, 138
  Time Man of the Year, 174
  typical evening, 144–45
  University of Missouri, attends, 10
  vibration monitoring system, 41
  victimized over divesture, 220–21
  Victorian home, 142, 144
  visits Boeing School of Aeronautics, 46–47
  Waltham Watch consulting agreement, 49–50
  Waltham Watch R&D study, 50
  work ethic, 127
Draper, Ivy Willard, 47, 48
Draper, James Stark, 11, 13, 142–43, 144, 145, 146
Draper, John, 145
Draper, Martha, 145, 218–19
Draper, Michael, 145
Draper, Ralph Clayton, 8, 40, 49
Draper notation, 129
Draperian guidance, 161
Dryden, Hugh L., 198, 201
Duffy, Robert A.
  comment on Draper's passing, 223
  on Draper's instrumentation notation, 28
  Draper's understanding of people, 141
  quote on development of single-degree-of-freedom gyro, 116
  remarks on assistance of MIT material scientists, 118
  reports on Doc's airsickness, 15
Dumaine, Frederick C., 49
Dummy Gun System project, 101
Dumont Laboratories, 83
Duolcam Co., 62
Dynamics Research Corporation, 187
Dyna-Soar, 198–99

E6B slide rule, 113
Eclipse-Pioneer Division, Bendix Corporation, 161

Index

Eden, Murray, 210
Edgerton, Harold E., xviii, 25
Eisenhower, Dwight D., 164, 192
El Dorado oil field, 10
Electro-Physical Laboratories, 32
Eller, Ernest M., 62
engine indicator, xviii, 17, 24–25, 44
engine knock. *See* detonation
Enterprise (CV 6), 70–72
Ethan Allen–class, 134
Everpin Technologies, 188
Executive Committee of the Corporation MIT, 215, 219
Eye in the Sky, 192

Fairchild Engine and Airplane Corporation, 196
Fairchild Semiconductor, 170
FD radar, 72
FEBE, 106, 113
    accuracy, 112
    influence on SPIRE, 2
    new gyros for, 116
    purpose, 111
    size, 112
    stable element, 128
Fenske, Merrill R., 117
Fertig, Kenneth, 161
Fire Control Branch (SP-23), SPO, 165, 166, 173
fire-control systems
    Polaris Mark 80, 171–72
    Polaris Mark 84, 172, 177
    Poseidon Mark 88, 177
first strike weapon. *See* counterforce weapon
Fitzgerald, Scott F., 11
Flickinger, Donald D., 205
Flight Test Center. *See* Instrumentation Laboratory MIT: Flight Test Center
FLIMBAL, 160–61, 163
floated gyroscope, 80, 117
Fluid Mechanics Laboratory MIT, 219
Fluorolube®, 118, 161
fly-by-wire guidance, 152
flying laboratory (Draper's), 35
flywheel damping, 194
Ford rangekeeper, 64
Forrester, Jay W., 96–97
Forter, Samuel A., 165, 177
Fowler, Sir Ralph Howard, 56
Fox and Hounds Restaurant, 127
Franklin Delano Roosevelt, 59

Freeman, Albert P., 109
From Polaris to Trident, 173

Gamow, George, 109–11
Gardner, Trevor, 149
General Electric Company
    anti-war protests, 218
    Corona subcontractor, 196
    Ordnance Systems plant, 170
    Polaris Mark 1 guidance, 168
    Polaris Mark 2 guidance, 173
"George." *See* autopilot: George
George Washington (SSBN 598), 134
Getting, Ivan A., 55
Gilbert, Ralph E., 109
Gilliland, Edwin R., 215
Gillmor, Reginald E.
    association with Elmer Sperry, 16
    employment history, 32
    establishes research lab, 15
    infrared research contract, 16
    Sperry Gyroscope president, 17
gimbal lock, 160
Goett, Harry J., 202
Gold, David, 166
Goodwin, Henry M., 17, 19, 20
gravity anomalies, 182
gravity stabilization, 190
Gray, Peter, 212
Grenier Field, 87
Grothe, Lester R., 154
Guggenheim Aeronautic Laboratory MIT, 21
Gugger, Edward, 40
"Guidance Mafia," 141
guidance systems
    Mark 1 (Polaris), 168
    Mark 2 (Polaris), 173–75
    Mark 3 (Poseidon), 177
    Mark 4, 178–80
    Mark 5 (Trident I), 178–83, 185
    Mark 6 (Trident II), 184–87
    Mark 6 Mod 1 (Life Extension program), 187–88
Gun Fire Control System Mark 57. *See* Mark 57 GFCS
Gun Fire Control System Mark 58. *See* Mark 58 GFCS
Gun Fire Control System Mark 63. *See* Mark 63 GFCS
Gunar, 80
Gustafson, Boyd E., 129

Gyronavigator, 134
gyroscopes
    2 FBG, 159
    10 FG, 123, 156, 159
    10 PIGA, 184
    16 PIGA-G, 161
    18 IRIG, 175
    25 IRIG, 168, 175
    25 PIG, 123
    25 series IRIGs, 153–54
    30-X-1, 117
    45 FG, 119
    in aircraft instruments, 32
    air-lubricated, gas spin, 125
    Autonetics: G2K, 133; G6B4, 161; G7B, 135
    in autopilots, 100–101
    characteristic time, 55
    drift, 104, 118, 159
    fiber optic, 163, 187
    fire control vs. inertial navigation, 118
    floated, 116–17, 135
    floated beryllium, 159
    gas floated, 163
    and gimbal lock, 160
    hermetically sealed, 118
    HIG-5, 118
    IFOG, 187
    improved for SINS, 132
    for inertial navigation, 104, 105
    integrating, 118, 123
    magnetic suspension system, 135
    reversible, 133
    Singer-Kearfott: MITA-4, 185; MITA-5, 85
    single-axis integrating, 124
    single-degree-of-freedom, 55, 161
    torque errors, 159
    for Trident I, 182
    tuned-rotor, 181
    two-degree-of-freedom, 161

Hacker, Barton C., 9, 13, 65
Hall, Elton C., 168, 169, 206–8
Hanscom Field, 1, 115
Hardison, Osborne B., 70–72
Hardy, Arthur C., 24
Hazen, Harold, 73, 74, 75, 77
Heather, John C. "Jack," 190–91, 194, 195, 196
Hendrick, David I., 69
Herman, Ernest E., 60, 67

heterogeneous engineer, 224–25
Hi-AC camera, 193, 195
Hill, Albert G., 220, 222
Hoag, David, 203, 204, 213
Honeywell Corporation, 185, 187, 188
Hood Building, 140
Hooker Chemical Company, 118
Houston, Forrest E., 129, 130–31
Hovorka, John, 129, 130
Hughes, Thomas P., 224
Hughes Aircraft Company, 185
Hume, Peter, 16
Hunsaker, Jerome C., 26–27, 41, 43, 137, 225
Hurse, John, 3

IBM 650 computer, 200
ICBM, 158
Idaho (BB 42), 71
Improved Accuracy Program (IAP), 183–84
IMU. *See* inertial measurement unit
inertial guidance
    for Atlas, 151
    correlated flight path, 152
    correlation trajectory, 156
    Delta guidance, 152
    fly-by-wire, 152
    gimbal-less system, 161
    Pace system, 199
    Polaris A3, 173
    Q Matrix, 152
    Q-guidance, 153
    strapdown system, 174
    velocity-to-be-gained vector, 152–53
    See also Instrumentation Laboratory
        MIT: inertial guidance
inertial measurement unit
    AIRS, 162
    and gimbal lock, 160
    gimbal-less, 160–61
    and gyro drift, 104
    Mark 3 guidance system, 177–78
    Mark 6 guidance system, 185
    new configurations, 184
    for satellite ascent stage, 190
    for Thor, 156
    weight-saving approaches, 168
inertial navigation
    for ballistic missiles, 123
    East Coast school of thought, 126
    gravity anomalies, 182
    and gyro drift, 118

N6A Autonavigator, 132
NAVAN cycle, 133
West Coast school of thought, 126
See also Instrumentation Laboratory MIT: inertial navigation
infrared communication, 16
Institute of Aeronautical Sciences (IAS), 26
Institute of Defense Analyses, 180
Instrumentation Laboratory MIT
 A1-C gunsight, 85, 87
 accelerometers, 121, 123, 153, 156, 159, 161, 168, 173, 178
 Agena spacecraft control, 192, 196
 AIRS IMU, 162
 annual budget (1951), 141
 anti-war protests, 211, 218–19
 Apollo guidance: design proposal, 204, 206; feasibility study, 202–3; NASA visit, 204–5; and presidential challenge, 203
 apparatus (early), 27
 ballistic missile guidance, 123
 beginnings, 24, 27, 227
 budget (1969), 227
 Corona guidance, 189
 and development of integrated circuits, 170
 digital computer study, 168
 divestiture, 220–21
 Doolittle flight, influence of, 46
 Draper's characterization of, 140
 Ducosyn, 135
 Dummy Gun System, 101
 employees: 1951, 141; 1969, 227; ca. 1956, 126–27
 engineering paradigm, 125
 Flight Test Center, 93, 115
 FLIMBAL, 159–60
 gyroscopes, 119, 156, 168, 175; development of, 109, 118; effort to improve, 153–54, 159; floated, 116–17, 135; floated beryllium, 159; gas floated, 163; for inertial navigation, 104, 105; Thor, 153; for Trident I, 182
 IAS meeting presentation, 27
 ICBM feasibility study, 150, 157, 159–60
 inertial guidance: Atlas guidance, 151, 154–55; Dyna-Soar, 199; IRBM requirements, 165; MIGIT, 175; Pace, 199; Polaris A3 design study, 172; Polaris contract, 167; Polaris Mark 1, 168; Polaris Mark 2, 173; Poseidon design, 176–77; Poseidon star tracker, 179; Sabre, 161; SLBM guidance, 180; strapdown system, 175; Titan guidance, 158; Thor guidance, 155
 inertial navigation: greatest contributions, 116; SINS development contract, 130; SINS Mark 4, 134; SINS study contract, 129; SINS technical support, 133; SPIRE, importance of, 7; SPIRE Jr. sensor redesign, 121; Stellar Bombing System, 104–5
 internships, 189–90
 JPL conflict, 202
 and Jupiter missile, 164
 Mars project, 200–201
 Missile Position Measurement System, 162
 Navy air intercept project, 102
 official name adopted, 79
 Pounds Panel victimization, 213
 and precision manufacturing, 154–55
 Project MAST, 128–29
 shipboard stabilization system, 164
 special adhesives, 118
 three-axis gyro stabilization, 192
 Tracking Control Project, 92
 visual aids, 194
integrated air defense system, 96
intercontinental ballistic missiles (ICBM), 123
interferometric fiber-optic gyroscope, 163
intermediate range ballistic missile (IRBM), 164
Itek Corporation, 193, 194, 196

Jarosh, Joseph J., 65
Jet Propulsion Laboratory (JPL), 202
John B. Nugent Medicinal Aid Foundation, 142
Johnson, Benjamin, 104
Johnson, Howard W., 34, 215–17, 220–22
Johnson, Lyndon B., 176
Jupiter missile, 164–65

K-18 gunsight, 91
Kearfott Div. General Precision Corporation, 178–79, 181
Kelly, Marvin J., 148
Kennedy, John F., 203
Key West Agreement, 164

Keys, Clement M., 32
Khrushchev, Nikita, 201
Kilby, Jack, 98, 99, 100
Killian, James R., 63, 98, 212, 220
  creates MIT Division of Defense, 100
  Draper's Mark 14 compensation, 65
  and federally sponsored research, 99
  proposes air defense laboratory, 99
  urges Atlas development, 164
Knockmeter, 44
Kraus, Sydney M., 45
K-Series navigation and bombing systems, 155

Lafayette-class submarines, 172, 177
Laird, Melvin, 179
Lancour, Joseph H., 45, 83–84
Laning, J. Halcome "Hal," 83–84, 151, 153, 166, 200–201
Law, John, 224
Lawrence, Anthony, 126
LDVC. *See* Apollo: Launch Vehicle Digital Computer
lead-computing gunsight. *See* Mark 14 gunsight
lead-computing sights, 60
Leavitt, William, 214, 217, 218
Lees, Sydney, 116
Leghorn, Richard S., 184
Leslie, Stuart, 223
Life Extension Program (Trident), 186–88
Lincoln Laboratory, MIT, 99, 211, 213, 220, 221
Link Trainer, 47
Lockheed Corporation, 175, 190, 191, 192
Luck, David G. C., 25

M-61 20-mm gun, 101
MacCloud, Frederick, 62
MacDonald, Duncan, 195
MacKenzie, Donald, 119, 124, 161, 162
Maclaurin Building, 211
magnetic dip, 37
magnetic torqueing, 178
magnetoreistive memory, 188
Mailman (warhead dispersal project), 175
Malcomson, Malcolm R., 190–91
Marchetti, John W., 96
Marine Stable System, 128
Mark 14 gunsight, 70
  British version, 57
  BuOrd interest in, 60
  design: concept sketch, 58; issues, 63
  importance of Sperry Gyroscope, 40
  lagging reticle, 71
  manufacturing: price, 65; production problems, 62; Sperry contract, 54; units produced, 65
  modified as bomb sight, 85
  Navy acceptance of, 62–63
  origin of Draper's concept, 52–53
  prototypes and working models, 55, 61
  reasons for success, 64–65
  testing: at Dahlgren, 60–61, 62; outshoots tracer control, 63–64; of prototype, 57; at rifle range, 55–56; Royal Navy firing tests, 57; on 20-mm gun, 64
  training and operation: how to lead, 66; manual, 66, 71; smooth tracking, 55
  upgrade requirements, 71
Mark 15 gunsight, 73
Mark 17 reentry vehicle, 178
Mark 33 director, 70, 72, 73
Mark 37 radar, 77
Mark 51 director, 73–77
Mark 57 GFCS, 76–77
Mark 58 GFCS, 77
Mark 63 GFCS, 77–78, 79, 92
Mark 64 GFCS, 80
Mark 84 fire-control system, 177
Mark 88 fire-control system, 177
Mars Probe, 200–201
Martin Company, 158, 199
Massachusetts Institute of Technology
  aeronautic program, 137–38
  anti-war protests, 211; collective guilt of war research, 221; debate on military research, 210; faculty statement against military research, 210; Pounds Panel, 212–15, 217; student strike, 210–11
  Board of Trustees, 2
  consulting fees and policy, 40, 43
  conversion of Special Laboratories, 219
  Division of Defense Laboratories, 100
  Division of Industrial Cooperation, 57
  Division of Industrial Cooperation and Research, 42, 43
  engine laboratory, xviii, 19, 21, 41, 44
  income from Mark 14, 65
  Project Lincoln, 100
  special laboratories budget, 211
  Technology Plan of 1918, 42
  Visiting Committee, 28

Massachusetts Institute of Technology–Sperry Apparatus for Measuring Vibration, 43
Massachusetts Institute of Technology–Sperry Detonation Indicator, 45
McKay, Walter, 36, 38, 142–43, 176
McNamara, Robert, 176, 199
Mechanical Integrating Accelerometer, 118–19
Mercury-Redstone rocket, 198
Meteorological Station, MIT, 35
Mexican arithmetic, 179
Microsyn, 60
Miles, Marion E., 60
Miller, Charles, 216, 220, 222
Millikan, Clark B., 147
Mindell, David, 64–65, 204, 205, 207
Minuteman missile, 112, 162, 163
Missile Position Measurement System, 162
Mitre Corporation, 220
Molella, Arthur, xvii
Morgan, Thomas A., 32
Morison, Elting, 212
Morse, Philip M., 23
MRAM. *See* magnetoresistive memory
MRVs. *See* multiple reentry vehicles
Mueller, Fritz, 118
Mueller integrating accelerometer, 119, *120*
multiple reentry vehicles (MRVs), 175, 176, 211
Multiscope Company, 173
Murphy, Marion E., 60, 67
Mustin, Lloyd M., 53
mutually assured destruction, 173
MX missile program, 162, 163

National Academy of Engineering, 223
National Advisory Council on Aeronautics (NACA), 25–26, 198
    Apollo: feasibility study, 202; guidance computer, 207; guidance contract, 204; funds: acoustic altimeter, 44; detonation detector 44; Mars Probe, 202
    mission of, 201
National Security Council, 164
Nautilus (SSN 571), 133
Navaho missile, 125, 132, 148
Naval Surface Warfare Center, 183
NAVAN cycle, 133
Navigation Data Computer (NAVDAC), 134
Nelkin, Dorothy, 222

New York Times, 214
Nixon, Richard M., 179
North American Aviation, 32, 111, 158
North Carolina (BB 55), 64
Northrop Corporation, 162, 163
Norton, Charles L., 22
Nortronics Div. Northrop Corporation, 161
November Action Committee (NAC), 218
Noyce, Robert, 179
Nugent, John B., 79

Oerlikon. *See* antiaircraft guns: 20-mm Oerlikon
Office of Naval Research, 129
Ohio-class submarines, 180–81, 184, 188
Olsen, Benjamin O., 191
on-orbit stabilization, 191
Orbiting Geophysical Observatory, 202–3
Our New Age, 31
OX-5 engine, 31, 35

Pace guidance system, 199
Palmer Engine Company, 40
passive gravity stabilization, 194
Peacekeeper. *See* MX missile program
pendulous integrating gyro accelerometer (PIGA), 18–19
Piland, Robert O., 204
Plan Position Indicator (PPI), 98
Polaris A1, 134, 178–79
Polaris A2, 134, 172
Polaris A3, 135, 172–73, 178
Polaris B3, 175, 176
Polaris missile (UGM-27), 132, 133
    fire control, 171–72
    guidance, 168–74
    mission, 178
    targeting 171
Polaris Missile Facility, 170
Polaris Steering Task Group, 134
Polaris submarines 136
    navigation equipment, 133–34
Poseidon C3, 176–79, 181
Pounds, William F., 192
Pounds Panel, 212–15, 217, 221
President's Board of Consultants on Foreign Intelligence Activities, 11
Prohibition, 11
Project Charles, 99
Project Feedback, 190
Project Lincoln, 99

Project MAST, 128
Project Mercury, 198
Project MX-402, 148
Project MX-1593, 147, 148
Putt, Donald L., 148

Q matrix, 171
Q-guidance, 154, 166, 176, 177, 200
Quarles, Donald A., 193

R-3350 engine, 113
Raborn, William F. "Red," 131, 164, 167
radio guidance
    Azusa (Atlas), 148
    Titan, 157, 158
    vulnerability, 158
Radio Corporation of America (RCA), 83
Ragan, Ralph, 165, 166
Ramo-Wooldridge Corporation, 153
Rand Corporation, 190
rangekeepers, 67
Raytheon Corporation, 187
Redman, Kent, 97–98
reentry vehicles, 178
Regulus cruise missile, 128
Rescuing Prometheus, 224
Research Steering Committee (NASA), 202
Review Panel of Special Laboratories.
    See Pounds Panel
Rivero, Horacio, Jr.
    BuOrd assignments, 59
    Draper disciple, 225–26
    Draper student, 53
    and Mark 14 gunsight, 58, 60
Roberts, Edward B., 18
Rogers Building, 211, 212
Rossby, Carl G., 35
Ruggles Orientator, 14–15
Ruin, Jack, 161

Sabre, 161
SAGE (Semi-Automatic Ground Environment System), 101, 220
Sage, Nathan M., 54, 62, 92–93, 141, 142
SAMOS, 193–94
Sattlemyer, John, 62
Sayre, Daniel C., 22, 35
Schlesinger, James, 183
Schriever, Bernard A., 150, 157, 158
Schuler principle, 108
Schuler tuning, 129
Schuyler, Garret L., 60

Schweidetzky, Walter, 152
Science, 213, 215
Science Action Coordinating Committee (SACC), 211, 216
Science Digest, 46
Scientists at War, 216
Seamans, Robert C., Jr.
    A1-C design, 86–87
    activities after leaving MIT, 115
    Collins supervisor, 115
    comments on Draper: behavior when thinking up new ideas, 142; habits during Dam Neck visit, 77, 79; request as Apollo crew member, 205; review of production models, 74; teaching philosophy, 138–39; test flight with Davis, 93–94; work ethic, 127
    defines Apollo guidance, 204
    as Draper's quintessential disciple, 226–27
    on Mark 14 manufacturing, 62
    on origins of floated gyro, 81
    Tracking Control Project, 92–93
Section T, National Defense Research Committee (NDRC), 74, 75
Seminar on Automatic Celestial and Inertial Long Range Guidance Systems, 110–11
Service, Robert W., 145
Servomechanism Laboratory, MIT, 68
Sevaried, Eric, 124
Shapiro, Asher H., 219
Sheehan Committee, 218
Shepard, Alan B., Jr., 198
Shumaker, Samuel, 76
Sims, William, 16
SINS, 129–36, 167, 171, 193
Sirbu, Marvin A., Jr., 215
Skunk Works, Lockheed Corporation, 196
Slater, John C., 22, 23, 28
Slater, John M., 125
Sloan, Alfred P., Jr., 17, 21
Smith, Gregory, 212
Smith, Levering, 165, 176, 178, 180
Smith, Thomas, 97
Smith, Victor C., 54
Söderqvist, Thomas, 110
Soviet Union, 191
Space Inertial Reference Equipment.
    See SPIRE
Space Task Group (STG), NASA, 202

Special Projects Office of the U.S. Navy (SPO)
  created, 33
  Mark 4 guidance nonessential, 180
  Mark 12 warhead studies, 175
  Polaris: A2 requirements, 134; A3, 175; A3 range goal, 172; B3 MIRVs, 133, 164, 176
  Poseidon: guidance, 176; reentry vehicles, 178
  SINS sea trials, 133
  Steller sighting skepticism, 178
Sperry, Elmer A., 16, 38, 39, 40
Sperry, Elmer A., Jr., 33, 38, 39–40
Sperry Gyroscope Company
  A-4 manufacture, 91
  contract for remote control director, 67
  divested from General Motors Corp., 32
  funds Draper turn indicator, 51
  instrument sales business, 44
  instruments for blind flying, 46
  Mark 14 gunsight: British order, 57; contract for, 63; problems producing, 62, 167; prototype production, 61; quantity produced, 295
  Mark 51 director order, 67
  refers Davis to RCA, 83
  selected to manufacture first A-1C gunsight, 90
  servomechanism agreement with MIT, 68
  SINS, 132; failure, 167; Gyronavigator, 134; Mark 3, 134
  vibration monitoring system, 42, 43
  visited by members of Tizard Mission, 57
Spilhaus, Athelstan F., 32, 33, 36
Spinardi, Graham, 173, 177
SPIRE, 1–7, 112, 119–22, 150, 226
SPIRE Jr., 121, 123
SPO. *See* Special Projects Office of the U.S. Navy
Sputnik, 191, 198, 200–202
St. Laurent–class destroyers (Canadian), 80
stable platform, 106–7
Stanford Pictorial, 11
star tracker, 104
  See also stellar sightings; stellar tracker
Stark, John, 8
Steering Task Group, SPO, 165
Steinhoff, Ernst, 141
Stellar Bombing System, 104–6

stellar sightings, 178
stellar tracker, 156
Stephen, Raymond J., 47
Stever, H. Guyford, 96
Stockbarger, Donald C., 14
Stockdale, Donald C., 16
strapdown guidance system, 174
Strategic Arms Reduction Treaty (START) II, 163
Strategic Missile Evaluation Committee, USAF, 149
Strategic Systems Project Office (SSPO), USN, 180, 186
Stratton, Julius A., 34, 212, 221
Strat-X, 180
Stuart, George R., 36
Stutz Bearcat, 11
Submarine Inertial Navigation System. *See* SINS
System 118P, 73

tacheometric directors, 73
Tartar missile, 102
Taylor, Charles Fayette "Fay"
  aids design of engine indicator, 25
  joins MIT faculty, 19
  offers Draper research assistant position, 20
  promotes detonation detector, 44–45
  recommends Draper to teach aircraft instrument course, 24
  replaced as head of aero program, 26
Taylor, Edward S.
  and Aero Engine Lab, 21
  coauthors paper with Draper, 24
  Gas Turbine Laboratory administrator, 140
  joins MIT faculty, 19
  and torsional vibration problem, 41
Technological Capabilities Panel, Scientific Advisory Committee, 164
Tennessee (SSBN 734), 186, 188
Terrier missile, 102
Texas Instruments Company, 170
Thor missile
  analog computer, 168
  Corona booster, 192
  guidance system, 155, 156, 165, 226; compared to Polaris, 168; gyros, 159 range, 155
three-axis gyro stabilization, 192
Time magazine, 198

Time Man of the Year, 174
Titan missile, 157–59
Titan II missile, 158, 159, 161, 223
Tizard, Sir Henry, 56
Tizard Mission. *See* British Technical and Scientific Mission to the United States
torsional vibration, 41
Towner, Winthrop H. "Win," 31
Tracking Control Project, 92
tracking instability, 87
Trageser, Milton, 200, 202
Trident I (C4), 180–84, 186
Trident II (C5), 184, 186
Trident submarines, 180–81
Trimble, William B., 26, 225
Tuve, Merle, 73
Two Ocean Navy, 59

UGM-27 missile. *See* Polaris missile (UGM-27)
Undersea Long-Range Missile System (ULMS), 180
Union of Soviet Socialist Republics (USSR), 191
    See also Soviet Union
Unistar Stellar System, 179, 180
United Shoe Company, 40

V-2 missile, 118
Valley, George E., 95–96
Valley Committee, 95–96
Van Allen, James, 174
Vandenberg, Hoyt S., 95, 98
"Vertical, Vertical, Who's Got the Vertical?," 109
vibration monitoring project, 41–42
Vidicon tube, 185
Viking missile, 148
Volstead Act, 11
von Karmon, Theodore, 95
von Neumann, John, 149
von Neumann Committee, 149–50

Wadleigh, Kenneth R., 218
Wallace, F. C., 57
Wallace, Robert F., 71
Walsh, John, 215
Walter, Hollis C., 68
Waltham Watch Company, 49–51
warhead delivery systems, 178
warhead dispersal systems, 175
Warner Brothers Paving Company, 40
Warrior System, 101
Weapons System 117L, 190, 192, 195
Weapons System Program, MIT, 189
Weapons Systems Section, Dept. of Aero. Engr. MIT, 150
Webb, James, 204
Weikert, John M., 85
Weisner, Jerome "Jerry," 96, 220
Weschler, Thomas, 138
Western Development Division, USAF, 150–51, 158
Weyerbacker, Ralph C., 41
Whirlwind computer, 96–98
Whirlwind engine, 19
Whitaker, Omar B., 61
Whitmore, William F., 165
Whittemore Building, 140
Wieser, Robert, 97
Willard, Ivy. *See* Draper, Ivy Willard
Williams, C. L., 44
Williamson, R. F., 10
Willis, H. Hugh, 45, 47, 54
Woodbury, Roger B., 3
Wright, Charles, 170
Wright Air Development Command, USAF, 151
Wright R-1840 engine, 41
Wrigley, Walter, 105, 106, 108, 125, 129, 190
WS-117L. *See* Weapons System 117L
Wurtman, Richard, 212

Young, Louis H., 24

About the Author

Thomas Wildenberg is an independent historian and scholar with special interests in aviators, naval aviation, and technological innovation in the military. He has written extensively about the U.S. Navy during the interwar period. His articles have appeared in several scholarly journals, including the *Journal of Military History*, *American Neptune*, and U.S. Naval Institute *Proceedings*. He is also the author of six books on U.S. naval history covering such varied topics as replenishment at sea, the development of dive bombing, and the history of the torpedo in the U.S. Navy. These include: *Gray Steel and Black Oil*, *Destined for Glory*, *Billy Mitchell's War with the Navy*, *Ship Killer*, and *Striking the Hornet's Nest*.

The Naval Institute Press is the book-publishing arm of the U.S. Naval Institute, a private, nonprofit, membership society for sea service professionals and others who share an interest in naval and maritime affairs. Established in 1873 at the U.S. Naval Academy in Annapolis, Maryland, where its offices remain today, the Naval Institute has members worldwide.

Members of the Naval Institute support the education programs of the society and receive the influential monthly magazine *Proceedings* or the colorful bimonthly magazine *Naval History* and discounts on fine nautical prints and on ship and aircraft photos. They also have access to the transcripts of the Institute's Oral History Program and get discounted admission to any of the Institute-sponsored seminars offered around the country.

The Naval Institute's book-publishing program, begun in 1898 with basic guides to naval practices, has broadened its scope to include books of more general interest. Now the Naval Institute Press publishes about seventy titles each year, ranging from how-to books on boating and navigation to battle histories, biographies, ship and aircraft guides, and novels. Institute members receive significant discounts on the Press' more than eight hundred books in print.

Full-time students are eligible for special half-price membership rates. Life memberships are also available.

For a free catalog describing Naval Institute Press books currently available, and for further information about joining the U.S. Naval Institute, please write to:

Member Services
U.S. Naval Institute
291 Wood Road
Annapolis, MD 21402-5034
Telephone: (800) 233-8764
Fax: (410) 571-1703
Web address: www.usni.org